Single Cell Analysis

Edited by
Dario Anselmetti

Related Titles

Yanagida, T., Ishii, Y. (eds.)

Single Molecule Dynamics in Life Science

2008
ISBN: 978-3-527-31288-7

Mirkin, C. A., Niemeyer, C. M. (eds.)

Nanobiotechnology II

More Concepts and Applications

2007
ISBN: 978-3-527-31673-1

Chalfie, M., Kain, S. (eds.)

Green Fluorescent Protein

Properties, Applications and Protocols

2005
ISBN: 978-0-471-73682-0

Niemeyer, C. M., Mirkin, C. A. (eds.)

Nanobiotechnology

Concepts, Applications and Perspectives

2004
ISBN: 978-3-527-30658-9

Laflamme, S. E., Kowalczyk, A. P. (eds.)

Cell Junctions

Adhesion, Development, and Disease

2008
ISBN: 978-3-527-31882-7

Single Cell Analysis

Technologies and Applications

*Edited by
Dario Anselmetti*

WILEY-VCH Verlag GmbH & Co. KGaA

The Editor

Prof. Dr. Dario Anselmetti
Fakultät für Physik
Universität Bielefeld
Universitätsstr. 25
33615 Bielefeld
Germany

All books published by Wiley-VCH are carefully produced. Nevertheless, authors, editors, and publisher do not warrant the information contained in these books, including this book, to be free of errors. Readers are advised to keep in mind that statements, data, illustrations, procedural details or other items may inadvertently be inaccurate.

Library of Congress Card No.: applied for

British Library Cataloguing-in-Publication Data
A catalogue record for this book is available from the British Library.

Bibliographic information published by the Deutsche Nationalbibliothek
The Deutsche Nationalbibliothek lists this publication in the Deutsche Nationalbibliografie; detailed bibliographic data are available on the Internet at http://dnb.d-nb.de.

© 2009 WILEY-VCH Verlag GmbH & Co. KGaA, Weinheim

All rights reserved (including those of translation into other languages). No part of this book may be reproduced in any form – by photoprinting, microfilm, or any other means – nor transmitted or translated into a machine language without written permission from the publishers. Registered names, trademarks, etc. used in this book, even when not specifically marked as such, are not to be considered unprotected by law.

Typesetting Thomson Digital, Noida, India
Printing betz-druck GmbH, Darmstadt
Binding Litges & Dopf GmbH, Heppenheim
Cover Design Adam Design, Weinheim

Printed in the Federal Republic of Germany
Printed on acid-free paper

ISBN: 978-3-527-31864-3

Foreword

In the history of science the development of new technologies, specifically technologies to collect new types of data, have often preceded breakthrough discoveries and induced paradigm changes. A point in case is astronomy. Over the centuries new devices for observing the sky – the first telescopes in the early seventeenth century, the horn reflector antenna of Penzias and Wilson that discovered the Cosmic Background Radiation, the Hubble space telescope with its multitude of discoveries in deep space – provided leaps in understanding and theory and created entirely new fields of science.

What is true for the large objects studied by astronomy is also true for small objects studied by physics, chemistry and biology. Many large advances in biology can be directly traced to the invention of new technologies or devices to collect data on biological samples and biomolecules. These include the first microscopes that revealed cells and later subcellular structures to the human eye, X-ray diffraction and NMR methods that solve the structure of complex biomolecules at atomic resolution, fluorescence imaging methods that detect, quantify and measure the dynamics of biomolecules in the living cell and mass spectrometry that identifies and quantifies the whole repertoire of proteins, lipids and metabolites in cells and tissues.

The ability to detect, identify, quantify and structurally analyze most or all of the biomolecules of a particular type culminated in the completion of the human genome sequence and that of hundreds of other species and is one of the foundations of molecular biology. As a consequence of the success of genome sequencing projects the interests of many biologists are now shifting from the detailed analysis of the structural, functional or catalytic properties of specific molecule towards the immensely challenging questions of how the molecules that constitute the cell are ordered and how they relate and interact with each other to carry out the myriads of complex functions of cells and organisms. These latter questions define the emerging field of systems biology.

A systems biologist equipped with a large budget might be tempted to assemble in his laboratory the most up to date and high-throughput data collection platforms of genomics, proteomics, metabolomics, lipidomics and so on that have been so successful for molecular biology and to start systematically measuring the molecules that define his system of interest. While he would likely collect a large volume of high-quality data, he would also likely fail miserably in his goal to reach a

Single Cell Analysis: Technologies and Applications. Edited by Dario Anselmetti
Copyright © 2009 WILEY-VCH Verlag GmbH & Co. KGaA, Weinheim
ISBN: 978-3-527-31864-3

comprehensive understanding of the system. The reason, of course, would not be incompetence, lack of resources or even the lack of powerful computers to crunch the data. The reason for failure would be that he was making the wrong types of measurements, those carried over from molecular biology research focused on the properties of molecules. Rather he might be more successful if he focused on those measurements that are relevant for systems biology and focused on the context in which those molecules operate, such as physical interactions, enzyme–substrate relationships, gene regulatory networks and their dynamic change.

Therefore, at this point, systems biology has to be considered substantially data-starved and also limited by the technologies required to collect the right types of data reliably, comprehensively and quantitatively. Measurements at the level of the single cell are, at the same time, among the most challenging and informative in systems biology. They are challenging because the specific cell to be measured has to be detected, isolated and in some instances lysed. Furthermore, the amount of each analyte present in or extracted from a single cell is minute, requiring analytical technologies of the highest possible sensitivity. Single cell measurements are informative because every cell analyzed produces a set of data that can be combined with data obtained from other individual cells to compute the distribution and statistical significance of the measured value over a population of cells, as opposed to the average value determined by classic methods in which large numbers of cells are analyzed in each assay. Knowledge of the distribution and statistical significance of values over a population of cells is significant for the classification of cells at an unprecedented resolution, for the detection of cellular variabilities representing different states (e.g. healthy or diseased, resting versus activated, etc.) and for the discrimination between deterministic and stochastic events in cells. It can therefore be expected that the ability to carry out extensive and robust measurements on single cells will catalyze significant conceptual advances in biology and medicine, the extent of which at this time can only be guessed. In addition to the fundamental advances they are expected to create, the nano- and microsystems required for single cell analysis also offer real practical advantages that can even be exploited for conventional assays. They allow the development of highly parallelized systems for high-throughput assays, the realization of substantial saving in analyte material and other consumables and dramatically increased speed of analysis.

The present volume describes exciting new technological developments that address the key issues related to single cell analysis. In a logical series of chapters a panel of leading researchers describe the state of the art of systems nanotechnology and single cell analysis. The chapters represent an interesting mix between topics of basic technology and prototype applications. The book is a highly welcome resource for beginner and advanced researchers alike who want to immerse themselves in the new and exciting field of systems nanotechnology. There is no doubt that the topics treated in this pioneering volume will be of lasting and growing significance for systems biology.

Basel, November 2008   *Ruedi Aebersold*

Contents

Foreword V
Preface XIII
List of Contributors XVII

Part I	**Single Cell Analysis : Imaging**
1	**Single Molecule Fluorescence Monitoring in Eukaryotic Cells: Intranuclear Dynamics of Splicing Factors** 1
	Ulrich Kubitscheck
1.1	Motivation 1
1.2	Experimental Approach 2
1.3	Single Particle Tracking within Living Cells 6
1.4	Pre-Messenger RNA Splicing 7
1.5	Intranuclear Splicing Factor Tracking 8
1.6	Intranuclear U1 snRNP Splicing Factor Binding 10
1.7	Events in Speckles 10
1.8	Intranuclear U1 snRNP Mobility 11
1.9	Perspectives of Single Molecule Microscopy 13
	References 15
2	**Gene Classification and Quantitative Analysis of Gene Regulation in Bacteria using Single Cell Atomic Force Microscopy and Single Molecule Force Spectroscopy** 19
	Robert Ros and Nicole Hansmeier
2.1	Introduction 19
2.2	AFM on Paracrystalline Cell Surface Layers of *C. glutamicum*: Protein Sequence Information and Morphology 20
2.3	Imaging of Living *C. glutamicum* Cells with Molecular Resolution: Genes, Transcriptional Regulation and Morphology 23
2.4	Single Molecule Force Spectroscopy on Specific Protein–DNA Complexes: Transcriptional Regulation in *S. meliloti* 25

Single Cell Analysis: Technologies and Applications. Edited by Dario Anselmetti
Copyright © 2009 WILEY-VCH Verlag GmbH & Co. KGaA, Weinheim
ISBN: 978-3-527-31864-3

2.5	Effector-Induced Protein–DNA Binding on the Single Molecule Level: Quorum Sensing in *S. meliloti* 29	
2.6	Conclusion 32	
	References 33	

3	**Cellular Cryo-Electron Tomography (CET): Towards a Voyage to the Inner Space of Cells** 39	
	Juergen M. Plitzko	
3.1	Introduction 39	
3.2	Tomography with the Electron Microscope – a Practical Perspective 42	
3.2.1	Sample Preparation 42	
3.2.2	Instrumental and Technical Requirements 48	
3.2.3	Alignment, Reconstruction and Visualization 54	
3.3	Molecular Interpretation of Cellular Tomograms 58	
3.4	Outlook: The Future is Bright 61	
	References 65	

Part II	**Single Cell Analysis: Technologies**	

4	**Single Cell Proteomics** 69	
	Norman J. Dovichi, Shen Hu, David Michels, Danqian Mao, and Amy Dambrowitz	
4.1	Introduction 69	
4.2	The Challenge 70	
4.3	Single Cell Proteomics: Mass Spectrometry 71	
4.4	Single Cell Separations 72	
4.5	Ultrasensitive Protein Analysis: Capillary Electrophoresis with Laser-Induced Fluorescence Detection 74	
4.6	Capillary Sieving Electrophoresis of Proteins from a Single Cancer Cell 75	
4.7	Cell Cycle-dependent Single Cell Capillary Sieving Electrophoresis 77	
4.8	Tentative Identification of Proteins in Single Cell Electropherograms 78	
4.9	Capillary Micellar and Submicellar Separation of Proteins from a Single Cell 79	
4.10	Two-Dimensional Capillary Electrophoresis of Proteins in a Single Cell 80	
4.11	Single Copy Detection of Specific Proteins in Single Cells 83	
4.12	Conclusion 85	
	References 87	

5	**Protein Analysis of Single Cells in Microfluidic Format** 91	
	Alexandra Ros and Dominik Greif	
5.1	Introduction 91	

5.2	Microfluidic Single Cell Analysis Concept	93
5.2.1	Single Cell Selection and Trapping	93
5.2.2	Single Cell Lysis	95
5.3	Single Cell Electrophoretic Separation and Detection of Proteins	96
5.3.1	Label-Based Fluorescence Detection	98
5.3.2	Label-Free Fluorescence Detection	99
5.3.2.1	UV-LIF in Quartz Microfluidic Devices	99
5.3.2.2	UV-LIF in PDMS Microfluidic Devices	99
5.3.2.3	Single Cell UV-LIF Electrophoretic Analysis	102
5.4	Future Directions in Single Cell Analysis	103
	References	104

6	**Single Cell Mass Spectrometry**	*109*
	Elena V. Romanova, Stanislav S. Rubakhin, Eric B. Monroe, and Jonathan V. Sweedler	
6.1	Introduction	109
6.2	Considerations for Single Cell Chemical Microanalysis using Mass Spectrometry	110
6.3	Mass Spectrometry as a Discovery Tool for Chemical Analysis of Cells	111
6.4	Single Cell Mass Spectrometric Applications	115
6.5	Subcellular Profiling	119
6.6	Imaging Single Cells with MS	121
6.7	Signaling Molecule Release from Single Cells	124
6.8	Future Developments	126
	References	126

7	**Single Cell Analysis for Quantitative Systems Biology**	*135*
	Luke P. Lee and Dino Di Carlo	
7.1	Introduction	135
7.2	Misleading Bulk Experiments	138
7.3	Common Techniques for High-Throughput and High-Content Single Cell Analysis	140
7.4	Improved Functionality for High-Throughput Single Cell Analysis	141
7.4.1	Microfluidic Techniques	141
7.4.2	Array-Based Techniques	144
7.4.3	High-Content Separation-Based Techniques	147
7.5	Example Studies Enabled by Microfluidic Cell Arrays	148
7.5.1	Pore-Forming Dynamics in Single Cells	148
7.5.1.1	Microfluidic Single Cell Arrays with Fluorescence Imaging	148
7.5.1.2	Toxin-Induced Permeability	148
7.5.1.3	Stochastic Model of Pore Formation	149
7.5.1.4	Amount and Size of Pores for Best Fit Models	151
7.5.1.5	Concerning the Pore Formation Mechanism of SLO	151

7.5.1.6	Conclusions on Pore-Forming Dynamics in Single Cells	*152*
7.5.2	Single Cell Culture and Analysis	*153*
7.5.2.1	Single Cell Trapping Arrays	*154*
7.5.2.2	Arrayed Single Cell Culture	*154*
7.5.2.3	Conclusions on Arrayed Single Cell Culture	*156*
7.6	Conclusions and Future Directions	*157*
	References	*158*
8	**Optical Stretcher for Single Cells**	*161*
	Karla Müller, Anatol Fritsch, Tobias Kiessling, Marc Grosserüschkamp, and Josef A. Käs	
8.1	Introduction	*161*
8.2	Theory, Methods and Experimental Setup	*163*
8.2.1	Fundamentals of Optical Stretching	*164*
8.2.1.1	Ray Optics	*165*
8.2.1.2	Resulting Forces	*167*
8.2.2	Microfluidics – Laminar Flow	*169*
8.3	Applications	*170*
8.3.1	Cancer Diagnostics	*171*
8.3.2	Minimally Invasive Analysis	*172*
8.3.3	Stem Cell Characterization	*173*
8.4	Outlook	*173*
	References	*174*

Part III	**Single Cell Analysis: Applications**	
9	**Single Cell Immunology**	*175*
	Ulrich Walter and Jan Buer	
9.1	Introduction	*175*
9.2	Single Cell Gene Expression Profiling	*175*
9.2.1	Single Cell (Multiplex) RT-PCR	*175*
9.2.2	Quantitative Single Cell Multiplex RT-PCR	*179*
9.3	Fluorescence-Activated Cell Sorting	*180*
9.4	Live Cell Fluorescence Microscopy	*183*
9.4.1	Confocal Laser Scanning Microscopy	*183*
9.4.2	Total Internal Reflection Fluorescence Microscopy	*185*
9.4.3	Förster Resonance Energy Transfer Imaging	*186*
9.4.4	Two-Photon Laser Scanning Microscopy	*186*
9.5	Other Techniques for Single Cell Analysis	*187*
9.5.1	Enzyme-Linked Immunospot Assay	*187*
9.5.2	In Situ Hybridization	*188*
9.5.3	Electron Microscopy	*188*
9.6	Conclusions and Outlook	*189*
	References	*189*

10	**Molecular Characterization of Rare Single Tumor Cells** *197*	
	James F. Leary	
10.1	Introduction *197*	
10.1.1	Importance of Rare Cells *197*	
10.1.2	Detection of Rare Tumor Cells *198*	
10.2	Finding Rare Event Tumor Cells in Multidimensional Data *199*	
10.2.1	Rare Event Sampling Statistics *200*	
10.2.2	High-Speed Sorting of Rare Cells *203*	
10.2.3	Sorting Speeds must be Fast Enough to be Practical *203*	
10.2.4	Limits in Sorting Speeds and Purities *204*	
10.3	Classification of Rare Tumor Cells *205*	
10.3.1	Using Classifiers to Sort Rare Tumor Cells *208*	
10.4	Molecular Characterization of Sorted Tumor Cell Cells *209*	
10.4.1	Model Cell Systems *209*	
10.4.2	Design of PCR Primers to Detect the PTEN Gene Region *209*	
10.4.3	Processing BT-549 Human Breast Cancer Cells *210*	
10.5	Detection of Mutated Sequences in Tumor Suppressor Genes *211*	
10.5.1	Detection of Mutations in Breast Cancer Tumor Suppressor Genes by High-Throughput Flow Cytometry, Single Cell Sorting and Single Cell Sequencing *211*	
10.5.2	Single Cell Sorting for Mutational Analysis by PCR *214*	
10.5.3	TA Cloning *215*	
10.5.4	Single Cell Analysis of Gene Expression Profiles *216*	
10.6	Conclusions and Discussion *219*	
	References *219*	
11	**Single Cell Heterogeneity** *223*	
	Edgar A. Arriaga	
11.1	Introduction *223*	
11.2	Measuring Heterogeneity using Single Cell Techniques *224*	
11.2.1	Optical Well Arrays *225*	
11.2.2	Capillary Electrophoresis Analysis of Organelles Released from Single Cells *225*	
11.3	Describing Cellular Heterogeneities and Subpopulations *227*	
11.4	Origins of Cellular Heterogeneity *228*	
11.5	Identifying Extrinsic and Intrinsic Noise Sources *230*	
11.5.1	Validation of the Flow Cytometry Measurements *230*	
11.5.2	Noise Dissection *232*	
11.5.3	Identification of Deviant Gene Products *232*	
11.5.4	Correlation of Gene Products with Potential Sources of Noise *232*	
11.6	Concluding Remarks *233*	
	References *234*	

12		**Genome and Transcriptome Analysis of Single Tumor Cells** *235*
		Bernhard Polzer, Claudia H. Hartmann, and Christoph A. Klein
	12.1	Introduction *235*
	12.2	Detection and Malignant Origin of Disseminated Cancer Cells *235*
	12.3	Methods for Amplifying Genomic DNA of Single Cells *237*
	12.4	Studying the Genome of Single Disseminated Cancer Cells *239*
	12.5	The Need for Higher Resolution: Array CGH of Single Cells *240*
	12.6	Studying the Gene Expression of Single Disseminated Cancer Cells *241*
	12.7	Combined Genome and Transcriptome Analysis of Single Disseminated Cancer Cells *246*
		References *246*

Index *251*

Preface

Single cell analysis (SCA) and systems nanobiology (SNB) are nowadays puzzling keywords in a scientific community that is more and more interested in tackling complex and transdisciplinary problems in biology – the science of life. However, why is the analysis of single cells so fascinating and important that biologist, chemists, physicists, engineers, biomathematicians and bioinformaticians contribute from different perspectives to achieve these goals? Is it the fascination of the smallest organizational system that – by definition – represents life and its related consciousness that such a small systems maintains a highly complex and hierarchical architecture of interconnected molecular networks? Is it the hope that understanding such a complex system will help us to understand the functioning of complete organs and will give us new approaches to cure diseases and to develop personalized medicine within the framework of systems biology? Or is it a means to identify the relevant mechanisms of complex processes in cells which have developed over millions of years with a limited set of molecules and which are responsible for their self-maintaining, efficient, robust and evolutionary properties? The answer to all these questions is yes. But what is the difference from the established biotechnological procedures and approaches that were based on investigating cellular ensembles and which were developed over recent decades, proving their validity and effectiveness in bringing breathtaking insights of cells and bringing their molecular blueprint to light?

This book gives you a number of possible answers to these questions. Of course it is only a snapshot of a rapidly growing field. Here, scientists of many origins share with you their results and view of a field which is at present in its infancy but is ready to develop into an equivalent and complementary partner to the existing approaches. A field that strives to extract quantitative information on cellular properties and information at different cell organizational levels (genome, proteome, metabolome, etc.) which are not ensemble-averaged. This is certainly the most obvious paradigm shift when you rationalize that, to date, proteome analyses have been based on 10^3–10^6 cells and rely on the fact that weak (but important) cellular signals and responses can be hidden within an unspecific cellular background of cells that do not behave in an identical way – they are not identical and are characterized by a heterogeneous

Single Cell Analysis: Technologies and Applications. Edited by Dario Anselmetti
Copyright © 2009 WILEY-VCH Verlag GmbH & Co. KGaA, Weinheim
ISBN: 978-3-527-31864-3

response. Moreover, novel single cell techniques will hopefully allow the investigation of individual cell cycle dependent effects, will give access to cellular variabilities during cell proliferation, will give access to cellular subpopulations and differentiation states and will allow an insight into the different and inhomogeneous cellular responses to external stimuli.

It will be highly interesting to see how parameters like robustness of a cell against external stimuli is a global property of the cellular ensemble and how it relates to the property of an individual cell.

A similar shift in paradigm happened in biophysics when, over the past 15 years, single molecule biophysics opened a new field of activity that allowed investigation of biomolecular processes and identification of physical mechanisms, with complementary approaches in mechanics (forces), optics (photonics), electrodynamics (charge transport) and thermodynamics (energies). Their novel insights into specific interactions, structure–function related binding mechanisms, molecular dynamics and the description within novel statistical mechanics models allowed the extraction of information on metastable transition states, molecular subpopulations and thermodynamically driven heterogeneous variabilities that are beyond the statistically averaged information provided by molecular ensembles in their initial and final state.

Single cell analysis will profit from such expertise in the sense that, beyond pragmatic benefits like saving material, time and resources (no cell cultivation and amplification needed), the novel highly parallelized and microchip-based single cell analysis approaches will allow new screening concepts and applications where only small amounts of cells are available (e.g. stem cell research).

In that sense, this book aims to cover some of the very prominent fields of activities that contribute to the area of single cell and subcellular analysis. General and more specialized readers will find a mixture of very recent results that are embedded in review and trend articles, where renowned researchers give their view of the state of the art of emerging and innovative analytical technologies as well as their motivations and visions of this lively area.

This book is structured in three main areas:

Part I   Single Cell Analysis: Imaging
Part II   Single Cell Analysis: Technologies
Part III   Single Cell Analysis: Applications

The three parts are preceded by an introductory feature article from Ruedi Aebersold, where the general topic of single cell analysis is put into the framework of molecular systems biology.

In Part One, where the book's focus is on SCA imaging, exemplarily the three major microscopy techniques – optical, electron, scanning probe microscopy – are highlighted with their emphasis to analyze individual cells. Namely, Ulrich Kubitschek reports on the investigation of intracellular protein dynamics with fluorescence microscopy at the single molecule level, Robert Ros and Nicole Hansmeier describe their microscopy efforts to map protein structures on bacteria and how single molecule force spectroscopy can nowadays be used to quantitatively affinity rank DNA–protein interactions from bacterial transcription regulation networks and Jürgen Plitzko gives an overview of the fascinating capabilities of 3D single cell electron tomography.

In Part Two, novel single cell analytical technologies are illustrated. Norman Dovichi and coworkers present their developments in single cell capillary electrophoresis (CE) and applications in cell development, oncology and neuroscience. Alexandra Ros and Dominik Greif demonstrate their achievements towards a label-free protein fingerprint of a single cell in a microfluidic chip format and Jonathan Sweedler and coworkers review current mass spectrometric approaches for examining single cells and highlight recent progress. Microfluidic cell isolation array structures for quantitative analysis of toxin pore formation and cell cycle dynamics in single cells are investigated by Luke Lee and Dino di Carlo, while Josef Käs and coworkers describe their optical stretcher technology that allows cancer diagnosis and stem cell characterization at the single cell level.

In Part Three, more application-driven examples are highlighted. Ulrich Walter and Jan Buer review how single cell analysis with various technologies contributed to a modern understanding of the immune system. Rare single tumor cells can nowadays be characterized by a variety of staining, detection, sorting and data analysis techniques. The chapter by James F. Leary gives an impressive overview about it and its consequences. Cellular variabilities and single cell heterogeneity is the topic of the contribution of Edgar A. Arriaga, where he investigates the diverse variations of isogenic cells, for example insulin-producing pancreas cells. Last but not least, Christoph Klein and coworkers review how single tumor cells can be analyzed at a genomic and transcriptomic level.

A number of different aspects of single cell analysis are gathered and described in this book. They are not complete, since this young and promising area of research has just left its infancy and started a very dynamic development. Although quite a number of possible applications and merits already can be identified, it will be interesting to see how the field develops when it gains momentum. Based on the pioneering and exciting work presented in this book (and others too, see e.g. the special issue on single cell analysis in *Analytical and Bioanalytical Chemistry* 387 (1), 2007) a more complete understanding of complex cellular processes will not only bridge different scientific disciplines in a transdisciplinary way from an analytical point of view, it will furthermore connect different perspectives from a reductionistic molecular view to a more holistic organism-oriented viewpoint. I am convinced that we can expect a wealth of new insights into the hidden mysteries of cells and their complex mechanisms in the future.

Let's get excited and inspired!

Bielefeld, November 2008 *Dario Anselmetti*

List of Contributors

Edgar A. Arriaga
University of Minnesota
Department of Chemistry
Minneapolis, MN 55455-0431
USA

Jan Buer
University Hospital of Essen
Institute of Medical Microbiology
Hufelandstrasse 55
45122 Essen
Germany

Dino Di Carlo
University of California
Biomolecular Nanotechnology Center
Berkeley Sensor & Actuator Center
Department of Bioengineering
Berkeley, CA 94720-1774
USA

Amy Dambrowitz
University of Washington
Department of Chemistry
Seattle, WA 98195-1700
USA

Norman J. Dovichi
University of Washington
Department of Chemistry
Seattle, WA 98195-1700
USA

Anatol Fritsch
University of Leipzig
PWM-Soft Matter Physics
Linnéstrasse 5
04103 Leipzig
Germany

Dominik Greif
Bielefeld University
Physics Faculty
Experimental Biophysics and Applied Nanoscience
33615 Bielefeld
Germany

Marc Grosserüschkamp
University of Leipzig
PWM-Soft Matter Physics
Linnéstrasse 5
04103 Leipzig
Germany

Nicole Hansmeier
Arizona State University
Department of Physics
Tempe, AZ 85287-1504
USA
and
Bielefeld University
Physics Faculty
Bielefeld
Germany

Single Cell Analysis: Technologies and Applications. Edited by Dario Anselmetti
Copyright © 2009 WILEY-VCH Verlag GmbH & Co. KGaA, Weinheim
ISBN: 978-3-527-31864-3

Claudia H. Hartmann
University of Regensburg
Department of Pathology
Division of Oncogenomics
Franz-Josef-Strauss-Allee 11
93053 Regensburg
Germany

Shen Hu
University of Washington
Department of Chemistry
Seattle, WA 98195-1700
USA

Josef A. Käs
University of Leipzig
PWM-Soft Matter Physics
Linnéstrasse 5
04103 Leipzig
Germany

Tobias Kiessling
University of Leipzig
PWM-Soft Matter Physics
Linnéstrasse 5
04103 Leipzig
Germany

Christoph A. Klein
University of Regensburg
Department of Pathology
Division of Oncogenomics
Franz-Josef-Strauss-Allee 11
93053 Regensburg
Germany

Ulrich Kubitscheck
Rheinische Friedrich-Wilhelms-
Universität Bonn
Institute for Physical and Theoretical
Chemistry
Department of Biophysical Chemistry
Wegelerstrasse 12
53115 Bonn
Germany

James F. Leary
Purdue University
Department of Basic Medical Sciences
and Weldon School of Biomedical
Engineering
Bindley Biosciences Center
Birck Nanotechnology Center
Oncological Sciences Center
West Lafayette, IN 47907-2057
USA

Luke P. Lee
University of California
Biomolecular Nanotechnology Center
Berkeley Sensor & Actuator Center
Department of Bioengineering
Berkeley, CA 94720-1774
USA

Danqian Mao
University of Washington
Department of Chemistry
Seattle, WA 98195-1700
USA

David Michels
University of California
Biomolecular Nanotechnology Center
Berkeley Sensor & Actuator Center
Department of Bioengineering
Berkeley, CA 94720-1774
USA

Eric B. Monroe
University of Illinois
Department of Chemistry
Urbana, IL 61801
USA

Karla Müller
University of Leipzig
PWM-Soft Matter Physics
Linnéstrasse 5
04103 Leipzig
Germany

Juergen M. Plitzko
Max-Planck-Institute of Biochemistry
Department of Molecular Structural Biology
Am Klopferspitz 18
82152 Martinsried
Germany

Bernhard Polzer
University of Regensburg
Department of Pathology
Division of Oncogenomics
Franz-Josef-Strauss-Allee 11
93053 Regensburg
Germany

Robert Ros
Arizona State University
Department of Physics
Tempe, AZ 85287-1504
USA

Alexandra Ros
Arizona State University
Department of Physics
Tempe, AZ 85287-1504
USA

Elena V. Romanova
University of Illinois
Department of Chemistry
Urbana, IL 61801
USA

Stanislav S. Rubakhin
University of Illinois
Department of Chemistry
Urbana, IL 61801
USA

Jonathan V. Sweedler
University of Illinois
Department of Chemistry
Urbana, IL 61801
USA

Ulrich Walter
University of Calgary
Faculty of Medicine
3330 Hospital Drive N.W.
Calgary
Alberta T2N 4N1
Canada

Part I
Single Cell Analysis: Imaging

1
Single Molecule Fluorescence Monitoring in Eukaryotic Cells: Intranuclear Dynamics of Splicing Factors
Ulrich Kubitscheck

1.1
Motivation

Many problems in single cell analysis involve very basic questions: where do proteins or larger molecular assemblies go? How do they move: by Brownian motion, freely or in some restricted manner, or by active transport along one of the cell's filament systems? Where are they captured, where do they bind and for how long? If binding, what are the interaction partners? Which factors are present at locations in large organelles, which are transiently bound, what is the sequence of molecular interactions? Especially the latter question is of great significance considering the functional role of large molecular complexes, which perform tasks like signal transduction, energy production, information processing, or protein formation and degradation. To approach these questions means developing an "intracellular biophysical chemistry." Currently there are only a small number of techniques available to approach these questions of intracellular protein dynamics. Light microscopy and in particular fluorescence microscopy is certainly one of the methods of choice; and it has reached a high level of maturation and sophistication. Especially the past 20 years have seen a storm of developments in quantitative fluorescence microscopic techniques, which was triggered by the perfection of microscope optics and light detectors, the widespread use of continuous wave and pulsed lasers as excitation sources, the introduction of elegant optical concepts, the availability of massive computing power to resolve complex image processing tasks and, last but not least, the introduction of genetically engineered autofluorescent protein conjugates. In the past few years tremendous progress has been made with regard to bringing optical microscope resolution almost to the ultimate level of molecular sizes with the introduction of stimulated emission depletion microscopy [1] and nonlinear structured illumination microscopy [2, 3]. But high-resolution methods are often not applicable or optimally suited to examine dynamical processes. However for such problems fluorescence techniques appear to be almost ideal. Probably the most well known is fluorescence recovery after photobleaching, abbreviated FRAP [4]. A further

Single Cell Analysis: Technologies and Applications. Edited by Dario Anselmetti
Copyright © 2009 WILEY-VCH Verlag GmbH & Co. KGaA, Weinheim
ISBN: 978-3-527-31864-3

technique, nowadays almost classic but nevertheless still rapidly expanding, is fluorescence correlation spectroscopy, FCS [5, 6]. A most recent and extremely powerful technique is single molecule tracking within cells [7]. Remarkably, monitoring single fluorescent molecules with a sufficiently high time resolution can provide real-time molecular views on biochemical processes within cells even *in vivo*. It is extremely fascinating and instructive to directly observe the motions and interactions of single protein molecules, ribonucleoprotein particles or oligonucleotides by state of the art light microscopy.

1.2
Experimental Approach

Fluorescence microscopic visualization of a single molecule in real time is relatively easy to achieve if some rules are observed. Methodological prerequisites for intracellular single molecule monitoring are the reduction of background fluorescence and the use of very low concentrations of the probe of interest, which should be in the picomolar range. Technical prerequisites are the use of laser light sources to yield the required irradiance of about $0.1\,\text{kW/cm}^2$, the optimization of light transmission in the detection pathway of the microscope and finally the use of fast camera systems of utmost sensitivity for signal detection. The identification and tracking of single molecule signals in usually noisy images is finally performed using sophisticated digital image processing. Figure 1.1 shows a schematic representation of the optical

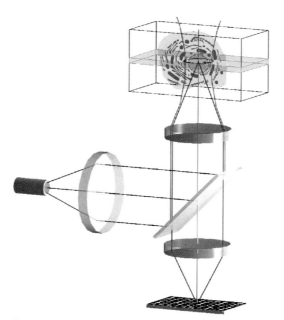

Figure 1.1 Principle optical setup of a single molecule microscope (for explanation see text).

beam path in a single molecule microscope. The principal idea is to use as few components as possible in order to optimize the light detection efficiency.

The illumination light is coupled via an optical fiber port into the epi-illumination light path of the microscope. In this manner the collimated light illuminates the image plane, which is conjugated to the object plane and located at the position of the field stop. The beam diameter at this plane divided by the magnification of the optical system determines the final extension of the illumination field with a Gaussian intensity profile. The incoming beam is reflected towards the sample by a dichromatic beam splitter. We use a fiber output diameter of 1 mm ($1/e^2$ diameter) and usually a $63 \times$ objective lens. This produces a Gaussian illumination pattern with a diameter of approximately 16 µm within the object plane. Currently, we study preferentially dynamic processes within cell nuclei; and therefore the illumination field size is adjusted approximately to the diameter of a single nucleus. The focal depth, that is the axial extension of the object area in focus, is less than 1 µm for high end objective lenses with a numerical aperture (NA) ≥ 1.3. Since the light collection efficiency of the objective is proportional to the square of the NA, the lens with the highest NA available should be chosen. When the laser is focused onto the back focal plane of the objective lens, a homogeneous illumination of the complete object field can be achieved. The fluorescence image is captured by the objective lens, transmitted by the dichromatic beam splitter and finally mapped by the tube lens onto the charged coupled device (CCD) of the camera. In order to achieve the highest frame rates and sensitivity available, a back-illuminated electron multiplying (EM) camera chip with 128×128 pixels is a suitable choice as imaging device. After careful optimization of all components in the detection path of the microscope, an overall detection efficiency in the range of 10% is achievable. Figure 1.2 shows more technical detail of the beam path within the microscope. We use Notch filters in the emission beam path instead of the commonly employed band pass filters.

These reject any residual excitation light with an optical density (OD) of 6 OD units which might pass the dichromatic mirror. Therefore, no Raleigh scattered light reaches the detector. The Notch filters are positioned within the infinity beam path in front of the tube lens. A set of three different Notch filters centered at 488, 532 and 633 nm is available and, depending on the employed laser excitation, the respective filter is inserted. The tube lens focuses the image but it is supplemented by a $4 \times$ magnifying lens system. This is required because our camera (iXon DV-860-BI; Andor Technologies, Belfast, Northern Ireland) has a pixel size of 24 µm. Used without a magnifier this would correspond to a pixel size in the object plane of 380 nm, which would not be sufficient for satisfactory detail when imaging diffraction-limited objects. Microscopic imaging of single fluorescent molecules produces diffraction-limited light spots in the image plane. The corresponding intensity pattern can be approximated by a two-dimensional Gaussian function with a full width at half maximum of approximately 250 nm, when light with a wavelength of 500 nm and an objective lens of NA 1.4 are used for imaging. The precise position of the molecule can therefore be obtained by a fitting process which yields the center of the light spot with a very high precision, but only if the single molecule signal is imaged with a good signal to noise ratio and sufficient geometric detail. The required

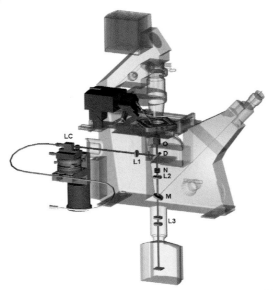

Figure 1.2 Details of the single molecule microscope. LC, fiber output; L1, tube lens in the illumination light path; L1, tube lens in the emission light path; L3, 4× magnifying lens system; O, objective lens; D, dichromatic beam splitter; N, Notch filters.

detail is obtained when a 4× magnifier is used, which reduces the object plane pixel size to well below the diffraction limit, namely to 95 nm. Then a diffraction-limited single molecule signal is spread over about 5 × 5 pixels, which is ideal for nanolocalization by the fitting process. The localization precision depends only on the signal to noise ratio (SNR) and the stability of the optical setup; values of 2–40 nm can be achieved [8–10].

Figure 1.3 gives a full schematic view of the complete setup, showing the details of the excitation laser setup. Our instrument allows the sequential use of up to three different laser lines for fluorescence excitation. The different laser beams are combined by two dichromatic beam splitters, then coupled into an acousto-optical tunable filter and focused into an optical mono-mode fiber. This delivers the light to the microscope. As outlined above, the triple dichromatic beam splitter reflects the three excitation lines onto the sample, while the three fluorescence bands are transmitted to the detection system via the Notch filters. This slightly unusual filter construction has the following purpose: by this means three fluorescence bands can be detected independently by a single EM CCD camera by simply alternatively triggering the AOTF driver governing the laser illumination. The camera generates the trigger signals whenever an image is acquired. Using a single camera for the imaging of the different fluorescence channels instead of using several cameras separated by additional emission beam path dichromatic beam splitters has the great advantage that alignment of the fluorescence images is not required, since registration is realized

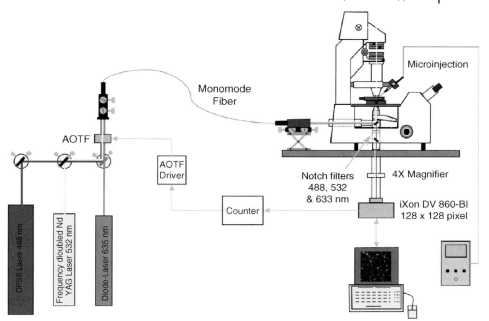

Figure 1.3 Sketch of the single molecule microscope including the excitation lasers. Three excitation laser lines were joined by two dichromatic beam splitters and passed through an acousto-optical tunable filter (AOTF), which allowed regulating the power for each laser separately within a few microseconds. Excitation light was directed to the inverted microscope by an optical mono-mode fiber, and coupled in at the epi-illumination port. The emitted fluorescence light was separated from the excitation by a triple dichromatic beam splitter, which reflected the three excitation lines onto the probe, while the three fluorescence bands were transmitted to the detection light pass by optionally insertable Notch filters with an absorbance of OD 6 at the laser lines. No further filters were used in the emission beam path. Hence, up to three fluorescence bands could be detected sequentially by a single electron-multiplying CCD camera by alternative triggering the AOTF driver via a programmable counter, which reacted to the camera-generated trigger signals. Fluorescent probe molecules could be microinjected into living cells.

in the very same beam path. Just minor lateral and axial chromatic aberration of the detection optics need be taken into account.

The most straightforward way to deliver a fluorescent probe molecule into living cells is by using a microinjection system. Intracellular fluorescent probe concentrations should be adjusted to <100 pM, since then only a few fluorescent molecule are present within the field of view. For extended observations, the molecules must be and move within the focal region of the objective lens. Signals from molecules leaving the focal plane rapidly vanish into the background noise, which means that the imaging of single point objects exhibits a three-dimensional resolution.

The observation of single molecules with high imaging frame rates allows the tracing of single molecule trajectories, if their movement during image acquisition is negligible. It has been shown that frame rates in the range of 300–400 Hz are sufficient to follow the pathways of protein molecules with molecular masses above 50 kDa in solution [11]. In that study it was shown for two model proteins – streptavidin and an IgG – that the quantification of diffusion based on tracing single protein molecules in aqueous buffer yielded the same results as the respective measurements using fluorescence correlation spectroscopy (FCS). Also, it was demonstrated that the high-speed video microscope employed could easily follow single fluorescent quantum dots. Hence, for molecules in the respective molecular mass and brightness range, single molecule tracking may be used as a complementary method to FCS for analyzing protein mobility. Tracer molecule mobility within biological cells is four- to 10-fold reduced, compared to mobility in aqueous solution [4]. Thus, Grünwald et al. showed that real-time visualization and tracking of protein molecules inside living cells is feasible using a state of the art imaging system.

1.3
Single Particle Tracking within Living Cells

Several recent studies demonstrated the power of single molecule tracing within living cells [12–16]. Of course, the technique is not restricted to imaging single molecules. Actually, it is even more straightforward to image other subdiffraction sized objects, like quantum dots, single viruses or single gene carriers, because they are usually moving more slowly and can often be labeled with a higher fluorophore to particle ratio [17–21]. The first successful observations of single proteins inside cells were performed using large autofluorescent proteins as probes for intracellular mobility [22]. The effective intracellular viscosity experienced by these probes was found to be about 15-fold higher than that of an aqueous solution, but with a much broader distribution of diffusion constants. The large range of protein mobility is obviously due to interactions with intracellular components or macromolecules and organelles forming barriers and obstacles.

When considering single molecules the tracking of molecules and particles is especially feasible within the cell nucleus, since the autofluorescence of the cell nucleus is significantly lower than that of the cytoplasm [13]. Single molecule observation within cell nuclei allows the analysis of nucleocytoplasmic transport and the characterization of intranuclear transport pathways, mobility restrictions and intranuclear binding processes. The past 10 years revealed that the nuclear architecture is much more complex than previously thought. The new view is that the cell nucleus is a highly organized organelle in which chromosomes occupy distinct territories and processes like DNA transcription and replication as well as mRNA splicing are carried out by supramolecular complexes which are organized in complex spatio-temporal patterns [23]. Even inert protein probes, which do not specifically interact with nuclear structures, do not perform free Brownian motion but rather exhibit various mobility modi (Grünwald, Martin, Buschmann, Leonhardt, Kubitscheck and Cardoso, unpublished data). Diffusion of such molecules is generally characterized by anomalous behavior, that is the effective mean square displacements do not depend linearly on time as for normal Brownian motion. The analysis of single molecule trajectories confirms the existence of very large restrictions on the mobility of large proteins within the cell nucleus, which is indicative of structural barriers and frequent binding–unbinding events to immobile or slowly moving supramolecular structures.

1.4
Pre-Messenger RNA Splicing

Molecules interacting specifically with further intranuclear components such as DNA, RNA or protein complexes exhibit a very different mobility in comparison to inert ones. In this case a detailed mobility analysis allows insight into the dynamics of the corresponding intranuclear processes.

Eukaryotic RNA transcripts go through several post-transcriptional modifications before their final transport into the cytoplasm [24]. Usually such pre-messenger RNAs have noncoding sequences – so-called introns – that must be removed to yield functional mRNA. This biochemical processing is designated as "splicing." It is a complex intranuclear process accomplished by pre-assembled complexes, the "spliceosomes," which comprise more than 70 different proteins. Many of these are contained in the uridine-rich small nuclear ribonucleoprotein particles (U snRNPs), which are classified as U1, U2, U5 and U4/U6 according to their small nuclear RNA (snRNA) content. With the exception of U6, the snRNAs are synthesized in the nucleus by RNA polymerase II and exported to the cytoplasm, where sets of common and specific proteins bind to the snRNAs [25]. After their assembly the U snRNPs are imported into the nucleus. Spliceosomes have been shown to be subcomplexes of huge multicomponent nuclear RNP complexes, so-called supraspliceosomes with geometric extensions of $50 \times 50 \times 35\,\text{nm}^3$ [26]. The group of proteins and RNPs involved in splicing is collectively designated as splicing factors. The intranuclear distribution of splicing factors is a prominent example of the high degree of spatio-temporal organization of the nuclear contents. Immuno-staining of cells with antibodies against splicing factors such as ASF/SF2 or U snRNPs produces images in which nuclei contain numerous tiny bright spots, the so-called "nuclear speckles", dispersed on a homogeneous background. These speckles are created by the enrichment of splicing factors in interchromatin granule clusters and perichromatin fibrils, which are designated as splicing factor compartments [27]. The mechanism of their formation and their function is still unresolved. They may represent sites at which splicing factors are reprocessed, stored or assembled together with other components of the transcription and RNA processing machinery into spliceosomes. It has been speculated that they arise by interaction with a putative karyoskeleton, or alternatively by self-assembly. The diffuse nucleoplasmic staining is probably due to unbound splicing factors and those actively involved in splicing. Of course, it is most interesting to study the dynamical behavior of splicing factors within and outside speckles in order to get insights into the different processes taking place in these intranuclear domains.

1.5
Intranuclear Splicing Factor Tracking

In a recent study we analyzed the mobility of native U1 snRNPs within live cells at the single particle level in order to yield insight into the dynamics of mRNA splicing [16]. U1 snRNPs were isolated from HeLa cell nucleus lysate by ultracentrifugation and then covalently labeled with the red fluorescent dye Cy5. The use of native RNPs excluded possible problems of particles that were assembled *in vitro* and might be incomplete or functionally imperfect. The Cy5-labeled U1 snRNPs were microinjected into the cytoplasm of living HeLa cells and were subsequently imported into the cell nuclei. Thereby the biochemical functions, integrity and structure of the nuclei remained as unaffected as possible. The concentration of the injection solution chosen was so low that single U1 snRNPs could be visualized and tracked within the cell nuclei with a spatial precision of roughly 30 nm by the single molecule microscope described above. In this manner movies from single U1 snRNPs with different time resolution ranging from 200 Hz to 5 Hz were recorded. The HeLa cells employed were transiently transfected with ASF/SF2-GFP in order to allow the study of the RNP dynamics within and outside nuclear speckles. Images of these reference structures were taken using GFP fluorescence excitation at 488 nm, while the single RNP trajectories were obtained switching laser excitation to 632 nm for excitation of Cy5. Automatic image analysis procedures on the basis of Diatrack 3.01 (SemaSopht, North Epping, Australia) were employed to identify single RNP signals and to reconstruct single particle trajectories (Figure 1.4).

These could be analyzed depending on their assignment to speckled or nucleoplasmic intranuclear domains. The greater fraction of the detected U1 snRNPs was immobile: roughly 80% of the U1 snRNPs identified in two subsequent frames did not move beyond the localization precision and therefore were classified as immobile. Such immobilization may be caused by binding to a very large and bulky structure with no or extremely low mobility. Evidently, being coupled to large intranuclear complexes is the standard state of U1 snRNPs.

This observation was in strong contrast to the results of a related single molecule tracking study, in which an inert probe molecule – streptavidin-cy5 – was tracked inside cell nuclei (D. Grünwald, R. Martin, V. Buschmann, H. Leonhardt, U. Kubitscheck and M.C. Cardoso, unpublished data). In that case, a significantly smaller fraction of molecules, namely about 20%, was immobile between successive frames. Furthermore, the duration of binding was significantly shorter in these experiments.

Presumably the observed immobilization of the U1 snRNPs was often caused by the involvement of the particles in ongoing splicing events. This assumption was suggested by comparing the results of a previous study on intranuclear U1 snRNP mobility in our laboratory, which was performed not on living but rather on digitonin-permeabilized cells [28]. Digitonin-permeabilized cells are mostly physiologically inactive and splicing has ceased or is to a very large extent reduced. Notably, in those cells a significantly smaller fraction of RNPs was found immobile, namely only about 20%, as in the case of the unfunctional tracer molecule streptavidin. It is well known

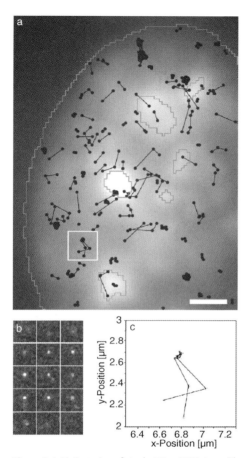

Figure 1.4 Trajectories of single U1 snRNPs in a living cell nucleus. (a) Trajectories of numerous single U1snRNPs extracted from a single molecule movie, which was recorded at 200 Hz. In the movie, mobile U1 snRNPs as well as transiently immobilized RNPs could well be distinguished within nucleoplasm and speckles. The black dots connected by lines represent the positions and molecular jumps as identified by Diatrack 3.0. The trajectories were plotted onto a gray-scale image of the splicing factor ASF/SF2-GFP, which was transiently expressed in the respective HeLa cell nucleus. Hence, the bright regions indicate nuclear speckles (dark gray lines). The white box marks a single trajectory which is detailed in (b) and (c). The white line indicates the estimated position of the nuclear envelope (bar: 2 μm). (b) The short image sequence of the trajectory marked in (a). The series demonstrates the excellent signal to noise ratio of the labeled U1 snRNPs. The sequence starts with the image in the upper left hand corner, and goes linewise to the lower right hand corner. Single frame integration time was 10 ms; single image size 1 μm^2. (c) Magnified view of the positions at which the U1 snRNP was observed, revealing that the RNP moved to a specific site, was bound there for 80 ms, and then left the site again; complete field size, 1 μm^2. The figure is modified from Ref. [16] and is shown with permission.

that pre-mRNA splicing, at least in part, is already occurring during transcription. This means that very large DNA–pre-mRNA–protein complexes are formed at the transcription sites, which would certainly represent large supramolecular complexes causing a gross immobilization of involved interaction partners – such as the U1 snRNPs.

1.6
Intranuclear U1 snRNP Splicing Factor Binding

The recorded movies also allowed a study of dissociation from the putative binding sites – simply by counting the number of frames after which the respective particles vanished from their fixed positions. Of course, the disappearance could in principle also be due to photobleaching of the Cy5-labeled RNPs, but it was found that the average photostability of the particles was not a limiting factor.

In the case of a simple bimolecular dissociation reaction one would expect a monoexponential decay of the number of complexes. In contrast to that, dissociation from the binding sites showed unexpected kinetics. Using movies taken with high frame rates to analyze the binding revealed short binding durations, while movies with low frame rate indicated long binding durations. In this manner dissociation times between 5 ms and 1400 ms were determined, ranging over three orders of magnitude. We interpret this result in the following manner. The complex dissociation kinetics most probably reflect the different ways in which U1 snRNPs interacted with other nuclear structures. We assume that we observed processes ranging from nonspecific interactions or "trapping" within a chromatin network – as in the case of nonfunctional streptavidin – and genuine splicing events to the assembly of spliceosomes before splicing and possibly the re-assembly of splicing factors. Furthermore, numerous quite diverse types of splicing reactions occur simultaneously within a cell nucleus, for example involving a few or numerous introns either in a single pre-mRNA reaction or in alternative splicing reactions. It is plausible that all of these display a different kinetics with regard to the duration of the involvement of different splicing factors. Therefore the observation of many different types of kinetics is probably not incidental but displays a fundamental property of intranuclear reaction kinetics. It is tempting to speculate that chemical kinetics occurring on many time scales is typical for the complex processes occurring in cell nuclei.

1.7
Events in Speckles

Speckles can *in vivo* be defined using cells expressing ASF/SF2-GFP conjugates. Then, distinct spots of strong GFP fluorescence mark the position of speckles, which appear quite contrasted in a confocal microscope. Acquisition of images upon green excitation in nonconfocal video images also allowed discrimination between speckles and nucleoplasmic spaces and created a basis for a separate analysis of U1 snRNP dynamics in speckles and in the nucleoplasm. Surprisingly, the dissociation kinetics did not differ significantly for binding within or outside speckles. This indicated that the increased splicing factor concentration in speckles is not due to a stronger binding of U1 snRNPs. Rather, two explanations for the higher concentrations of splicing factors within speckles remain: (i) the on rate of the binding process is increased, for example by a facilitated accessibility of the speckle binding sites, (ii) the concentration of binding sites within the speckles is higher than in the remaining nucleoplasmic space, while the interaction is of similar nature. The data presently available do not allow discrimination between the two options. Above, we presented arguments that at least part of the observed binding was related to splicing during ongoing transcription. Therefore it can be assumed that mRNA transcription and splicing also takes place in speckles, because the processes causing U1 snRNP immobilization are identical inside and outside the speckles. In summary, the reason for the complex immobilization pattern cannot yet be explained. Nevertheless, it is clear that the study of complex physiological processes such as transcription and splicing require further live cell measurements.

1.8
Intranuclear U1 snRNP Mobility

Obviously, single particle tracking is ideal for recording molecular trajectories (Saxton and Jacobson, 1997). Above, we discussed that a frame rate of 350 Hz is sufficient to track single protein molecules in buffer solution. Therefore, a repetition rate of 200 Hz should be high enough to record real-time U1 snRNP trajectories in the cellular interior. There exist several approaches to analyze single molecule tracks.

The trajectory for each U1 snRNP was determined by Diatrack as a set of coordinates $\{x_i, y_i\}$, where $1 \leq i \leq N$ with N denoting the number of observations. From each trajectory a total of $(N-1) \cdot N/2$ square displacements, $r^2(t_{lag})$, was determined between two positions separated by a time lag, t_{lag}, with $t_{lag} = n\,(t_{ill} + t_{delay})$. t_{ill} denotes the single frame integration time, t_{delay} the delay time between two successive frames and n the difference between the frame numbers. By averaging the square displacements for equal to t_{lag} the mean square displacements (MSD), $\langle r^2(t_{lag}) \rangle$ can be calculated. In the case of two-dimensional Brownian motion the diffusion coefficient D is related to the MSD by:

$$\langle r^2(t_{lag}) \rangle = 4D\,t_{lag} \tag{1.1}$$

Thus, a linear relationship between MSD and lag time indicates unrestricted Brownian motion and can be used to derive diffusion coefficients from single or many trajectories. But if the observed molecular motion is not based on free diffusion but on confined diffusion or directed flow, the relation between MSD(t_{lag}) and t_{lag} is no longer linear. Analysis of trajectories by Equation 1.1 is also not appropriate, when the particles have a different mobility, possibly due to interactions with further molecules, or when particles change their mode of motion along a trajectory. In such cases the data can be analyzed by so-called jump distance distributions. This analysis considers the probability $p(r,t)dr$ that a particle starting at some arbitrary position is encountered within a distance between r and $r + dr$ from the starting point at a given, later time t. For particles diffusing in two dimensions the jump distance distribution can be written as:

$$p(r, t) \, dr = \frac{1}{4\pi D t} e^{-r^2/4Dt} 2\pi \, r \, dr \qquad (1.2)$$

Experimentally this probability distribution can be approximated by a normalized frequency distribution obtained by counting the jump distances within respective intervals $[r, r + dr]$ covered by single particles after a given number of frames respective to the lag times. The essential point in the jump distance distribution analysis is that subpopulations of jumps can be determined by curve fitting.

So far, the mobility of single molecules within cell nuclei cannot be characterized by simple Brownian motion. This was demonstrated by quite a number of previous studies using FCS and single particle tracking [13, 22, 29]. In our study on U1 snRNP dynamics, we had to discriminate one immobile and at least two mobile fractions of molecular jumps. Hence, a satisfactory fit of the data required three terms like that given in Equation 1.2 (Figure 1.5).

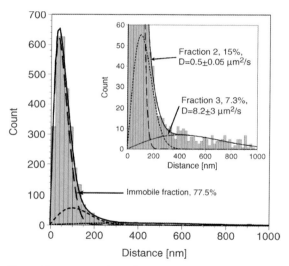

Figure 1.5 Motion analysis of single U1 snRNPs in successive frames. Distribution of the distances covered by single, nucleoplasmic U1 snRNPs, which were identified and tracked in successive frames of movies acquired at a frame rate of 100 Hz. A minimum number of three different diffusion terms was needed for a satisfactory fit of the data. The greatest fraction of jumps (long dashes, 77.5%) corresponded to immobile particles. These performed only a seeming movement due to the finite localization precision. Two mobile jump fractions, f_2 and f_3, corresponded to fast diffusional motion with coefficients of $D_2 = 0.51 \pm 0.05 \, \mu m^2/s$ (short dashes, 15%) and $D_3 = 8.2 \pm 3 \, \mu m^2/s$ (dotted line, 7.3%). The sum of all three fractions excellently described the data (full line). The insert shows the same plot with a different scaling of the ordinate to show the great number of large jumps corresponding to D_3. The figure was taken from Ref. [16] and is shown with permission.

The dissection of the distribution into three fractions represented a minimum number, meaning that three fractions were sufficient to fit the jump distance distribution. Certainly the fit could have been optimized by using four or five fractions, or even a continuous distribution. Furthermore it is important to note that the different jump fractions do not correspond to different particles with distinct mobilities. Rather – as could be noted in many trajectories – single particles switch their mode of motion along their trajectory, for example from mobile to bound or back. This is no surprising observation considering the complex intranuclear dynamics. Hence the fractions identified in the jump distance histograms also represent different modes of motion of possibly identical particles. Actually, we suppose that distinct mobility fractions do not exist, but rather that the U1 snRNP mobility can be characterized by motions which range from immobility up to dynamics which can be characterized by a diffusion coefficient of $8\,\mu m^2/s$.

The mobile U1 snRNPs showed diffusion coefficients varying from $0.5\,\mu m^2/s$ to $8.0\,\mu m^2/s$. A diffusion coefficient of $8\,\mu m^2/s$ is four- to fivefold lower than that expected for a 240-kDa particle such as U1 snRNP in aqueous solution. A fourfold reduction in mobility was found in several previous mobility studies of tracer molecules within cell nuclei, as already stated. Therefore we suppose that this diffusion coefficient corresponds to the motion of uncomplexed U1 snRNPs moving within nuclei, as in a solution with an effective viscosity of 5 cPoise. It has been shown that U snRNPs are central components of pre-formed complexes designated as spliceosomes and even supraspliceosomes, which contain four native spliceosomes with a total mass of 21 MDa and geometric dimensions of $50 \times 50 \times 35\,nm^3$ [26]. For such complexes, the Stokes–Einstein law predicts a diffusion constant of $D \approx 2\,\mu m^2/s$ in a solution of 5 cPoise. Obviously this value perfectly lies within the range of diffusion constants determined for U1 snRNPs. However huge complexes such as supraspliceosomes would probably not move in an *unrestricted* manner within the molecular crowded and entangled intranuclear space. They would suffer multiple collisions per unit time with chromatin or other molecular complexes, which would slow it down even further. Exactly this has been observed for large 2 MDa dextran molecules. Their mobility was dependent on the concentration of intracellular obstacles [30]. Furthermore it has been shown that chromatin regions represent a significant obstruction for the accessibility of large probe molecules [31]. Hence jumps corresponding to a mobility as low as $D = 0.5\,\mu m^2/s$ could well correspond to U1 snRNPs contained in supraspliceosomes moving in a strongly hindered manner through a crowded nucleoplasm.

1.9
Perspectives of Single Molecule Microscopy

Single molecule visualization and tracking is especially advantageous in single cell analysis for a number of reasons. Many signaling processes in biological systems are triggered by a very small number of effector molecules which can easily escape the classic methods of detection. Single molecule microscopy requires exceedingly low probe concentrations and hence is ideally suited to study such events. In addition the use of low probe concentrations guarantees that the physiological processes in investigation are only minimally disturbed. Processes like RNA splicing or transport through the nuclear pore complex take place on length scales between molecular distances (0.1–1.0 nm) and the optical microscope resolution limit (>250 nm). Actions on such "intermediate-sized" length scales cannot directly be studied with conventional microscopy. Furthermore they usually involve large supramolecular complexes containing numerous protein molecules and RNPs, which makes their reconstitution *in vitro* very difficult. Single molecule detection within live cells with a spatial precision in the nanometer range combined with the simultaneous use of multiple fluorescent labels and exploitation of single molecule energy transfer is a suitable way to approach such complex problems.

The molecular dynamics during ongoing live processes within cell nuclei have often been measured by photobleaching or photoactivation techniques or FCS. In order to correctly account for the mobility data obtained, binding processes are postulated [32]. Alternatively anomalous diffusion models can also adequately describe the data [29]. In contrast, binding events do not have to be *assumed* in single molecule tracking, but rather can be directly *observed* and discriminated from molecular movements. It is possible not only to observe distinct binding events but also to measure the duration of individual interactions. Dissociation data like that discussed above for U1 snRNPs can usually be obtained only by a direct preparation of the molecular complexes in the initial, associated state. This often presents a problem which is very difficult or impossible to solve, especially when working *in vivo*. Obviously single molecule detection elegantly solves this problem, because the individual events can be aligned in time a posteriori. The observation of individual molecular interactions does not require synchronization of the molecular ensemble. Finally in contrast to conventional techniques, such as FCS or photobleaching techniques, different forms of mobility in heterogeneous systems may well be discriminated by single molecule tracking. It is especially well suited for dynamic processes with diffusion coefficients <10 $\mu m^2/s$.

Two methodological drawbacks can currently be identified. As a video microscopy the technique lacks a true optical sectioning capacity. Although the imaging of single particles provides an intrinsic three-dimensional resolution, molecules in the out of focus regions are excited and drastically reduce the signal to noise ratio of the molecules in focus and thus lower the respective localization precision. Furthermore the reference structures, which must be imaged in order to locate the dynamic single molecule processes within a cellular context, appear blurred in current applications. However an elegant solution to this problem can be imagined. The sample may be illuminated using a light sheet perpendicular to the observation axis. This yields an intrinsic optical sectioning effect [33–35]. This clever illumination principle can be incorporated without major problems in a high-speed video microscope with single molecule sensitivity. Finally a bottleneck is still represented by the image processing of single molecule movies. Inherently noisy single molecule signals must be identified in thousands of images and the true trajectories must be constructed out of the single, fitted positions. Obviously this is often not possible in an unambiguous manner and sophisticated numerical approaches must be applied to extract a maximum of information from the data. A very small number of commercial program packages are currently available for this task, but none of them is able to cope with the data in a really satisfactory manner. The development of suitable modern image analysis techniques to this problem would really produce a significant progress for the field.

The conventional way to analyze intracellular mobility is by FRAP, which measures bulk mobility on a spatial scale of several microns within a time window of up to 50 s. Currently, FCS is increasingly used to study intracellular dynamics at selected focal positions. Single molecule tracking quantifies the mobility of individual molecules on length scales from 20 nm up to several microns within time windows from milliseconds to seconds. It is not at all straightforward to extrapolate from single molecule data to the results of FCS or FRAP measurements and vice versa. We think that it would be most fruitful to carefully apply all three techniques in a complementary manner and to correlate the distinct results with each other. To accomplish this a full simulation of molecular mobility on the basis of the single molecules would be required, taking reactions, interactions and geometrical restrictions into account. Such a simulation does not yet exist but is urgently needed in order to obtain a full view of the dynamic processes inside live cells.

Numerous interesting problems in biology and biophysics have been studied by single molecule techniques during the past few years [36, 37] but application to living cell systems is still in its infancy [38]. Single molecule microscopy and single particle tracking permit a completely new and fascinating view to intracellular dynamics. For the first time ever, this technique provides a direct, real-time insight into molecular processes in living cells with almost molecular resolution.

Acknowledgements

I gratefully acknowledge financial support by the Volkswagen Foundation, the German Research Foundation (DFG) and the Rheinische Friedrich-Wilhelms-Universität Bonn. I am indebted to Reiner Peters for a successful and long-standing cooperation; and I thank David Grünwald, Thorsten Kues and Jan Peter Siebrasse for a very successful cooperation during recent years. Finally, I thank Werner Wendler for preparing Figures 1.1 and 1.2, Andreas Hoekstra for preparing a first draft of Figure 1.3 and Jan Peter Siebrasse for a critical reading of the manuscript.

References

1 Donnert, G., Keller, J., Medda, R., Andrei, M.A., Rizzoli, S.O., Luhrmann, R., Jahn, R., Eggeling, C. and Hell, S.W. (2006) Macromolecular-scale resolution in biological fluorescence microscopy. *Proceedings of the National Academy of Sciences of the United States of America*, **103**, 11440–11445.

2 Heintzmann, R., Jovin, T.M. and Cremer, C. (2002) Saturated patterned excitation microscopy – a concept for optical resolution improvement. *Journal of the Optical Society of America*, **19**, 1599–1609.

3 Gustafsson, M.G. (2005) Nonlinear structured-illumination microscopy: wide-field fluorescence imaging with theoretically unlimited resolution. *Proceedings of the National Academy of Sciences of the United States of America*, **102**, 13081–13086.

4 Verkman, A.S. (2002) Solute and macromolecule diffusion in cellular aqueous compartments. *Trends in Biochemical Sciences*, **27**, 27–33.

5 Enderlein, J., Gregor, I., Patra, D., Dertinger, T. and Kaupp, U.B. (2005) Performance of fluorescence correlation spectroscopy for measuring diffusion and concentration. *Chemphyschem*, **6**, 2324–2336.

6 Haustein, E. and Schwille, P. (2007) Fluorescence correlation spectroscopy: novel variations of an established technique. *Annual Review of Biophysics and Biomolecular Structure*, **36**, 151–169.

7 Kubitscheck, U. (2006) Fluorescence microscopy: single particle tracking, in *Encyclopedic Reference of Genomics and Proteomics in Molecular Medicine* (eds D. Ganten and K. Ruckpaul), Springer.

8 Yildiz, A., Forkey, J.N., McKinney, S.A., Ha, T., Goldman, Y.E. and Selvin, P.R. (2003) Myosin V walks hand-over-hand: single fluorophore imaging with 1.5-nm localization. *Science*, **300**, 2061–2065.

9 Kubitscheck, U., Kückmann, O., Kues, T. and Peters, R. (2000) Imaging and tracking of single GFP molecules in solution. *Biophysical Journal*, **78**, 2170–2179.

10 Thompson, R.E., Larson, D.R. and Webb, W.W. (2002) Precise nanometer localization analysis for individual fluorescent probes. *Biophysical Journal*, **82**, 2775–2783.

11 Grunwald, D., Hoekstra, A., Dange, T., Buschmann, V. and Kubitscheck, U. (2006) Direct observation of single protein molecules in aqueous solution. *Chemphyschem*, **7**, 812–815.

12 Harms, G.S., Cognet, L., Lommerse, P.H., Blab, G.A., Kahr, H., Gamsjager, R., Spaink, H.P., Soldatov, N.M., Romanin, C. and Schmidt, T. (2001) Single-molecule imaging of l-type Ca(2 +) channels in live cells. *Biophysical Journal*, **81**, 2639–2646.

13 Kues, T., Peters, R. and Kubitscheck, U. (2001) Visualization and tracking of single protein molecules in the cell nucleus. *Biophysical Journal*, **80**, 2954–2967.

14 Yang, W., Gelles, J. and Musser, S.M. (2004) Imaging of single-molecule translocation through nuclear pore complexes. *Proceedings of the National Academy of Sciences of the United States of America*, **101**, 12887–12892.

15 Kubitscheck, U., Grunwald, D., Hoekstra, A., Rohleder, D., Kues, T., Siebrasse, J.P. and Peters, R. (2005) Nuclear transport of single molecules: dwell times at the nuclear pore complex. *The Journal of Cell Biology*, **168**, 233–243.

16 Grunwald, D., Spottke, B., Buschmann, V. and Kubitscheck, U. (2006) Intranuclear binding kinetics and mobility of single U1 snRNP particles in living cells. *Molecular Biology of the Cell*, **17**, 5017–5027.

17 Seisenberger, G., Ried, M.U., Endress, T., Buning, H., Hallek, M. and Brauchle, C. (2001) Real-time single-molecule imaging of the infection pathway of an adeno-associated virus. *Science*, **294**, 1929–1932.

18 Babcock, H.P., Chen, C. and Zhuang, X. (2004) Using single-particle tracking to study nuclear trafficking of viral genes. *Biophysical Journal*, **87**, 2749–2758.

19 Shav-Tal, Y., Darzacq, X., Shenoy, S.M., Fusco, D., Janicki, S.M., Spector, D.L. and Singer, R.H. (2004) Dynamics of single mRNPs in nuclei of living cells. *Science*, **304**, 1797–1800.

20 Michalet, X., Pinaud, F.F., Bentolila, L.A., Tsay, J.M., Doose, S., Li, J.J., Sundaresan, G., Wu, A.M., Gambhir, S.S. and Weiss, S.

(2005) Quantum dots for live cells, in vivo imaging, and diagnostics. *Science*, **307**, 538–544.

21 Bausinger, R., von Gersdorff, K., Braeckmans, K., Ogris, M., Wagner, E., Brauchle, C. and Zumbusch, A. (2006) The transport of nanosized gene carriers unraveled by live-cell imaging. *Angewandte Chemie (International Edition in English)*, **45**, 1568–1572.

22 Goulian, M. and Simon, S.M. (2000) Tracking single proteins within cells. *Biophysical Journal*, **79**, 2188–2198.

23 Misteli, T. (2005) Concepts in nuclear architecture. Bioessays: News and Reviews in Molecular, Cellular and Developmental Biology, **27**, 477–487.

24 Darzacq, X., Singer, R.H. and Shav-Tal, Y. (2005) Dynamics of transcription and mRNA export. *Current Opinion in Cell Biology*, **17**, 332–339.

25 Will, C.L. and Luhrmann, R. (2001) Spliceosomal UsnRNP biogenesis, structure and function. *Current Opinion in Cell Biology*, **13**, 290–301.

26 Medalia, O., Typke, D., Hegerl, R., Angenitzki, M., Sperling, J. and Sperling, R. (2002) Cryoelectron microscopy and cryoelectron tomography of the nuclear pre-mRNA processing machine. *Journal of Structural Biology*, **138**, 74–84.

27 Lamond, A.I. and Spector, D.L. (2003) Nuclear speckles: a model for nuclear organelles. *Nature Reviews. Molecular Cell Biology*, **4**, 605–612.

28 Kues, T., Dickmanns, A., Lührmann, R., Peters, R. and Kubitscheck, U. (2001) High intranuclear mobility and dynamic clustering of the splicing factor U1 snRNP observed by single particle tracking. *Proceedings of the National Academy of Sciences of the United States of America*, **98**, 12021–12026.

29 Wachsmuth, M., Waldeck, W. and Langowski, J. (2000) Anomalous diffusion of fluorescent probes inside living cell nuclei investigated by spatially-resolved fluorescence correlation spectroscopy. *Journal of Molecular Biology*, **298**, 677–689.

30 Seksek, O., Biwersi, J. and Verkman, A.S. (1997) Translational diffusion of macromolecule-sized solutes in cytoplasm and nucleus. *Journal of Cell Biology*, **138**, 131–142.

31 Gorisch, S.M., Richter, K., Scheuermann, M.O., Herrmann, H. and Lichter, P. (2003) Diffusion-limited compartmentalization of mammalian cell nuclei assessed by microinjected macromolecules. *Experimental Cell Research*, **289**, 282–294.

32 Phair, R.D. and Misteli, T. (2001) Kinetic modelling approaches to in vivo imaging. *Nature Reviews. Molecular Cell Biology*, **2**, 898–907.

33 Voie, A.H., Burns, D.H. and Spelman, F.A. (1993) Orthogonal-plane fluorescence optical sectioning: three-dimensional imaging of macroscopic biological specimens. *Journal of Microscopy*, **170**, 229–236.

34 Huisken, J., Swoger, J., Del Bene, F., Wittbrodt, J. and Stelzer, E.H. (2004) Optical sectioning deep inside live embryos by selective plane illumination microscopy. *Science*, **305**, 1007–1009.

35 Engelbrecht, C.J. and Stelzer, E.H. (2006) Resolution enhancement in a light-sheet-based microscope (SPIM). *Optics Letters*, **31**, 1477–1479.

36 Zlatanova, J. and van Holde, K. (2006) Single-molecule biology: what is it and how does it work? *Molecular Cell*, **24**, 317–329.

37 Greenleaf, W.J., Woodside, M.T. and Block, S.M. (2007) High-resolution, single-molecule measurements of biomolecular motion. *Annual Review of Biophysics and Biomolecular Structure*, **36**, 171–190.

38 Peters, R. (2007) Single-molecule fluorescence analysis of cellular nanomachinery components. *Annual Review of Biophysics and Biomolecular Structure*, **36**, 371–394.

39 Saxton, M.J. and Jocobsen, K. (1997) Single-particle tracking: applications to membrane dynamics. *Annual Review of Biophysics and Biomolecular Structure*, **26**, 373–399.

2
Gene Classification and Quantitative Analysis of Gene Regulation in Bacteria using Single Cell Atomic Force Microscopy and Single Molecule Force Spectroscopy

Robert Ros and Nicole Hansmeier

2.1
Introduction

Atomic force microscopy (AFM) [1] offers fascinating possibilities to resolve biologically relevant surface structures with molecular or even submolecular resolution under ambient as well as liquid environments. For imaging, the molecules are usually immobilized on flat substrates [2, 3], but measurements on fixed or living cells have also been reported [4–7].

AFM belongs to the continuously increasing family of the scanning probe microscopes, which all go back to the scanning tunneling microscope (STM) invented by Binnig *et al.* [8]. In the case of AFM a tiny tip with a radius in the range of 10 nm mounted on a microfabricated cantilever moves line by line over the surface of interest, while the interaction forces between the surface and the tip are probed (Figure 2.1a). Nowadays various modes of operation are available, differing for example in the applied lateral and vertical forces and the contrast mechanism. For soft biological samples, dynamic modes are often applied, like the tapping mode where the cantilever oscillates, therefore avoiding the destruction of the fragile structures [9, 10].

Furthermore AFM is not only an imaging tool; it also allows the quantitative characterization of biomolecular interactions and intramolecular forces on the level of single molecules. The technique can be applied to a remarkable range of interactions; from the binding of complex biological molecules like antibodies [11–13], proteoglycans [14, 15], cytochromes [16], chaperones [17], selectines [18] and protein–DNA interactions [19–21] to small bioorganic or organic compounds like peptides [22] and supramolecular systems [23–25]. The binding affinities of the probed complexes can differ several orders of magnitude. For example the seminal early force spectroscopy works on streptavidin/avidin–biotin interactions [26, 27] yield a dissociation constant K_D in the range of 10^{-15} M, whereas for weak calexarene-ion complexes one finds a K_D of 10^{-5} M [23]. The mechanic behavior of single molecules could also be addressed with this method. Examples include the mechanics of single DNA strands [28–31], polymers [32] and proteins [33, 34].

Single Cell Analysis: Technologies and Applications. Edited by Dario Anselmetti
Copyright © 2009 WILEY-VCH Verlag GmbH & Co. KGaA, Weinheim
ISBN: 978-3-527-31864-3

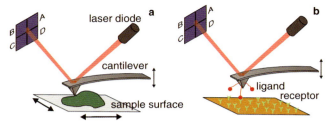

Figure 2.1 Atomic force microscopy and single molecule force spectroscopy. (a) Schematic representation of AFM. The tiny tip is moved line by line over the sample surface while the deflection of the cantilever is detected by the position of a reflected laser beam. (b) For single molecule force spectroscopy experiments on ligand–receptor pairs the binding partners are respectively attached to the surface and the tip and the cantilever is cycled up and down. The reflected laser beam indicates the force acting on the molecular complex.

For single molecule force spectroscopy experiments on ligand–receptor systems one binding partner is connected to the tip of a soft AFM cantilever and the other is bound to a surface (Figure 2.1b). The ligand and/or receptor molecules are usually linked via polymeric tethers to the tip and to the surface, respectively [12, 19, 22, 35, 36]. In order to obtain force–distance curves the AFM tip or the surface is cycled up and down while measuring the force acting on the cantilever. From these curves rupture or unbinding forces are analyzed for various pulling velocities.

In this chapter we report on different examples where AFM was used to compare protein sequences with molecular structures, the influence of gene regulation on the protein extraction of surface proteins and examples for the quantitative analysis of specific protein–DNA interactions in the field of transcriptional regulation.

2.2
AFM on Paracrystalline Cell Surface Layers of *C. glutamicum*: Protein Sequence Information and Morphology

One remarkable feature of prokaryotic cell envelopes is the presence of monomolecular paracrystalline protein 2D lattices, so-called cell surface layers (short S-layers) [37]. These S-layers consist of (glyco-)proteins with molecular masses ranging from 40 kDa to 200 kDa depending on the respecific bacterial genera. Because of their location S-layers are involved in the interaction between bacteria and their environment. Therefore diverse functions have been attributed to the S-layers of individual bacterial species, including protection of the cell from hostile factors, serving as molecular sieves and mediation of attachment to host tissues [38]. Important features of S-layers especially in regard to bio- and nanotechnological applications are: (a) pores with identical size and morphology in the nanometer range passing through the S-layers, (b) functional groups on the surface and in the pores are aligned in well defined positions and orientations and (c) isolated S-layer subunits can recrystallize by self-assembly processes on diverse surfaces [38].

Based on this repetitive physicochemical surface properties of the S-layers down to the subnanometer scale, a broad bio- and nanotechnological application potential exists for those S-layers [39]. The diameters of the pores are in the range 2–8 nm and therefore usable as ultrafiltration membranes or lithographic masks [39]. Further applications of S-layers are bioanalytical sensors, enzyme and affinity membranes, immunoassays and electronic or optical devices. Several excellent reviews can be found in the literature dealing with this topic [39–42].

For all these applications the design of lattices with accurately defined physicochemical properties is essential. One strategy to obtain such defined lattices is to take a look at the natural diversity of S-layer properties of bacterial communities. Therefore, in a comparative study the S-layer proteins of 28 different *Corynebacterium glutamicum* strains were analyzed by the use of classic biological techniques (SDS-PAGE, PCR, sequencing) and AFM. The investigation of the extracted S-layer proteins from these 28 different *C. glutamicum* strains by SDS-PAGE showed variation in their molecular masses in the range 55–66 kDa [43]. Based on these mass variations, significant differences could be expected in the corresponding genes or their physicochemical properties. Using PCR techniques and subsequently sequencing the corresponding genes of the 28 different *C. glutamicum* strains revealed high gene variability. This variability is also reflected in the deduced amino acid sequences of the 28 *C. glutamicum* S-layer proteins, which showed calculated identities between 69% and 98% [43].

To assess whether the detected sequence differences of the *C. glutamicum* S-layer proteins are reflected by their morphology, AFM high-resolution imaging was performed. S-layer proteins of the 28 *C. glutamicum* isolates were extracted from cells with 2% SDS, precipitated by centrifugation and adsorbed on amino-functionalized mica surfaces. The S-layers of the different isolates adsorbed to silanized mica as mono-, double or multilayers and were finally imaged in the absence of buffers. High-resolution topographs (~1 nm) revealed two different surface types of the *C. glutamicum* S-layer, one flower-shaped and one triangular (Figure 2.2a, b). Mostly surfaces with flower-shaped hexameric S-layer cores are exposed. This depends on the physicochemical properties the flower-shaped S-layer side which does not adsorb 6nto the hydrophilic mica [44]. However the triangular S-layer side nicely adsorbs to the mica and is therefore only detectable on the border regions (Figure 2.1). With the exception of four *C. glutamicum* strains, S-layers sheets of all analyzed *C. glutamicum* strains could be isolated and displayed a common hexagonal lattice symmetry (Figure 2.2) [43], in which monomers from hexameric cores are connected to six other cores [44]. Applying an AFM nano-dissection analysis, a 3D reconstruction of the *C. glutamicum* S-layer architecture could be calculated and the S-layer was classified as an M_6C_3 layer type [44].

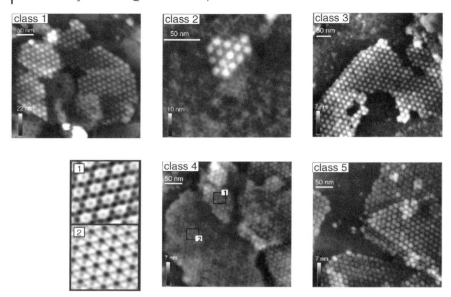

Figure 2.2 AFM images of C. glutamicum S-layers from five different sequence similarity-based classes adsorbed to aminofunctionalized mica. All strains display hexagonal lattices with flower-shaped bottom surfaces and triangular top surfaces, but with different unit cell dimensions. Fourier transformations of the bottom (a) and the top (b) sides of the S-layer of C. glutamicum ATCC 19240 are shown. AFM images represent S-layers of each class: class 1 S-layer of ATCC 17966, class 2 S-layer of ATCC 22243, class 3 S-layer of ATCC 13058, class 4 S-layer of ATCC 19240 and class 5 S-layer of ATCC 14751. Graph adapted from Hansmeier et al. [43].

Interestingly, detailed analysis of the 24 C. glutamicum S-layers showed beside the common hexagonal lattice symmetry small but significant differences in their unit cell dimension, specifying the distance between the centers of two S-layer hexamers. The unit cell dimension of the 24 C. glutamicum S-layers varies between $a = b = 15.2 \pm 0.25$ nm and $a = b = 17.4 \pm 0.2$ nm (Figure 2.3) [43].

Based on the protein sequence data, a grouping of the analyzed C. glutamicum strains in dependence on their S-layer identity was performed and a division into five classes was conducted (Figure 2.3). A correlation between the AFM-based classification of the S-layers and the protein sequence-based classification is striking, although S-layer proteins from two isolates (C. glutamicum ATCC 31830 and ATCC 13744) do not fit well into this classification (Figure 2.3). Therefore, class 3 might need to be further subdivided [43]. Consequently, the sequence variations of the S-layer proteins are directly displayed in morphological differences of the S-layer unit cell dimensions.

In this respect this interdisciplinary combinatory analysis of classic biological methods and high-resolution imaging techniques by AFM is the first step into the direction of protein engineering to design S-layers with well defined dimensions for nano- and biotechnological applications.

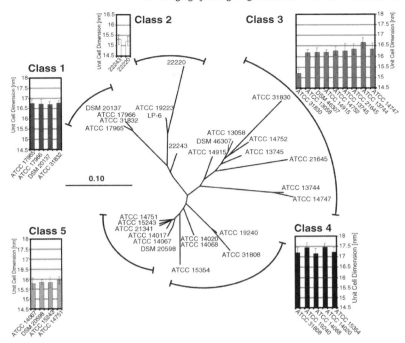

Figure 2.3 Correlation of phylogenetic and AFM-based classification of the S-layer proteins from 28 *C. glutamicum* strains. The dendrogram is constructed with the CLUSTALX program under the usage of the BLOSUM 62 matrix with the neighbour-joining method and depicted the relationship among the different S-layer proteins from the 28 *C. glutamicum* strains. Scale bar: 0.1% amino acid substitution. Diagrams showed measurement of the unit cell dimension of imaged S-layers. Each bar represents 180 independent measurements (three dimensions of 20 hexagons originating from three different images). Graph adapted from Hansmeier et al. [43].

2.3
Imaging of Living *C. glutamicum* Cells with Molecular Resolution: Genes, Transcriptional Regulation and Morphology

It has been calculated that a rod-shaped bacterium of average size with a generation time of 20 min has to synthesize around 500 S-layer subunits per second to cover the cell surface completely [37]. Accordingly S-layer genes are expressed at extremely high levels and S-layer proteins constitute 10–15% of the cell's total protein content. The high-level expression of S-layer genes is not only based on strong promoters and efficient transcription but also on highly stable mRNA molecules [37]. However S-layer proteins are seldom found in culture supernatants, suggesting that their synthesis is tightly regulated.

First hints regarding the regulation of S-layer gene expression in *C. glutamicum* were published by Chami et al. [45], who showed that *C. glutamicum* cells grown on solid medium are completely covered with an S-layer, whereas cells cultivated in liquid medium are only partially covered by an ordered lattice. Subsequent studies have demonstrated that the amount of S-layer protein synthesized by *C. glutamicum* is dependent on the carbon source of the growth medium [46]. On the basis of these observations, a relationship between S-layer formation, carbohydrate metabolism and the physiological status of the cell in general has been suggested [46].

To gain insight into transcriptional regulation of S-layer formation in *C. glutamicum*, a combinatory approach of classical biological techniques (RACE-PCR, DNA affinity purification assay, RT-PCR) and high-resolution AFM imaging techniques were applied. By using a classic RACE-PCR the promoter of the S-layer gene was mapped, enabling the search for regulatory DNA-binding proteins interacting with this DNA region by a so-called DNA affinity purification assay. With this assay a putative upstream of the S-layer gene DNA-binding LuxR-type transcriptional regulator was identified [47].

The role of the identified transcriptional regulator in S-layer formation was deduced from the characterization of this regulator mutant *C. glutamicum* strain Intcg2831 in comparison to negative and positive *C. glutamicum* control strains by AFM. Therefore the *C. glutamicum* cells were washed with 20 mM Tris/HCl buffer solution to avoid medium contaminations before they were adsorbed onto glass plates for AFM imaging in tapping mode. This easy sample preparation procedure helped in combination with mild imaging conditions to avoid artefacts during the measurements. The resulting AFM images showed highly ordered hexagonal lattice patches, which covered nearly the complete cell surfaces of the two S-layer-containing *C. glutamicum* strains, the positive control strains (Figure 2.4a, b). In contrast the cell surface of the negative control is as expected devoid of any ordered structure that resembles hexagonal S-layers (Figure 2.4c). Interestingly in comparison to both control strains, the cell surface of *C. glutamicum* regulator mutant Intcg2831 displayed very small patches of the hexagonal S-layer (Figure 2.4d). Those S-layer patches covered only about 1% of the cell surface of *C. glutamicum* mutant strain, which is a sign that the inactivated regulator Cg2831 has an activating function on the S-layer production in *C. glutamicum*. Furthermore it could be determined that these small S-layer patches were primarily located at the bacterial cell poles, which could be a hint for the starting point of S-layer subunit extrusion and assembly.

Further validation of the AFM analyses with quantitative expression analysis of the S-layer proteins in the mutant compared to the wild-type background could definitely determine that the identified LuxR-type regulator Cg2831 plays an important role as transcriptional activator of S-layer gene expression, resulting in an enhanced transcription of the S-layer gene and the formation of an ordered S-layer lattice on the surface of *C. glutamicum* cells. Finally, also the last open question respectively the carbohydrate dependency of the S-layer regulation could be elucidated, because recent analysis of the transcriptional regulator Cg2831, also now known under the name RamA, revealed whose activating function in dependency on the carbon source of the growth medium and showed the direct linkage between this regulator and the carbohydrate metabolism in *C. glutamicum* [48].

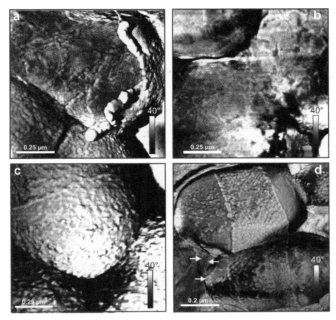

Figure 2.4 Atomic force microscopy of living C. glutamicum cells. AFM phase images recorded in tapping mode showed cell surfaces from: (a, b) S-layer carrying C. glutamicum strains acting as positive control, (c) S-layer devoid C. glutamicum strain as negative control and (d) C. glutamicum regulator mutant (Intcg2831). In contrast to the control strains, the cells of regulator mutant C. glutamicum Intcg2831 displayed only marginal S-layer patches at the bacterial cell poles. Graph adapted from Hansmeier et al. [47].

These experiments nicely prove that AFM is able to resolve subcellular protein surfaces down to the level of individual protein monomers on single living cells, giving access to a direct test of genetic and protein variability.

2.4
Single Molecule Force Spectroscopy on Specific Protein–DNA Complexes: Transcriptional Regulation in *S. meliloti*

The soil bacterium *Sinorhizobium meliloti* is capable of fixing molecular nitrogen in a symbiotic interaction with plants of the genera *Medicago*, *Melilotus* and *Trigonella*. It has the ability to produce two acidic exopolysaccharides (EPSs), succinoglycan (EPS I) and galactoglucan (EPS II). EPS I is required for invasion of *Medicago sativa* root nodules by *S. meliloti*, but can be replaced by EPS II [49–51]. Biosynthesis of EPS II is directed by the 30 kb *exp* gene cluster that comprises 22 genes organized in four operons. One of the corresponding regulatory gene on this gene cluster is *expG* [52, 53]. Extra copies of *expG* stimulate transcription of the *expA*, *expD* and *expE* operons [52]. Furthermore *expG* is required for the stimulation of these operons under phosphate-limiting conditions indicating that the protein ExpG acts as a transcriptional

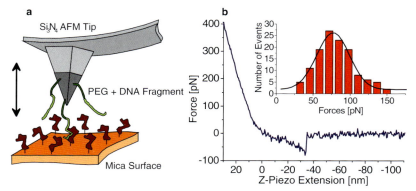

Figure 2.5 Force spectroscopy measurements. (a) The experimental setup consists of a Si_3N_4 AFM tip with DNA fragments attached via poly (ethylene glycol) spacer molecules and a flat mica surface on which the $(His)_6ExpG$ proteins are immobilized. Cycling the tip and sample between approach and retract with constant velocities results in a series of force-distance measurements. (b) A typical force–distance curve (only retractive part displayed). An unbinding event can be identified by a certain distance from the point of contact due to the length of the polymer linker and the stretching of this linker directly prior to the point of bond rupture. Rupture forces of a given series are combined to form a histogram (small insert) with an almost Gaussian distribution. The mean value of the Gaussian equals the most probable unbinding force, with statistical errors given by standard deviation ($2\sigma/N^{-1/2}$ for 95.4% confidence level). Graph adapted from Bartels et al. [19].

activator of *exp* gene expression [52, 54]. ExpG binds to the three different promoter fragments *expA1*, *expG/expD1* and *expE1* in a sequence-specific manner.

AFM-based force spectroscopy allows the investigation of protein–DNA interaction on a single molecule basis. Therefore the DNA fragment was attached to the tip via a polymer spacer while the protein was immobilized on the surface by a short linker molecule (Figure 2.5a). When the tip was approached to and retracted from the surface, the flexibility of the polymer chain allowed the DNA molecules to access the binding pockets of immobilized proteins. By plotting the force acting on the AFM tip against the vertical position (given by the extension of the piezo), unbinding events can be identified by a characteristical stretching of the polymer spacer before the point of bond rupture (where the tip snaps back to zero force). A typical force–distance curve is shown in Figure 2.5 b. The rupture forces from multiple approach/retract cycles under a single retract velocity were combined in a histogram. The mean value of the nearly Gaussian distribution was taken as the most probable rupture force.

The total unbinding probability (events/cycles) usually amounted to 15%. When free DNA fragments were added to the buffer solution in an excess of approximately 60:1 regarding the proteins, a distinct reduction in unbinding probability was observed for all three DNA target sequences (shown in Figure 2.6, a–c for the *expA1/A4* fragment) [19]. Moreover when both tip and sample were washed with buffer solution and reinstalled with the original, competitor-free buffer in place, the system could be reactivated to almost its full former unbinding probability. The distinct influence of the competitor fragments clearly indicates the specificity of the binding process [19].

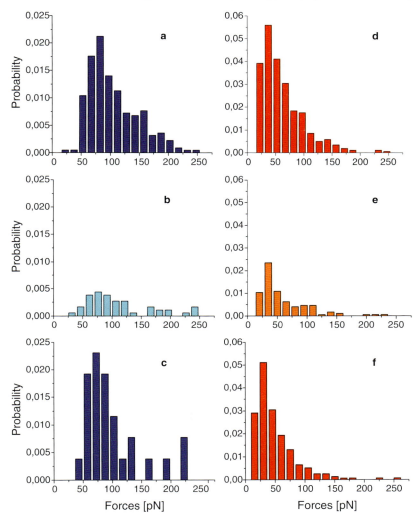

Figure 2.6 Single molecule competition experiments. Force spectroscopy measurements were performed under alternating buffer conditions with a single tip and sample system for each series. (a–c) In this series, the *expA1/A4* DNA fragment was attached to the tip. Unbinding events were first measured in the original binding buffer solution (a). When free *expA1/A4* fragments were inserted into the buffer at an excess of 60:1, the total unbinding probability was reduced (b). By exchanging the buffer back to the original conditions, the system was reactivated after 20 min (c). (d–f) A series with the *expG1/G4* fragment bound to the tip reveals a reduction in unbinding probability from the original binding buffer (d) to the application of free $(His)_6$ExpG protein as a competitor in a ratio of 1:1 (e). By a buffer exchange back to the original buffer solution, the system was reactivated after 85 min (f). Graph adapted from Bartels *et al.* [19].

In an experiment with the *expG1/G4* DNA, free (His)$_6$ExpG proteins were added to the buffer as an alternative competitor. Even at a ratio as small as approximately 1:1 regarding the DNA density on the tip (which was calculated based on measurements for PEG-bound ligands on aminosilane monolayers by [55]), a distinct reduction in unbinding probability was observed. Again, the system could be reactivated to almost its former unbinding probability when buffers are changed back to the original conditions (Figure 2.6, d–f) [19]. The experiments demonstrate that ExpG binds to promoter regions in the *exp* gene cluster in a sequence-specific manner.

Thermodynamical and structural information concerning the binding can be obtained by measuring the most probable unbinding forces for the three DNA–protein complexes in dependence on the loading rate by varying the retract velocity. For each DNA target sequence, typically 150–300 unbinding events (from 1000–2000 approach/retract cycles) were recorded at seven to nine different retract velocities ranging from 10 nm/s to 8000 nm/s, while the approach velocity was kept constant at 1000 nm/s. These resulted in loading rates in the range from 70 pN/s to 79 000 pN/s.

The results for the three DNA target sequences are shown in Figure 2.7. When the unbinding forces are plotted against the corresponding loading rates on

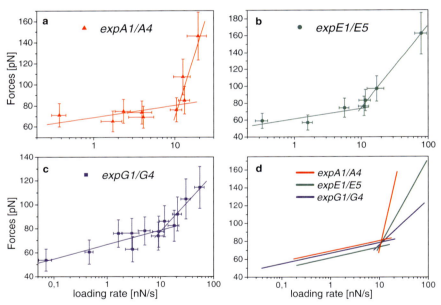

Figure 2.7 Dynamic force spectroscopy. Loading rate dependent measurements of the unbinding forces are displayed for complexes formed by the (His)$_6$ExpG protein and each of its three DNA target sequences (a–c). Two regions can be distinguished in every dataset. In the higher loading rate region, when each dataset is subjected to a linear fit, the slopes differ widely from each other (see (d) for comparison of the fits), and the individual DNA fragment in use can be identified by this behavior. In the lower loading rate region, however, the different protein–DNA complexes share a single slope under a linear fit within the error margin. This corresponds to a joint natural thermal off-rate of $k_{off} = (1.2 \pm 1.0) \times 10^{-3}\,s^{-1}$, derived by extrapolating the line fit to the state of zero external force. Graph adapted from Bartels *et al.* [19].

a logarithmic scale, two regions with different slopes emerge for all DNA fragments. In both regions the experimental data can be fitted to a function according to [56]:

$$F = \frac{k_B T}{x_\beta} \ln \frac{x_\beta r}{k_B T k_{off}}$$

where F is the most probable unbinding force, $k_B T = 4.114$ pN nm (at 298 K) the thermal energy, x_β is a molecular length parameter along the reaction coordinate (see below), r is the loading rate and k_{off} is the thermal off-rate under zero load. In the lower region ($r < 11\,000$ pN/s), the slopes corresponding to the three fragments do not differ within the error margin when this fit is applied to each dataset (see Figure 2.7). According to the theoretical model behind the formula used [56, 57], this slope can be attributed to the last potential barrier in the energy landscape of the system. In this case, the natural thermal off-rate k_{off} can be derived by extrapolating the linear fit to the state of zero external force. We obtain a medium off-rate $k_{off} = (1.2 \pm 1.0) \times 10^{-3}$ s^{-1} for all three DNA target sequences, which corresponds to a mean life time $\tau = (13.9 \pm 11.6)$ min for the bound protein–DNA complex. The molecular parameter x_β defines the distance between the minimum of the potential well of the bound state and the maximum of the energy barrier separating the bound state from the free state along the reaction coordinate. This is often interpreted as the depth of the binding pocket (e.g. [57]). For the lower region, $x_\beta = (7.5 \pm 1.0)$ Å for all protein–DNA complexes can be deduced.

In the upper region ($r > 11\,000$ pN/s), different values for the slopes corresponding to the individual DNA fragments can be found. In accordance with [57] this can be attributed to a second energy barrier in the system, with different properties for the three DNA target sequences. We measure a molecular parameter $x_\beta = (2.0 \pm 0.6)$ Å for the protein–DNA complex with *expG1/G4*, $x_\beta = (0.97 \pm 0.06)$ Å with *expE1/E5* and $x_\beta = (0.39 \pm 0.14)$ Å with *expA1/A4*. These differences could be attributed to the non-identical nucleotides of the three target sequences in the binding region [19].

These experiments prove that AFM based single molecule force spectroscopy is a powerful investigative tool to study sequence-specific protein–DNA interactions on the molecular scale.

2.5
Effector-Induced Protein–DNA Binding on the Single Molecule Level: Quorum Sensing in S. meliloti

Quorum sensing (QS) is a form of population density-dependent gene regulation controlled by low molecular weight compounds called autoinducers, which are produced by bacteria. QS is known to regulate many different physiological processes, including the production of secondary metabolites, conjugal plasmid transfer, swimming, swarming, biofilm maturation and virulence in human, plant and animal pathogens [58]. Many QS systems involve N-acyl homoserine lactones (AHLs) as signal molecules (for a review, see Ref. [59]). These AHLs vary in length, degree of substitution and saturation of the acyl chain. Bacterial cell walls are permeable to AHLs, either by unassisted diffusion across the cell membrane (for shorter acyl chain length) or active transport (possibly for longer acyl chain length) [60]. With the increase in the number of cells AHLs accumulate both intracellularly and extracellularly. Once a threshold concentration is reached, they act as co-inducers, usually by activating LuxR-type transcriptional regulators.

In *S. meliloti* Rm1021, a QS system consisting of the AHL synthase SinI and the LuxR-type AHL receptors SinR and ExpR was identified [61]. SinI is responsible for production of several long-chain AHLs (C_{12}–HL to C_{18}–HL) [62]. The presence of a second QS system, the Mel system, controlling the synthesis of short-chain AHLs (C_6–HL to C_8–HL) was suggested [62]. In addition to SinR, five other putative AHL receptors, including ExpR, were identified [63]. As originally described for the model QS LuxI/LuxR system of *Vibrio fischeri* [64] and demonstrated for the TraR–AHL complex of *Agrobacterium tumefaciens* whose crystal structure was resolved [65] it is assumed that the LuxR-type regulators of *S. meliloti* are activated by binding of specific AHLs. Once activated, the expression of target genes is regulated by binding upstream of the promoter regions of these genes [66]. The first target genes identified for the *S. meliloti* Sin system were the *exp* genes mediating biosynthesis of the exopolysaccharide galactoglucan. The expression of the *exp* genes not only relies on a sufficient concentration of Sin system-specific AHLs but also requires the presence of the LuxR-type AHL receptor ExpR [63, 67]. ExpR is highly homologous to the *V. fischeri* LuxR. Activated LuxR-type regulators usually bind to a consensus sequence known as the *lux* box, typically located upstream of the promoters of its target genes [66]. A *lux* box-like sequence (TATAGTACATGT) 70 bp upstream of *sinI* which constitutes a putative binding site for the LuxR-type regulators ExpR was identified [68].

To investigate the ExpR–DNA interaction on a single molecule basis, the binding partners were covalently bound to the sample surface and to the AFM tip as described before. The total dissociation probability (events/cycles) remained below 0.5% for the bare protein–DNA system in buffer solution, the histogram consisting of scattered events (Figure 2.8a) [68]. The background signal, that is a series of measurements with a functionalized surface and an AFM tip without any DNA but prepared as normal in all other respects reveal no dissociation events. The profile changed drastically when AHL was added to give a final concentration of 10 µM (oxo-C_{14}–HL in the case of Figure 2.8b). The total dissociation probability increased to 8–10%, and the dissociation forces form a distribution of almost Gaussian shape. Data from this experiment indicate that ExpR binds to DNA even in the absence of any effector due to unspecific attraction (e.g. electrostatic forces) but the probability of binding is highly increased in the presence of a proper effector [68].

AFM force spectroscopy proved to be a sensitive tool to determine the influence of different effector molecules. Six of the seven effectors tested stimulated protein–DNA binding, namely C_8–HL, C_{10}–HL, C_{12}–HL, oxo-C_{14}–HL, $C_{16:1}$–HL and C_{18}–HL. Only C_7–HL caused no noticeable increase in activity when added to the buffer solution (Figure 2.9a) [68].

The different effectors showed a distinct influence on the kinetics and structure of the ExpR–DNA binding. Most off-rates were close to $k_{off} = 2\,\text{s}^{-1}$ which corresponds

2.5 Effector-Induced Protein–DNA Binding on the Single Molecule Level: Quorum Sensing in S. meliloti

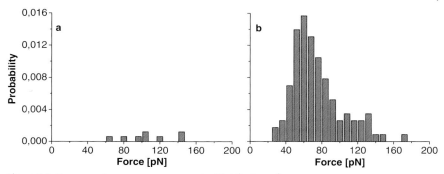

Figure 2.8 Force spectroscopy measurements. Distribution of dissociation forces for the DNA-(His)$_6$ExpR complex (a) without effector (at v = 2000 nm/s) and (b) after adding 10 μM of oxo-C$_{14}$-HL. Graph adapted from Bartels et al. [68].

to a mean lifetime of approximately 0.5 s for the bound protein–DNA complex. The notable exceptions are C$_8$–HL which caused an off-rate of $k_{off} = 0.5$ s^{-1} ($\tau = 2.3 \pm 0.8$ s) and C$_{12}$–HL with $k_{off} = 5.4$ s^{-1} ($\tau = 0.19 \pm 0.04$ s). The measured reaction length x_β indicates three different states as well: the long-chain AHLs (oxo-C$_{14}$–HL, C$_{16:1}$–HL, C$_{18}$–HL) center around $x_\beta \sim 4.0$ Å, while the short-chain AHLs (C$_8$–HL, C$_{10}$–HL) tend to higher values at $x_\beta \sim 5.5$ Å and C$_{12}$–HL has a strikingly low reaction length of $x_\beta \sim 3$ Å [68].

Why is C$_7$–HL not able to stimulate protein–DNA interaction (Figure 2.9a)? One likely explanation would be that it does not bind to the ExpR protein. But this is not the case. Even after C$_7$–HL was removed from both the sample surface and the AFM tip by multiple washing steps with AHL-free buffer solution over the course of 1 h, addition of C$_{12}$–HL yielded very few dissociation events (Figure 2.9b) [68]. It seems that most proteins retained a C$_7$–HL effector which inhibited activation by C$_{12}$–HL. This suggests that, although C$_7$–HL binds to the ExpR protein, it is not able to change its conformation into an active state. This is a further indication that the chemical structure of a particular effector has a strong influence on the sterics of the ExpR protein and thereby on the kinetics of possible protein–DNA interactions [68].

The aforementioned experiment suggests a long lifetime of the C$_7$–HL effector protein complex, resulting in an effective inhibition of activation of ExpR by C$_{12}$–HL. We performed an additional experiment in which C$_{12}$–HL was the first AHL added, which showed its usual degree of activity with a fresh protein sample (Figure 2.9c) [68]. As in the previous experiment, sample surface and AFM tip were washed multiple times over the course of 1 h before the addition of the second AHL, C$_7$–HL. Although the binding probability was marginally reduced after the washing step (Figure 2.9d), the system still showed a considerable degree of activity in the presence of C$_7$–HL. In contrast to the lifetime of the protein–DNA complex (which is $\tau = 185 \pm 35$ ms in the case of C$_{12}$–HL), the lifetime of the protein–effector bond seems to be much longer. ExpR–DNA kinetics can therefore be regarded as independent from AHL–ExpR kinetics, indicating that only the structural change

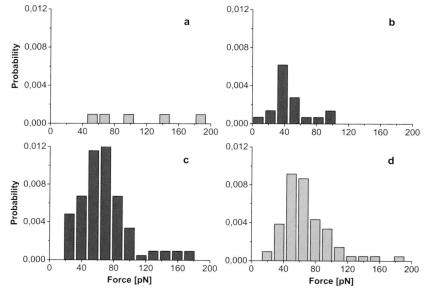

Figure 2.9 Stability of the protein–effector bond. (a) In the presence of C_7–HL, no binding is observed. (b) After the sample was washed multiple times with buffer solution over the course of 1 h and C_{12}–HL was added, the protein–DNA complex is still inactive. The reverse process was investigated with a new tip and sample surface. (c) In the presence of C_{12}–HL, the protein–DNA complex shows its usual degree of activity. (d) C_7–HL was added after the sample was washed multiple times over the course of 1 h. Activity is only slightly reduced. Obviously C_7–HL is not able to displace C_{12}. Both experiments indicate a high stability of the protein–effector bond. Graph adapted from Bartels et al. [68].

of the protein induced by a particular AHL effector governed the properties of ExpR–DNA interaction in the DFS experiments [68].

Dynamic force spectroscopy based on AFM provided detailed information about the molecular mechanism of protein–DNA binding upon activation by effector molecules. These single molecule experiments revealed that the mean lifetime of the bound protein–DNA complex varies depending on the specific effector molecule. Small differences between individual effectors can induce a pronounced influence on the structure of protein–DNA interaction.

2.6
Conclusion

In this chapter we present four different examples for the use of AFM in microbiology research, where lateral resolution in the nanometer range, piconewton force sensitivity and measurements under liquid environments are essential. This demonstrates that AFM is a powerful tool for protein engineering and for the direct investigation of genetic and protein variability of bacterial surface proteins on the level of single cells. Furthermore, it allows the investigation of sequence specific protein–DNA interaction on the level of single molecules elucidating kinetic constants and energy landscape parameters.

Acknowledgements

We thank D. Anselmetti for fruitful discussions and manuscript reading, F.W. Bartels, B. Baumgarth, T. Damberg, A. Fuhrmann, M. McIntosh and P. Plattner for laboratory assistance and gratefully acknowledge our collaboration partners A. Pühler, J. Kalinowski, A. Becker, N. Sewald. The work reviewed was funded by Sonderforschungsbereich 613 of the Deutsche Forschungsgemeinschaft (DFG) and DFG project HA 8530).

References

1 Binnig, G., Quate, C.F. and Gerber, C. (1986) Atomic Force Microscope. *Physical Review Letters*, **56**, 930–933.
2 Hansma, H.G. and Hoh, J.H. (1994) Biomolecular imaging with the atomic force microscope. *Annual Reviews of Biophysics and Biomolecular Structure*, **23**, 115–139.
3 Engel, A. and Müller, D.J. (2000) Observing single biomolecules at work with the atomic force microscope. *Nature Structural Biology*, **7**, 715–718.
4 Ohnesorge, F.M., Hörber, J.K.H., Häberle, W., Czerny, C.P. and Binnig, G. (1998) AFM review study on pox viruses and living cells. *Biophysical Journal*, **73**, 2183–2194.
5 Rotsch, C., Jacobson, K. and Radmacher, M. (1999) Dimensional and mechanical dynamics of active and stable edges in motile fibroblasts investigated by using atomic force microscopy. *PNAS*, **96**, 921–926.
6 Henderson, R.M. and Oberleithner, H. (2000) Pushing, pulling, dragging, and vibrating renal epithelia by using atomic force microscopy. *American Journal of Physiology. Renal Physiology*, **278**, F689–F701.
7 Dufrene, Y.F. (2001) Application of atomic force microscopy to microbial surfaces: from reconstituted cell surface layers to living cells. *Micron*, **32**, 153–165.
8 Binnig, G., Rohrer, H., Gerber, C. and Weibel, E. (1982) Tunneling through a controllable vacuum gap. *Applied Physics Letters*, **40**, 178–180.
9 Anselmetti, D., Lüthi, R., Meyer, E., Richmond, T., Dreier, M., Frommer, J. and Güntherodt, H.-J. (1994) Imaging biological materials with dynamic force microscopy. *Nanotechnology*, **5**, 87–94.
10 Hansma, P.K., Cleveland, J.P., Radmacher, M., Walters, D.A., Hillner, P.E., Bezanilla, M., Fritz, M., Vie, D., Hansma, H.G., Prater, C.B., Massie, J., Fukunaga, L., Gurley, J. and Elings, V. (1994) Tapping mode atomic force microscopy in liquids. *Applied Physics Letters*, **64**, 1738–1740.
11 Hinterdorfer, P. and Dufrene, Y.F. (2006) Detection and localization of single molecular recognition events using atomic force microscopy. *Nature Methods*, **3**, 347–355.
12 Ros, R., Schwesinger, F., Anselmetti, D., Kubon, M., Schäfer, R., Plückthun, A. and Tiefenauer, L. (1998) Antigen binding forces of individually addressed single-chain Fv antibody molecules. *Proceedings of the National Academy of*

Sciences of the United States of America, **95**, 7402–7405.

13 Schwesinger, F., Ros, R., Strunz, T., Anselmetti, D., Güntherodt, H.-J., Honegger, A., Jermutus, L., Tiefenauer, L. and Plückthun, A. (2000) Unbinding forces of singel antibody-antigen complexes correlate with their thermal dissociation rates. *Proceedings of the National Academy of Sciences of the United States of America*, **97**, 9972–9977.

14 Dammer, U., Popescu, O., Wagner, P., Anselmetti, D., Güntherodt, H.-J. and Misevic, G.N. (1995) Binding strength between cell adhesion proteoglycans measured by atomic force microscopy. *Science*, **267**, 1173–1175.

15 Garcia-Manyes, S., Bucior, I., Ros, R., Anselmetti, D., Sanz, F., Burger, M.M. and Fernandez-Busquets, X. (2006) Proteoglycan mechanics studied by single-molecule force spectroscopy of allotypic cell adhesion glycans. *Journal of Biological Chemistry*, **281**, 5992–5999.

16 Bonanni, B., Kamruzzahan, A.S.M., Bizzarri, A.R., Rankl, C., Gruber, H.J., Hinterdorfer, P. and Cannistraro, S. (2005) Single Molecule Recognition between Cytochrome C 551 and Gold-Immobilized Azurin by Force Spectroscopy. *Biophysical Journal*, **89**, 2783–2791.

17 Vinckier, A., Gervasoni, P., Zaugg, F., Ziegler, U., Lindner, P., Plückthun, A. and Semenza, G. (1998) Atomic force microscopy detects changes in the interaction forces between GroEL and substrate proteins. *Biophysical Journal*, **74**, 3256–3263.

18 Fritz, J., Katopodis, A.G., Kolbinger, F. and Anselmetti, D. (1998) Force-mediated kinetics of single P-selectin/ligand complexes observed by atomic force microscopy. *Proceedings of the National Academy of Sciences of the United States of America*, **95**, 12283–12288.

19 Bartels, F.W., Baumgarth, B., Anselmetti, D., Ros, R. and Becker, A. (2003) Specific Binding of the Regulatory Protein ExpG to Promoter Regions of the Galactoglucan Biosynthesis Gene Cluster of *Sinorhizobium meliloti* - A Combined Molecular Biology and Force Spectroscopy Investigation *Journal of Structural Biology* **143**, 145–152.

20 Baumgarth, B., Bartels, F.W., Anselmetti, D., Becker, A. and Ros, R. (2005) Detailed studies of the binding mechanism of the *Sinorhizobium meliloti* transcriptional activator ExpG to DNA. *Microbiology*, **151**, 259–268.

21 Kuhner, F., Costa, L.T., Bisch, P.M., Thalhammer, S., Heckl, W.M. and Gaub, H.E. (2004) LexA-DNA bond strength by single molecule force spectroscopy. *Biophysical Journal*, **87**, 2683–2690.

22 Eckel, R., Wilking, S.D., Becker, A., Sewald, N., Ros, R. and Anselmetti, D. (2005) Single Molecule Experiments in Synthetic Biology – A New Approach for the Affinity Ranking of DNA-binding Peptides. *Angewandte Chemie-International Edition*, **44**, 3921–3924.

23 Eckel, R., Ros, R., Decker, B., Mattay, J. and Anselmetti, D. (2005) Supramolecular chemistry at the single molecule level. *Angewandte Chemie-International Edition*, **44**, 484–488.

24 Auletta, T., de Jong, M.R., Mulder, A., van Veggel, F.C.J.M., Huskens, J., Reinhoudt, D.N., Zou, S., Zapotoczny, S., Schönherr, H., Vancso, G.J. and Kuipers, L. (2004) Beta-cyclodextrin host-guest complexes probed under thermodynamic equilibrium: thermodynamics and AFM force spectroscopy. *Journal of the American Chemical Society*, **126**, 1577–1584.

25 Zapotoczny, S., Auletta, T., de Jong, M.R., Schönherr, H., Huskens, J., van Veggel, F.C.J.M., Reinhoudt, D.N. and Vancso, G.J. (2002) Chain length and concentration dependence of beta-cyclodextrin-ferrocene host-guest complex rupture forces probed by dynamic force spectroscopy. *Langmuir*, **18**, 6988–6994.

26 Florin, E.-L., Moy, V.T. and Gaub, H.E. (1994) Adhesion forces between individual

ligand-receptor pairs. *Science*, **264**, 415–417.

27 Lee, G.U., Chrisey, L.A. and Colton, R.J. (1994) Direct measurement of the forces between complementary strands of DNA. *Science*, **266**, 771–773.

28 Rief, M., Clausen-Schaumann, H. and Gaub, H.E. (1999) Sequence-dependent mechanics of single DNA molecules. *Nature Structural Biology*, **6**, 346–349.

29 Krautbauer, R., Clausen-Schaumann, H. and Gaub, H.E. (2000) Cisplatin changes the mechanics of single DNA molecules. *Angewandte Chemie-International Edition*, **39**, 3912–3915.

30 Krautbauer, R., Fischerländer, S., Allen, S. and Gaub, H.E. (2002) Mechanical Fingerprints of DNA Drug Complexes. *Single Molecules*, **3**, 97–103.

31 Eckel, R., Ros, R., Ros, A., Wilking, S.D., Sewald, N. and Anselmetti, D. (2003) Identification of Binding Mechanisms in Single Molecule - DNA Complexes. *Biophysical Journal*, **85**, 1968–1973.

32 Rief, M., Oesterhelt, F., Heymann, B. and Gaub, H.E. (1997) Single molecule force spectroscopy on polsaccharides by atomic force microscopy. *Science*, **275**, 1295–1297.

33 Rief, M., Gautel, M., Oesterhelt, F., Fernandez, J.M. and Gaub, H.E. (1997) Reversible unfolding of individual titin immunoglobulin domains by AFM. *Science*, **276**, 1109–1112.

34 Marszalek, P.E., Lu, H., Li, H., Carrion-Vazquez, M., Oberhauser, A.F., Schulten, K. and Fernandez, J.M. (1999) Mechanical unfolding intermediates in titin modules. *Nature*, **402**, 100–103.

35 Hinterdorfer, P., Baumgartner, W., Gruber, H., Schilcher, K. and Schindler, H. (1996) Detection and localization of individual antibody-antigen recognition events by atomic force microscopy. *Proceedings of the National Academy of Sciences of the United States of America*, **93**, 3477–3481.

36 Hinterdorfer, P., Kienberger, F., Raab, A., Gruber, H.J., Baumgartner, W., Kada, G., Riener, C., Wielert-Badt, S., Borken, C. and Schindler, H. (2000) Poly(ethylene glycol): an ideal spacer for molecular recognition force microscopy/spectroscopy. *Single Molecules*, **1**, 99–103.

37 Sleytr, U.B. (1997) Basic and applied S-layer research: an overview. *FEMS Microbiology Reviews*, **20**, 5–12.

38 Beveridge, T.J., Pouwels, P.H., Sara, M., Kotiranta, A., Lounatmaa, K., Kari, K., Kerosuo, E., Haapasalo, M., Egelseer, E.M., Schocher, I., Sleytr, U.B., Morelli, L., Callegari, M.L., Nomellini, J.F., Bingle, W.H., Smit, J., Leibovitz, E., Lemaire, M., Miras, I., Salamitou, S., Beguin, P., Ohayon, H., Gounon, P., Matuschek, M., Sahm, K., Bahl, H., GrogonoThomas, R., Dworkin, J., Blaser, M.J., Woodland, R.M., Newell, D.G., Kessel, M. and Koval, S.F. (1997) Functions of S-layers. *FEMS Microbiology Reviews*, **20**, 99–149.

39 Sleytr, U.B., Bayley, H., Sara, M., Breitwieser, A., Kupcu, S., Mader, C., Weigert, S., Unger, F.M., Messner, P., JahnSchmid, B., Schuster, B., Pum, D., Douglas, K., Clark, N.A., Moore, J.T., Winningham, T.A., Levy, S., Frithsen, I., Pankovc, J., Beale, P., Gillis, H.P., Choutov, D.A. and Martin, K.P. (1997) Applications of S-layers. *FEMS Microbiology Reviews*, **20**, 151–175.

40 Sara, M., Pum, D., Schuster, B. and Sleytr, U.B. (2005) S-Layers as Patterning Elements for Application in Nanobiotechnology. *Journal of Physics-Condensed Matter*, **5**, 1939–1953.

41 Sleytr, U.B., Egelseer, E.M., Ilk, N., Pum, D. and Schuster, B. (2007) S-Layers as a basic building block in a molecular construction kit. *FEBS Journal*, **274**, 323–334.

42 Sleytr, U.B., Huber, C., Ilk, N., Pum, D. and Schuster, B. (2007) S-layers as a tool kit for nanobiotechnological applications. *FEMS Microbiology Letters*, **267**, 131–144.

43 Hansmeier, N., Bartels, F.W., Ros, R., Anselmetti, D., Tauch, A., Pühler, A. and Kalinowski, J. (2004) Classification of

hyper-variable Corynebacterium glutamicum surface-layer proteins by sequence analyses and atomic force microscopy. *Journal of Biotechnology*, **112**, 117–193.

44 Scheuring, S., Stahlberg, H., Chami, M., Houssin, C., Rigaud, J.L. and Engel, A. (2002) Charting and unzipping the surface layer of *Corynebacterium glutamicum* with the atomic force microscope. *Molecular Microbiology*, **44**, 675–684.

45 Chami, M., Bayan, N., Dedieu, J.-C., Leblon, G., Shechter, E. and Gulik-Krzywicki, T. (1995) Organization of the outer layers of the cell envelope of *Corynebacterium glutamicum:* A combined freeze-etch electron microscopy and biochemical study. *Biology of the Cell*, **83**, 219–229.

46 Soual-Hoebeke, E., D'Auria, C., Chami, M., Baucher, M.F., Guyonvarch, A., Bayan, N., Salim, K. and Leblon, G. (1999) S-layer protein production by *Corynebacterium* strains is dependent on the carbon source. *Microbiology*, **145**, 3399–3408.

47 Hansmeier, N., Albersmeier, A., Tauch, A., Damberg, T., Ros, R., Anselmetti, D., Pühler, A. and Kalinowski, J. (2006) The S-layer gene *cspB* of *Corynebacterium glutamicum* is transcriptionally activated by a LuxR-type regulator and located on a 6-kb genomic island from the type strain ATCC13032. *Microbiology*, **152**, 923–935.

48 Cramer, A. and Eikmanns, B.J. (2007) RamA, the transcriptional regulator of acetate metabolism in *Corynebacterium glutamicum*, is subject to negative autoregulation. *Journal of Molecular Microbiology and Biotechnology*, **12**, 51–59.

49 Glazebrook, J. and Walker, G.C. (1989) A novel exopolysaccharide can function in place of the calcofluor-binding exopolysaccharide in nodulation of alfalfa by *Rhizobium meliloti*. *Cell*, **56**, 661–672.

50 Wang, L.X., Wang, Y., Pellock, B. and Walker, G.C. (1999) Structural characterization of the symbiotically important low-molecular-weight succinoglycan of *Sinorhizobium meliloti*. *Journal of Bacteriology*, **181**, 6788–6796.

51 Gonzalez, J.E., Reuhs, B.L. and Walker, G.C. (1996) Low molecular weight EPS II of *Rhizobium meliloti* allows nodule invasion in *Medicago sativa*. *Proceedings of the National Academy of Sciences of the United States of America*, **93**, 8636–8641.

52 Rüberg, S., Pühler, A. and Becker, A. (1999) Biosynthesis of the exopolysaccharide galactoglucan in *Sinorhizobium meliloti* is subject to a complex control by the phosphate-dependent regulator PhoB and the proteins ExpG and MucR. *Microbiology*, **145**, 603–611.

53 Becker, A., Rüberg, S., Küster, H., Roxlau, A., Keller, M., Ivashina, T., Cheng, H.-P., Walker, G.C. and Pühler, A. (1997) The 32-kilobase *exp* gene cluster of *Rhizobium meliloti* directing the biosynthesis of galactoglucan: genetic organization and properties of the encoded gene products. *Journal of Bacteriology*, **179**, 1375–1383.

54 Astete, S.G. and Leigh, J.A. (1996) *mucS*, a gene involved in activation of galactoglucan (EPS II) synthesis gene expression in *Rhizobium meliloti Molecular Plant-Microbe Interactions* **9**, 395–400.

55 Hinterdorfer, P., Schilcher, K., Baumgartner, W., Gruber, H. and Schindler, H. (1998) A mechanistic study of the dissociation of individual antibody-antigen pairs by atomic force microscopy. *Nanobiology*, **4**, 177–188.

56 Evans, E. and Ritchie, K. (1997) Dynamic strength of molecular adhesion bonds. *Biophysical Journal*, **72**, 1541–1555.

57 Merkel, R., Nassoy, P., Leung, A., Ritchie, K. and Evans, E. (1999) Energy landscapes of receptor-ligand bonds explored with dynamic force spectroscopy. *Nature*, **397**, 50–53.

58 Gonzalez, J.E. and Marketon, M.M. (2003) Quorum sensing in nitrogen-fixing

rhizobia. *Microbiology and Molecular Biology Reviews*, **67**, 574–592.

59 Fuqua, C., Parsek, M.R. and Greenberg, E.P. (2001) Regulation of gene expression by cell-to-cell communication: acyl-homoserine lactone quorum sensing. *Annual Review of Genetics*, **35**, 439–468.

60 Pearson, J.P., Delden, C.V. and Iglewski, B.H. (1999) Active efflux and diffusion are involved in transport of *Pseudomonas aeruginosa* cell-to-cell signals. *Journal of Bacteriology*, **181**, 1203–1210.

61 Marketon, M.M. and Gonzalez, J.E. (2002) Identification of two quorum-sensing systems in *Sinorhizobium meliloti*. *Journal of Bacteriology*, **184**, 3466–3475.

62 Marketon, M.M., Groquist, M.R., Eberhard, A. and Gonzalez, J.E. (2002) Characterization of the *Sinorhizobium meliloti sinR/sinI* locus and the production of novel N-acyl homoserine lactones. *Journal of Bacteriology*, **184**, 5686–5695.

63 Pellock, B.J., Teplitski, M., Boinay, R.P., Bauer, W.D. and Walker, G.C. (2002) A LuxR homolog controls production of symbiotically active extracellular polysaccharide II by *Sinorhizobium meliloti*. *Journal of Bacteriology*, **184**, 5067–5076.

64 Hanzelka, B.L. and Greenberg, E.P. (1995) Evidence that the N-terminal region of the *Vibrio fischeri* LuxR protein constitutes an autoinducer-binding domain. *Journal of Bacteriology*, **177**, 815–817.

65 Vannini, A., Volpari, C., Gargioli, C., Muraglia, E., Cortese, R., De Francesco, R., Neddermann, P. and Di Marco, S. (2002) The crystal structure of the quorum sensing protein TraR bound to its autoinducer and target DNA. *The EMBO Journal*, **21**, 4393–4401.

66 Stevens, A.M., Dolan, K.M. and Greenberg, E.P. (1994) Synergistic binding of the *Vibrio fischeri* LuxR transcriptional activator domain and RNA polymerase to the *lux* promoter region. *Proceedings of the National Academy of Sciences of the United States of America*, **91**, 12619–12623.

67 Marketon, M.M., Glenn, S.A., Eberhard, A. and Gonzalez, J.E. (2003) Quorum sensing controls exopolysaccharide production in *Sinorhizobium meliloti*. *Journal of Bacteriology*, **185**, 325–331.

68 Bartels, F.W., McIntosh, M., Fuhrmann, A., Metzendorf, C., Plattner, P., Sewald, N., Anselmetti, D., Ros, R. and Becker, A. (2007) Effector-Stimulated Single Molecule Protein-DNA Interactions of a Quorum-Sensing System in *Sinorhizobium meliloti*. *Biophysical Journal*, **92**, 4391–4400.

3
Cellular Cryo-Electron Tomography (CET): Towards a Voyage to the Inner Space of Cells
Juergen M. Plitzko

3.1
Introduction

"One picture is worth 10 000 words"; this slogan from former times depicts clearly the fact that human beings are, by and large, visually centered. It definitely holds true for scientists as well, no matter which branch they belong to. A long hold dream, for example, of biologists is the ability to "zoom" in on very fine details of living matter, literally in one go, from the complete organism to one single cell and beyond, mainly with the help of one single microscope. However, today, we have to utilize different microscopes operating at different resolution levels to make this dream halfway come true. Therefore researchers use different probing signals, for example, photons, electrons and X-ray quanta to cover the different length scales and to visualize the gamut of organic and cellular functions. The cornucopia of available imaging methods and techniques in the twentyfirst century is enormous, and we have already travelled a long way to reach our ultimate goal – the voyage to the inner space of cells – but we are clearly not there, at least not yet. At the level of a single cell we might think that we already know a great deal, but at the supramolecular level the cell is still an uncharted territory, and only a few "hot spots" have been studied in greater detail. Electron microscopy for example is an established tool in structural research, despite its relatively young lifetime (roughly 70 years have passed since it first came into the world [1, 2] compared to the hundreds of years old "methuselah" light microscopy (Figure 3.1).

Biologists have used the electron microscope practically from its beginning and nowadays the textbook "picture" of a cell at the organelle level is a direct result out of it. They have been ingenious in designing and implementing suitable preparation techniques for the soft and very frail living matter to be studied in the inhospitable environment of an electron microscope: ultra-high vacuum and a constant electron "bombardment." While the first condition literally sucks out every trace of liquid and almost 70% of a cell is built up of water, electrons cruising at a fraction of the speed of light are as harmful as any electromagnetic radiation is at one point, thus making radiation damage a big challenge (for a retrospective, see Ref. [3]). Moreover living

Single Cell Analysis: Technologies and Applications. Edited by Dario Anselmetti
Copyright © 2009 WILEY-VCH Verlag GmbH & Co. KGaA, Weinheim
ISBN: 978-3-527-31864-3

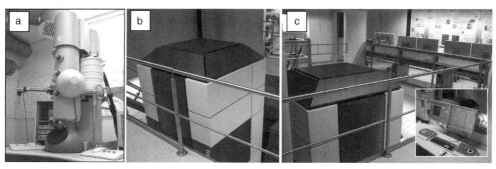

Figure 3.1 Dedicated cryo-transmission electron microscopes. (a) Tecnai F30 Helium ("Polara") microscope from FEI Company, Eindhoven, The Netherlands as installed at the MPI of Biochemistry in Martinsried (near Munich), Germany. This cryo-EM comprises a manual operated multispecimen stage (six samples) and it can be used at liquid nitrogen or helium temperature. (a) and (c) latest generation of computer controlled TEM dedicated for cryo-EM, combining the latest technology in automation and robotics (especially for sample mounting). A microscope in a box, the worldwide first Titan Krios (FEI Company, Eindhoven, The Netherlands) as installed at the MPIB in Martinsried, Germany. Both microscopes operate at 300 kV acceleration voltage.

matter is primarily built up of carbonaceous compounds with very lightweight elements (e.g. carbon, oxygen, hydrogen, nitrogen, etc.) and it is easily understandable that the "fast" electrons do not necessarily interact strongly with these atoms (very small elastic cross-sections) and molecules constituting the biological structures. This directly results in a very low or barely visible contrast in single electron micrographs. Of course, solutions have been found over the years to offer a partial remedy to these problems; negative staining and metal shadowing to enhance the contrast were introduced in the late 1950s and early 1960s. However the problem of dehydration (severely altering the structure under investigation) and radiation damage were addressed in the 1970s with low-dose imaging techniques and vitrification, the embedding of cells or macromolecules in vitrified (amorphous) ice in the 1980s [4–6]. During the 1980s the term "cryo" entered the field of electron microscopy, since ice-embedded samples have to be kept "cold" (namely at the temperature of liquid nitrogen −196 °C) at all times, even during investigation within the electron microscope. Building on this basis, over the past 20 years cryo-electron microscopy (cryo-EM) has become one major tool in structural research of unstained (and thus unaltered) biological substances in a hydrated frozen state. Cryo-EM comprises three major branches: cryo-electron crystallography, cryo-electron microscopy of purified single particles (where "particles" stand for proteins and macromolecular complexes) and cryo-electron tomography (the three-dimensional investigation of any nonperiodic – "pleiomorphic" – object) [7, 8].

Cryo-electron tomography (CET) for the study of cellular structures at macromolecular resolution is based on the principle of any "tomographic" investigation technique: the acquisition of images from different viewing angles of a three-dimensional (3D) object and the subsequent reconstruction of that particular

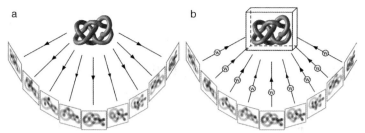

Figure 3.2 The figure shows a single axis tilt series data acquisition scheme. The object is represented by a pleiomorphic object (in this case a knot) to emphasize the fact that electron tomography can retrieve 3D information from nonrepetitive structures. (a) A set of 2D projection images is recorded while the specimen is tilted incrementally around an axis perpendicular to the electron beam, and projection images of the same area are recorded on a CCD camera at each tilt angle. Tilt range is typically ±70° with tilt increments of 1.5–3.0°. (b) The backprojection method explains the principle of 3D reconstruction in a fairly intuitive manner. For each weighted projection, a backprojection body is calculated, and the sum of all projection bodies represents the density distribution of the original object – the tomogram.

structure at hand (Figure 3.2). This is known to everyone nowadays from computer-assisted tomography (CT) or magnetic resonance imaging (MRI) commonly used in almost every large hospital around the world for diagnosing internal medical problems, ranging from simple bone or ligament fractures to highly advanced investigations of cancerous tumors, for example in combined brain mapping and brain surgery procedures [9, 10].

The idea of using a tomographic acquisition technique in electron microscopy is "quite" old. Already in 1968 and in the early 1970s two groups independently worked on this method, namely Roger Hart at the Lawrence Livermore National Laboratory (California, USA) and Walter Hoppe at the Max-Planck Institute of Biochemistry (Martinsried, Germany) [11, 12]. They reported for the first time on experiments with the transmission electron microscope (TEM), rotating the specimen holder, the recording of micrographs from different views and the subsequent combination of all acquired images to obtain a three-dimensional structure. A third group, namely Aaron Klug and David DeRosier at the MRC Laboratory of Molecular Biology (Cambridge, UK) circumvented the laborious data acquisition procedure, but showed for the first time a 3D reconstruction of the tail of a T4 bacteriophage, literally based on a single electron micrograph [13]. In any case, all three research groups did start the field of 3D electron microscopy nowadays called 3DEM. However it took quite some time, mainly due to instrumental and technical restraints, before the first cryo-electron tomogram of a eucaryotic cell was shown [14] (Figure 3.3). Since then cryo-ET or CET has become increasingly important and its impact can be seen in the rising amount of publications addressing CET or simply utilizing its capabilities for cellular studies with almost macromolecular resolution ([67]; [75]; [15–19]; Figure 3.4).

With this chapter we are trying to explain in detail this method to shed light on several key issues almost intrinsically linked to it and to teach the reader in the major

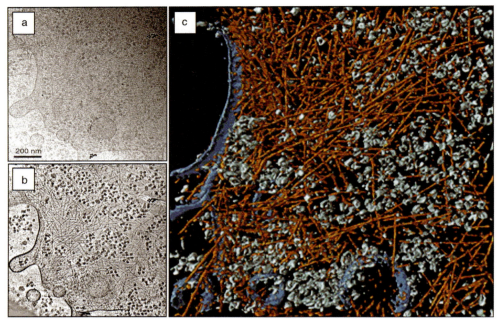

Figure 3.3 First electron tomographic investigation of a eukaryotic cell; the slime mould *Dictyostelium discoideum* embedded in vitrified ice. (a) Cryo-electron micrograph at 0° tilt (conventional 2D projection) of an approximately 200 nm peripheral region of the cell. (b) x–y slice of a tomographic reconstruction from a complete tilt-series (120 images). (c) Visualization by segmentation. Large macromolecular complexes, for example ribosomes, are shown in green, the actin filament network in orange-red and the cells' membrane in blue. Cryo-tomograms of *D. discoideum* cells grown directly on carbon support films have provided unprecedented insights into the organization of actin filaments in an unperturbed cellular environment [14].

concepts of tomographic imaging with the electron microscope for structural biology. Moreover, we want to provide an outlook into future developments especially regarding hybrid approaches, where several methods are combined to work in unison for the one goal, which we have already stated – a voyage to the inner space of a cell [20, 21].

3.2
Tomography with the Electron Microscope – a Practical Perspective

3.2.1
Sample Preparation

Ahead of every good menu lies the recipe and most important the suitable ingredients. However, the outcome is highly dependent on the "cook" and furthermore on

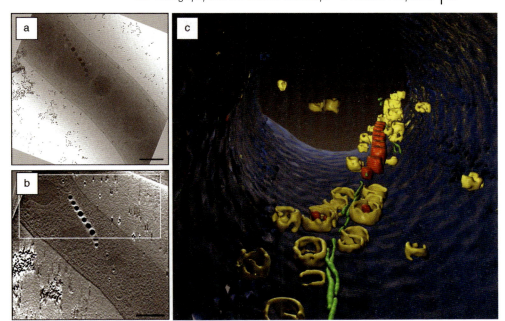

Figure 3.4 Cellular cryo-electron tomography of the magnetotactic microorganism *Magnetospirillum griphiswaldense*. The entire bacterium is oriented like a compass needle inside the magnetic field in its search for optimal living conditions. The miniature cellular compass is made by a chain of single nano-magnets, called magnetosomes. (a) The 2D image represents one projection (at 0°) from an angular tilt-series. (b) x–y slices along the z axis through a typical 3D reconstruction (tomogram). (c) Surface-rendered representation of the inside of the cell showing the membrane (blue), vesicles (yellow), magnetite crystals (red) and a filamentous structure (green). Until now, it was not clear how the cells organise magnetosomes into a stable chain, against their physical tendency to collapse by magnetic attraction. However biochemical analysis revealed a protein responsible for the chain formation and the 3D investigation a cytoskeletal structure which aligns the magnetosomes like pearls on a string [17].

the quality of all the trimmings. In cryo-electron tomography it is quite similar; ahead of any successful CET study is the sample and the quality of the sample preparation. Therefore here we will focus on cryo-sample preparation suitable for cellular tomography.

The most common procedure to get a vitrified or frozen hydrated specimen is called "plunge freezing" (sometimes also addressed as "shock or rapid freezing") [4, 5]. The major requirement is to circumvent any crystalline ice formation, which would have adverse effects in the imaging process and which could potentially result in catastrophic damage to the cells to be frozen. The latter contrary effect is mainly due to the all too well known anomaly of water: its volume increase during crystallization, which understandably disrupts the internal cellular integrity. The whole process of freezing therefore has to be done as quickly as possible to give the present water literally no time to start to crystallize. Typically this is done with a

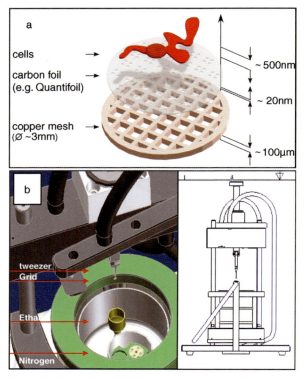

Figure 3.5 Plunge freezing instrumentation. (a) Typical arrangement of cells cultured on TEM grid just prior to plunging. The schematic indicates the dimensions. (b) Computer-aided design (CAD) image of a home-built plunge freezing apparatus (MPI of Biochemistry, Martinsried (near Munich), Germany; courtesy of R. Gatz). The small reservoir (yellow) in the middle of the dewar (green) contains the liquid ethane, which is chilled by a surrounding bath of liquid nitrogen. The forceps which hold the grid are attached to a weighted arm. When the arm is released by means of a foot-trigger, the grid is gravity-plunged into the ethane. The cross-sectional sketch (b, left) illustrates the "guillotine-like" arrangement.

"plunger," a guillotine-like apparatus which makes use of the earth gravitational force (Figure 3.5). While holding a standard EM grid with an inverted tweezer, this assembly is released into a liquid ethane bath, which is cooled down with liquid nitrogen to approximately −180 °C (to prevent solidification of the liquid ethane). Liquid ethane is commonly used because it has somehow higher cooling rates than pure liquid nitrogen and thus it boosts the success rate for vitrification. Before the final plunging process takes place the sample solution, typically cells in suspension or within their buffer solution, are placed with a microliter pipette on the carbon coated side of an EM grid. An amount of 1–4 μL is normally sufficient; however this droplet is visible with the "naked eye" and thus far too big to be directly imaged in a frozen state by transmission electron microscopy. Therefore, a fraction of this solution has to be removed as gently as possible. This is typically done with a filter paper, in the case of cells, brought to the backside of the EM grid. This way the excess solution will be

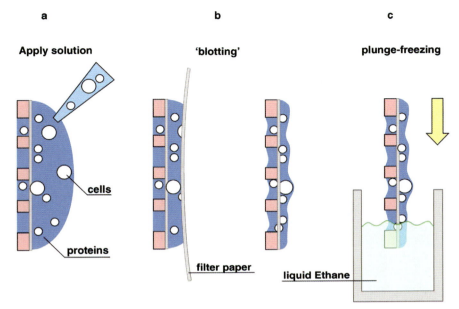

Figure 3.6 Steps to be taken in the plunge freezing process. (a) Apply the solution containing protein complexes or cells (TEM grid illustrated in cross-section). To remove the excess solution and thus adjust the ice layer thickness, (b) blotting is done with a filter paper, carefully brought to the front- or backside of the TEM grid. After blotting the remaining solution on the grid is rapidly plunged (c) into liquid ethane and afterwards kept at all times at liquid nitrogen temperature to prevent contamination, crystallization or smelting of the amorphous (vitrified) ice layer.

soaked away by the filter paper, leaving behind a very thin aqueous film. The whole process is called "blotting" and it is done from the backside for cellular cryo-preparations ("backside blotting") to minimize the risk of damaging the cells (e.g. the pressure applied with the blotting paper). Directly after the blotting the tweezer-EM grid assembly is released and rapidly falls down, driven by gravitational force, into a tiny liquid ethane dewar for the final step: vitrification of the sample (Figure 3.6). This frozen hydrated specimen is then transferred into a cryo-storage box for later investigation in the electron microscope. There are two necessary pre-preparation steps for successful sample preparation. Typically the carbon coating of the EM grid is hydrophobic, making it quite difficult to place an aqueous droplet of cellular solution on top of it. Therefore, the EM grids are first "glow discharged" to make them hydrophilic. Second, to assist the alignment of low-contrast electron micrographs (we will cover that topic in the next paragraph) a solution of gold nanoparticles (typically spherical particles with a diameter of 10 nm) is applied on top of the carbon foil. One might also add that gold solution to the sample itself, but either way, it has to be done to facilitate the very crucial alignment procedure of the angular series of images required for tomography. "Glow discharging" is achieved by exposing the EM grids to a "plasma" for a couple of seconds. In this way the "passivation layer" of the carbon coating, typically a layer of hydroxyl groups, is removed and the carbon is turned to a more hydropilic state.

The whole procedure of plunging sounds simple and rather straightforward; however there are several environmental and other factors which can influence the outcome severely, for example the salt concentration of the applied sample solution. As stated already crystals do have adverse effects on the imaging process in the TEM and salty solutions tend to form salt crystals during plunging. So the salt content should be kept as low as possible, quite difficult if one aims for example to study halophilic bacteria cells. Of course a crucial factor is not only the salt concentration but also the concentration of cells within the solution. If there are too many cells present, they literally "stick" together within the vitrified ice layer and thus hamper the tomographic acquisition process. Closely neighboring cells overlay each others if viewed from different angles and thus obscure or even completely block finer details within the cell of interest. So, cells should be distributed evenly and at best separated. Another essential requirement lies in the size of the cells. If they exceed a diameter of more than 500 nm, they will be almost too thick to let for example 300 keV electrons pass through for the image acquisition, especially at higher tilt angles (see below). They can be of course several microns long, but in the direction of the incoming/probing electron beam they should be below this critical threshold value of 500 nm. This is also slightly dependent on the "crowding" of the interior of the cell type to be studied. This "macromolecular crowding" is a real problem in both eucaryotic and procaryotic cells [22]. Even procaryotes, cells without a nucleus, are not just a "bag of (freely diffusing) enzymes," they are densely packed with molecules of different sizes, literally touching each other [23]. A prime example of macromolecular crowding is the nucleus itself; typically several microns thick, thus definitely too thick to be penetrated by electrons and definitely very dense to be studied as a whole entity by electron tomography. Additionally one has to keep in mind that the "freezing depth" of the standard plunge freezing procedure is limited to roughly $10\,\mu m$; and beyond that thickness we again face the difficulty of ice crystal formation or even incomplete freezing of the specimen.

For "thick" cells (like mammalian cells) we cannot just prescribe a diet to get them thinner, we have to find other means of a suitable preparation or to choose suitable genetically modified "miniature" versions of their pristine counterparts. Larger cells or even tissue for example can be addressed with high-pressure freezing and subsequent cryo-sectioning [24, 25]. Cryo-sectioning works in principle like standard sectioning with a microtome; the material of interested is embedded in a support matrix and with the help of a diamond or glass knife thin slices, respectively sections are generated (Figure 3.7). At room temperature, samples are typically embedded in epoxy resins and sections down to 50 nm in thickness can be fabricated and directly placed on an EM grid. For cryo-sections the support matrix has to be vitrified ice, which has to be kept "cold" (at liquid nitrogen temperature) at all times to prevent ice crystal formation or even smelting of the sample. A normal plunge frozen sample cannot be used for sectioning, first it is too small and the whole "sandwich" (sample material in ice, carbon and copper) does exhibit different materials to the cutting knife, with different material and more important with different cutting properties. Therefore one is using high-

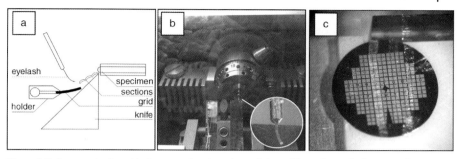

Figure 3.7 Cryo-sectioning. (a) Cross-sectional schematic of the cryo-sectioning process with a cryo-ultra-microtome. (b) View into the cryo-ultramicrotome during sectioning. The inset shows a magnified view of the trimmed copper tube and the "ribbon" formed after several cutting cycles which becomes very visible in (c) image of a cryo-section placed on a TEM grid (images courtesy of A. Leis and M. Gruska, MPI of Biochemistry, Martinsried, Germany).

pressure frozen samples, where only the sample embedded in the vitrified ice will be sectioned. For high-pressure freezing (HPF) the cell solution is forced into a copper capillary (15 mm copper tube, inner diameter of approximately 0.3 mm). In a next step an apparatus is utilized applying a high pressure typically in the range of 2000 bar in a few milliseconds and as soon as the pressure is established cooling of the specimen takes place, for example with a jet of pressurized liquid nitrogen. The increase in pressure lowers the freezing rate and the lowest values are observed at around 2000 bar, where the depth of vitrification increases up to 10 times as compared to freezing at ambient pressures [26]. The high-pressure freezing device is commercially available and rather easy to use. The successful use of the cryo-ultramicrotome, however, is highly dependent on the "cook," typically a highly trained specialist. Since everything has to be done at liquid nitrogen temperature (either in the gaseous or liquid phase) one just cannot pick the fabricated sections with a tweezer or let them float on a water surface, as normally done with epoxy resin embedded samples. For cryo-sections one uses an eyelash within the cryo-chamber to place them on top of a standard EM grid. Moreover vitrified ice is an insulator and the whole low temperature atmosphere is additionally prone to electrostatic charging, thus it is even more difficult for reproducible handling the sections under these conditions.

Before cutting sections, the copper jacket has to be removed. This is done with a diamond trimming device to form a square/block face of about 150–200 µm. Afterwards the sections can be cut in a serial fashion and, due electrostatic charging, the single sections tend to stick together to form a ribbon of consecutive cuts which then can placed with the already mentioned eye lash on top of a standard EM grid. Almost a whole scientific field has evolved around cryo-sectioning in recent years to analyze and overcome the difficulties and to lessen the amount of potential cutting artefacts (deformation, crevasses, chatter, knife marks, etc.). Therefore we will not continue to elaborate on cryo-sectioning but instead refer to numerous publications about cryo-sectioning and related cutting artefacts where everything is addressed exhaustively. It is just noteworthy that it is much easier to obtain almost artefact free thin sections (below 50 nm), which are of course not really ideal to study by

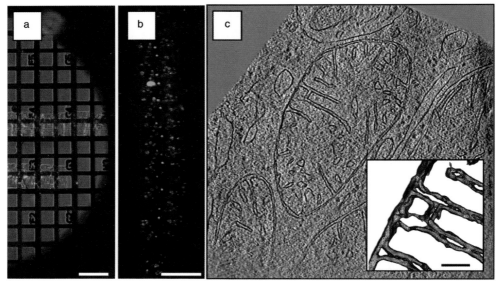

Figure 3.8 Example for TOVIS: fluorescence images of vitrified HL-1 cardiomyocytes after staining with Mitotracker Green FM and vitreous sectioning with a nominal microtome thickness setting of 150 nm. (a) superimposed fluorescence and phase contrast micrographs of ribbons of vitreous sections on a 200 mesh copper finder grid (scale bar: 150 μm). (b) fluorescence micrograph of a ribbon of sectioned HL-1 cardiomyocytes. Serially sectioned mitochondria visible along the length of the ribbon (scale bar: 50 μm). (c) digital x–y slice from the reconstructed tomogram. Three mitochondria as well as endoplasmic reticulum are especially prominent. Inset shows a surface rendering of the "H" connection (scale bar 60 nm; [67]).

tomography. The thicker the sections get (above 150 nm), the more artefacts will be present. However, both, thin or thick sections do not lie flat on the grid surface (another difficulty) which will hamper the acquisition of a tomographic tilt series. The two-dimensional microscopy of sections is named cryo-electron microscopy of vitrified sections (CEMOVIS) and the three-dimensional counterpart is called tomography of vitrified sections (TOVIS; Figure 3.8).

It has been shown that with cryo-sectioning real "big stuff" can be investigated and that for example the nucleus is not anymore an impossible object to be studied with almost molecular resolution tomography [75]. In any case, while plunge freezing is fairly simple and widespread the amount of laboratories producing cryo-sections is steadily increasing and it is not really astonishing since many secrets lie ahead of us to be discovered within mammalian cells and tissue samples.

3.2.2
Instrumental and Technical Requirements

Having now concluded with the most critical part in any transmission electron microscopy study, sample preparation, we will now focus on the instrumental and

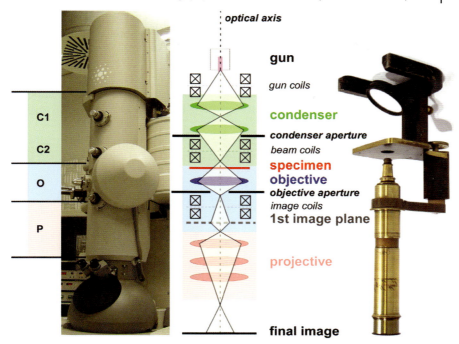

Figure 3.9 Major electron optical components of a transmission electron microscope. Representative image of a TEM (Tecnai F30 "Polara") comprising a field emission gun (FEG) a double condenser lens system (C1, C2), objective lens system (O) and a projective system (P). Noteworthy are the three different deflector coils, especially the beam and image coils, which are crucial for the automation in electron tomography (beam tilt used for "auto"-focusing) and beam shift/image shift are used in conjunction for beam and image centering ("auto"-tracking).

technical requirements to perform electron tomography on cellular samples. In this rather short chapter we are not able to give the reader a complete manual in microscope operation, nor a step by step guide how to produce a tomographic tilt series. Nevertheless we aim to give a comprehensive summary of the basic concepts of 3D microscopy.

In the beginning of course is the instrument, a transmission electron microscope (TEM), which is built up of an electron source and three major lens systems: the condenser, the objective and the projective lens system, quite analogous to a conventional light microscope if oriented upside down (Figure 3.9). The most important lens is the objective, which almost exclusively determines the quality of the recorded image. Its characteristics, mainly the geometric (spherical) and chromatic aberrations, influence the transfer of electrons and together with the illumination source (thermionic or field emission) and the chosen focus determines the achievable resolution. This is elegantly summarized in the contrast transfer function (CTF), which can be calculated and measured precisely and which enables us to determine suitable experimental settings before we even start an image acquisition.

The sample to be investigated is located within the objective lens system. This is a really "tiny" space (a couple of millimeters) and this opening directly affects the magnitude of the spherical aberration coefficient (C_s). In simple terms one could say the bigger this gap, the higher is the C_s and the higher the resulting contrast, but the lower the resulting resolution. However, the sample holder with the specimen is located inside the objective lens system and therefore the gap dimensions cannot be too small, especially if one would like to tilt the holder, for example for an angular acquisition.

Today's microscopes are operated typically in an acceleration range between 120 kV and 300 kV. The voltage used to accelerate the electrons gives them their speed (e.g. approximately two-thirds of the speed of light at 300 keV) and this is directly related to their penetration depth, which is typically expressed with the mean free path (the average distance covered by a particle between subsequent impacts). One has to keep in mind that electrons interacting with the sample atoms can be scattered elastically (no loss in energy but change in direction) and inelastically (loss of energy due to interaction with core shell electrons). While the first type of electrons carry the main portion of structural information, inelastically scattered electrons form a strong background and furthermore contribute to a blur of the image. The inelastic mean free path of vitrified ice at 300 kV at liquid nitrogen temperature is around 300 nm. Thus for specimen with a thickness above 300 nm multiple scattering is inevitable and especially for biological specimen (mainly built up of light atoms) inelastic scattering is much stronger than elastic scattering. Now it is quite understandable why we cannot investigate micron-thick samples with a standard TEM operated at 300 kV and why we are restricted to this 500 nm threshold value. To lessen the extent of "image blurring" (or in other words to "sharpen" the image) due to inelastic scattered electrons, nowadays microscopes for cellular tomography are equipped with so-called energy filters, an additional instrument attached to the microscope (either in-column or post-column) which separates electrons with different energies, making use of the fact that electrons with different kinetic energies will travel on different circles if exposed to a homogeneous magnetic field (Lorentz force). With the help of a simple slit aperture at the end of a magnetic prism, only the electrons with none or only a minor energy loss are selected to contribute to the final image. This procedure is called zero-loss filtering and nowadays used in almost every cellular cryo-electron tomography study [27, 28] (Figure 3.10).

Unlike CT or MRI methods where the patient remains still and the detector and probing device X-rays or a strong magnetic field) is slowly rotated around him, electron tomography tilts the sample incrementally around the probing electron beam, while for every angular increment an image is recorded. The tilting device is normally located outside the EM column and it is called the goniometer tilt stage and in some instances the "compustage," because of its ability to be directly computer controlled. This tilting is exclusively mechanical and it is limited by the holder–sample geometry to $\pm 70°$ due to the limited spacing within the objective lens polepieces, the slab geometry of the holder and moreover the planar geometry of the object itself. The missing angular region can be nicely illustrated in reciprocal space (respectively Fourier space), where a wedge shaped "blind" region is formed

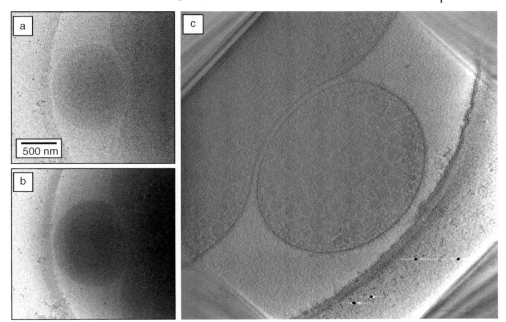

Figure 3.10 Influence of zero-loss filtering on the investigation of ice-embedded thick specimens (*Thermoplasma acidophilum*; images courtesy of C. Kofler, MPI of Biochemistry, Martinsried, Germany). Unfiltered (a) and zero-loss (b) filtered images of a *Thermoplasma acidophilum* cell. The specimen thickness is almost 500 nm, which is greater than the mean free path of 300 keV electrons in ice at liquid nitrogen temperature. The slit width was 20 eV. (c) shows a section through a tomographic reconstruction based on the filtered image set.

the so-called "missing wedge". However, the requirement for a distortion-free 3D reconstruction, all projections of the sample over the complete ±90° tilt range [29], is not fulfilled, thus leading to imperfections in the reconstructed object. To reduce the missing information space in single-axis tomography, the acquisition of a second single-axis tilt series, where the tilt axis is rotated in-plane by 90° can offer some remedy. Dual-axis tilting will reduce the "missing wedge" to a "missing pyramid" and thus increase the amount of information up to almost 20% [30, 31].

One can easily imagine that mechanical imperfections of the tilting device result in an imprecise tilting which then add up to displacements in the micron range. This is not really ideal while trying to image nanometer-sized structures within the cellular context. At the moment the best current stages can tilt and move the sample within 0.5 μm accuracy and this gets normally worse when tilting above ±45°. To partially compensate for the still large displacement, automation is mandatory during the acquisition of a tilt series composed out of typically 100 micrographs and therefore the complete computer control of the microscope and the tilting device is absolutely mandatory.

When electron tomography was invented the researches did not have the luxury of a computer controlled microscope and therefore they had to do everything manually, a very time-consuming and laborious procedure. Moreover they had to record the images on negative plates (film) and for the subsequent image processing they had to scan the images before they even could start the reconstruction. Nowadays we have digital cameras (charge couple device (CCD) cameras) for recording and for fast online processing during the acquisition process. These cameras entered the field of electron microscopy in the late 1980s [32] and the first computer-controlled microscopes were introduced around the same time. So it is not really surprising that it took quite some time from the initial experiments to the first successful applications in cellular cryo-electron tomography.

For this reason we have to explain a little about the background for the automation involved in the acquisition of a tilt series, especially if one investigates ice-embedded specimens at cryo-conditions. When biological specimens are irradiated by the electron beam in the electron microscope, the specimen structure is damaged as a result of ionization and subsequent chemical reactions. Ionization occurs if energy is deposited in the specimen as a result of inelastical scattering events. This can primarily induce the heating of the sample in the irradiated area, radiochemical decomposition of water (radiolysis) as well as secondary chemical reactions (breaking of chemical bonds, formation of new molecules or even radicals). Unlike in materials science, where beam damage is almost negligible, radiation damage in life sciences is the fundamental limitation in any cryo-EM investigation.

The amount of damage is proportional to the applied dose and therefore we have to minimize the exposure time to the area of interest quite drastically. This is typically done with so-called "low-dose" acquisition schemes, where only during the time of recording electron micrographs is the electron beam allowed to illuminate the specimen [33, 34]. In all other cases the beam remains blanked. Additionally the microscope can be preset in different states, to reduce the amount of time one needs for changing the magnification or adjusting other electron optical parameters during the process of screening, focusing and the final image acquisition. During the acquisition of a tilt series in electron tomography, where typically 100–200 images are recorded, it is essential that the total applied dose stays below a critical value, which is given by the specific sample at hand. Most biological materials can tolerate an exposure of no more than $10-100\,e^-/\text{Å}^2$, at which point major changes will have become evident in the structure of the specimen. According to the dose-fractionation theorem [35], the integrated dose of a conventional two-dimensional image is likewise sufficent for a three-dimensional reconstruction, if the resolution and the statistical significance between the two are identical. It is therefore feasible to distribute the total applied dose among as many statistically "noisy" 2D images as possible. A total applied dose of $50\,e^-/\text{Å}^2$ distributed among 100 2D images over an angular range would correspond to $0.5\,e^-/\text{Å}^2$ per image. Thus the resulting single images will be of a very noisy quality, which directly demands highly sensitive CCD cameras for image recording.

Automated tomography comprises three major steps for every tilt angle: tracking, focusing and final image acquisition [33, 34, 36–39] (Figure 3.11). While tracking compensates for lateral movements (xy) of the area of interest during tilting, focusing compensates the movement in the direction of the incoming beam (z). In any case,

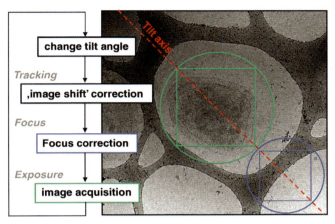

Figure 3.11 A automated data collection scheme for the "full tracking/full focusing" case. For tracking, focusing and the final image acquisition the same magnification is used and the beam is adjusted in a way that the areas where tracking, focusing and exposure is done do not overlap (blue and green circle). Using very short exposure times in combination with high binning values (4×8 or even 8×8) the exposure to the sample in auto-tracking and auto-focusing can be minimized (image adapted from Ref. [68]).

the offset between the object and the tilt axis can be minimized by accurately setting the eucentric height. However, due to the mechanical imperfections of the goniometer, it is very difficult to tune the eucentric height to zero and therefore "auto"-tracking and "auto"-focusing procedures are essential.

Under strict low-dose conditions, the tracking and focusing steps are normally carried out adjacent to the area of interest, to reduce the dose that is delivered to the specimen area. These areas are positioned along the tilt axis with respect to the actual acquisition area to determine the correct shift and focus levels prior to the final image recording. During the tracking step a micrograph with a very low exposure time at a higher binning mode of the CCD camera is acquired, which is correlated to a micrograph from the previous tilt. Utilizing cross-correlation or by a manual inspection of prominent features (e.g. gold beads), the shift relative to the preceding micrograph can be determined. Depending on the signal to noise ratio (SNR) of the image, the cross-correlation algorithm might fail and has to be supported by suitable image processing routines, for example, band-pass filtering or the use of a Hanning window. This is especially the case at higher tilt angles where the projected thickness increases. Owing to the planar (slab) geometry of the sample–holder arrangement, the transmission path of the electrons will increase. Simple geometrical calculations show that the transmission path of the electrons for a specimen initially 100 nm thick doubles at 60° (200 nm), is almost triple at 70° (290 nm) and the projected thickness increases more than five times at 80° (590 nm). Thus at very high tilt angles the image contrast is dramatically reduced, owing to the increase in multiple scattering. There are two possible ways to account for this effect: a tilting scheme with a finer sampling at higher tilt angles and an increase in exposure time for higher tilt angles to obtain a similar SNR within the projections of a tilt series ("Saxton scheme"; [40]).

Today's acquisition schemes and software packages can be grouped in roughly three major domains: the almost classic "full tracking/full focusing", the "pre-calibrated" and the "on the fly or dynamical prediction" procedure. Naturally, there are various derivatives and hybrids of these three different acquisition models, especially modified to serve specific purposes and applications. The full tracking/full focusing scheme (as described above) is the most common, due to the fact that it can be literally used universally, for example with almost any kind of sample and microscope and holder system. For the other schemes, we refer to the original publications [41, 42].

3.2.3
Alignment, Reconstruction and Visualization

In order to obtain the 3D image from a set of acquired projections two initial and one final step have to be carried out: first, the individual micrographs need to be aligned to a common coordinate system and, second, the actual 3D reconstruction of the tomographic volume. The third step is the visualization and interpretation of the tomogram. In this context one has to understand that cellular tomograms are, by and large, 3D images of the entire proteome of the cell and contain an imposing amount of information. Although one can see almost "everything," identifying what one actually sees may be difficult in the crowded macromolecular environment where complexes literally touch each other. Therefore it is quite understandable that the visualization and interpretation of low signal to noise cryo-tomograms is not only quite difficult and tedious but also crucial for an unbiased view in the inner space of a cell.

The compensation of the specimen movement during the data acquisition process will keep the feature focused in most cases but the compensation for the xy displacement is not sufficient for a subsequent reconstruction, which makes a second, more precise alignment necessary. Primarily the alignment procedure has to determine the angle of the tilt axis and the lateral shifts. Other changes, such as magnification changes or image rotation due to large defocus changes during the acquisition, have to be determined and compensated as well if present in the recorded tilt series.

Alignment of the individual micrographs in cellular tomography is normally done by high density markers, so-called fiducial markers. Typically, spherical gold beads with a diameter of ~ 10 nm are added to the sample solution or directly onto the carbon foil prior to the vitrification process. They can be easily recognized within a single 2D micrograph as "black dots," due to their high Z number and thus their pronounced elastic scattering. The coordinates of the markers on each projection are determined manually or automatically. To minimize the alignment error as a function of the lateral translations and the tilt axis angle, an alignment model can be calculated based on least squares procedures [43]. Their locations are then adjusted to a 3D coordinate system. Apart from translations, this procedure for the determination of a common origin often accounts for possible image rotations or magnification changes [43, 44]. However, the larger the number of determined parameters

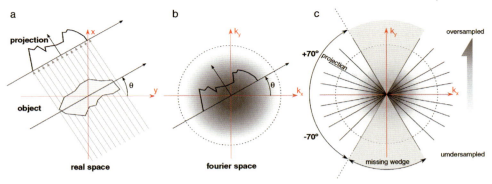

Figure 3.12 "Projection slice theorem". (a) An object is shown in real space (x, y) at the origin and one of its projection images is presented formed by tilted parallel beams. (b) The Fourier transform of the projection is a section through the origin of the Fourier space (k_x, k_y) tilted by θ. (c) illustration of the sampling "density" problem in Fourier space. The large number of sampling points at low frequencies (darker area) are in contrast to the periphery (high frequencies), very only few sampling points are included in the single projections (brighter area). This sampling imbalance results in "blurred" reconstructions, which can be overcome by weighting schemes. The sectors in the Fourier domain, which remain unsampled owing to the limited tilt range (here ±70°), create the "missing wedge".

gets, the more gold beads need to be located to achieve a sufficient significance level during minimization of the residual.

Although the first practical formulation for applied tomography was achieved in the 1950s [45], Johan Radon first outlined the mathematical principles behind the technique in 1917 [46] (see English translation in Ref. [47]). In practice the reconstruction from projections is aided by the understanding of the relationship of a projection in real space and Fourier space (Figure 3.12). The "projection-slice theorem" states that a two-dimensional projection at a given angle is a central section through the three-dimensional Fourier transform of this object. If a series of projections is acquired at different tilt angles, each projection corresponds to part of an object's Fourier transform, thus sampling the object over the full range of frequencies in a central section. The shape of most objects is only partially described by the frequencies in one section but, by taking multiple projections at different angles, many sections can be sampled in Fourier space. This will describe the Fourier transform of an object in many directions, allowing a fuller description of an object in real space. In principle a sufficiently large number of projections taken over all angles provides a complete description of the object. Therefore tomographic reconstruction is possible from an inverse Fourier transform of the superposition of a set of Fourier transformed projections: an approach known as direct Fourier reconstruction. This was the approach formulated by Bracewell [45] and was used for the first tomographic reconstruction from electron micrographs [13].

The direct interpretation of the projection-slice theorem would suggest a reconstruction algorithm based in Fourier space [48]. Despite the fact that the first 3D reconstruction by DeRosier and Klug [13] was carried out in Fourier space, it is common to use real space-based reconstruction algorithms, because the practical implementation of Fourier space reconstruction is not as simple as an inverse transform. By far the most widely utilized algorithm is weighted backprojection [49]. Nowadays, algebraic reconstruction techniques (ART) are also well established, despite the fact that they were initially subject to criticism ([50]; for a detailed review of reconstruction algorithms we refer to Frank, 2006) [7, 8]. The theory of backprojection relies on a simple principle: a point in space may be uniquely described by any three "rays" passing through that point. However the object increases in complexity more "rays" are required to describe it uniquely. A projection of an object is the inverse of such a "ray" and describes some of the complexity of the object at hand. Therefore by inversing the projection, smearing out the projection into an object space at the angle of projection, one generates a "ray" that uniquely describes an object in the projection direction, a process known as backprojection. With sufficient projections, from different angles, superposition of all the backprojected "rays" reproduces the shape of the original object: a reconstruction technique known as direct backprojection. Direct backprojection is used for reconstruction in classic computer-assisted tomography (CT; [51]); and this was the technique used by Hart [11] for the "polytropic montage". However reconstructions made by direct backprojection are exceptionally blurred, showing distinct enhancement of low frequencies, while fine spatial details are reconstructed poorly. This problem is an effect of the uneven sampling of spatial frequencies in the series of original projections. In 2D each of the acquired projections is a line intersecting the center of the Fourier space. Assuming a regular sampling of Fourier space in each projection, far more sampling points are located in the center of Fourier space than in the periphery. The outcome of this is an "undersampling" of the high spatial frequencies and an "oversampling" of the low spatial frequencies of the object, which subsequently results in a "blurred" reconstruction. Therefore to remove the blurring in real space and to restore the correct "frequency balance" in Fourier space, one has to apply weighting schemes. The weighted backprojection consists of two steps: first, the aligned projections are weighted in Fourier space by a function that characterizes the different sampling density in Fourier space. The second step is the backprojection of the weighted micrographs into the reconstruction body, most frequently by trilinear interpolation.

The visualization and interpretation of tomograms at the ultrastructural level requires decomposition of a tomogram into its structural components, for example, the segmentation of intracellular membranes or the assignment of organelles. Manual assignment has been commonly used, since human pattern recognition is often superior to the available segmentation algorithms. In principle however, machine-based segmentation should be more objective and thus unbiased. Continuous structures are relatively easy to recognize and delineate, in spite of the low signal to noise ratio present in cryo-tomograms. For example, visualizing the organization of the cytoskeleton in both *Spiroplasma melliferum* and *Dictyostelium discoideum* was possible at the level of individual filaments, without the need for extensive post-processing [14, 18] (see Figure 3.3).

Instead of addressing the ultrastructure [52], cryo-electron tomography provides the basis for interpreting tomograms even at the molecular level. However analysis and three-dimensional visualization are hampered by a very low SNR. In order to increase the SNR, so-called denoising algorithms have been developed (reviewed in Ref. [53]). These algorithms aim to identify noise and remove it from the tomogram, but in practice they also remove a certain fraction of the signal, resulting in data with reduced information but higher SNR.

Although averaging can obviously not be applied to tomograms of unique structures such as individual cells or organelles, such tomograms may nevertheless contain multiple copies of components such as ribosomes, chaperones, or proteases. Small regions of the tomogram containing isolated complexes can be extracted from the tomogram *in silico* and these subtomograms can be subjected to classification against a library of known structures. The subtomograms can then be aligned to a common orientation and averaged within the appropriate class. The result should be an average with better signal to noise and higher resolution. The original low resolution tomographic image can be replaced by the average or by the higher resolution template itself, if available. The result is a "synthetic" tomogram with a much improved, local signal to noise ratio. Such a procedure was used in a tomographic study of enveloped *Herpes simplex* virions [19] and to also visualize nuclear pore complexes in intact nuclei [16, 54] (Figure 3.13).

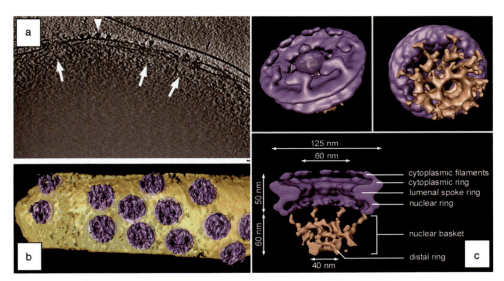

Figure 3.13 Cryo-ET in combination with the single particle approach of transport-competent *Dictyostelium discoideum* nuclei. (a) Three-dimensional reconstruction of the peripheral rim of an intact nucleus. x–y slice of 10 nm thickness along the z axis through a typical tomogram. Side views of nuclear pore complexes (NPCs) are indicated by arrows. Ribosomes connected to the outer nuclear membrane are visible (arrowheads). Inset displays a phase-contrast image and the corresponding fluorescence image. (b) Surface-rendered representation of a segment of nuclear envelope (NPCs in blue, membranes in yellow). (c) Structure of the *Dictyostelium* NPC after classification and averaging of subtomograms. Cytoplasmic face of the NPC (upper left); the cytoplasmic filaments are arranged around the central channel. Nuclear face of the NPC (upper right); the distal ring of the basket is connected to the nuclear ring by the nuclear filaments. Cross-sectional view of the NPC (bottom). The dimensions of the main features are indicated. All views are surface-rendered (nuclear basket in brown; [54, 16]).

Averaging of cryo-ET data offers some real advantages that are worth the additional efforts, at least in some cases. The most important reason is the ability to image nonpurified samples. Cryo-ET is currently the only technique that can be used for quaternary structure determination of fragile or even transient complexes. The degree of alteration that biological macromolecules undergo under phsiological conditions is largely undetermined since there is no existing imaging technique that could resolve them in the context of a cell. Apart from this, averaging of subtomograms also offers one principal advantage compared to averaging from projections: It is fundamentally easier to determine the orientations of an individual copy from 3D data than from single 2D projections.

3.3
Molecular Interpretation of Cellular Tomograms

In tomographic reconstructions of vitrified samples, the macromolecular content of an organelle or cell is present in its native state, thereby making interpretation at the molecular level less problematic. Although the resolution of an individual tomogram may be limited, the advantage is one can see everything in its native context. There is a vast amount of information. Tomograms are 3D images of the entire proteome and they should ultimately enable us to map the spatial relationships of macromolecules in an unperturbed cellular environment. However retrieving this information is confronted with major problems. First, although one can see almost "everything," identifying what one can see may be difficult in the crowded macromolecular environment where complexes literally touch each other [21]. Furthermore, because it is not possible to tilt a full 180°C, the tomographic reconstructions are distorted by missing data, resulting in a nonisotropic resolution. There are essentially only two options for identifying macromolecules in tomograms: specific labeling or pattern recognition methods where complexes are matched against a library of known structures. Of course, the two approaches are not mutually exclusive.

Proteins exposed on the surface of cells or organelles can be labeled with antibodies or specific ligands bound to gold nanoparticles. Such labels provide indicators for the presence of a specific molecule within a broad molecular landscape. Intracellular labeling is more problematic and requires innovative approaches. These approaches could be based on non-invasive genetic manipulations generating covalent fusions with a protein such as metallothionien, which has the potential to bind heavy metals such as gold [55]. Ideally we would like to introduce a label in a time resolved experiment identifying a particular event and to subsequently remove the background as done with the ReAsH compound in fluorescence microscopy [56]. However achieving labeling which can be statistically quantified is a challenging task. It is also hard to imagine that labeling can be developed such that it becomes a high-throughput technology capable of mapping entire proteomes. In order to identify every molecule of interest, the whole procedure of labeling as well as data acquisition and reconstruction of tilt series needs to be repeated. Moreover the unique nature of

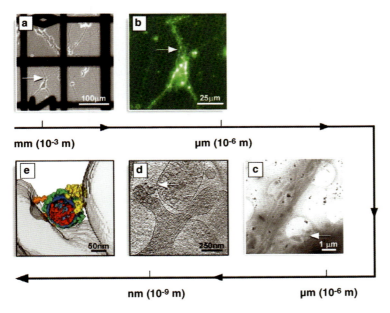

Figure 3.14 Cryo-correlative microscopy: Illustrative example of a light optical and electron optical correlation where light/fluorescence microscopy was used to guide a subsequent cryo-electron tomography investigation. (a) Phase-contrast image and (b) fluorescence image of neurons (neurons were labeled with FM1–43). The arrow points to the spot with a high FM1–43 fluorescence. (c) Cryo-EM image of neuronal processes surrounding the area where the tomogram was taken (arrow). (d) Tomographic slice showing two neuronal processes, an extracellular vesicle connected to one of the processes and to a protrusion from the other one. Additionally an endocytotic invagination on the protrusion is visible. (e) Surface-rendering showing the extracellular vesicle (blue), neuronal processes (gray) and some of vesicle-bound molecular complexes. The vesicle is shown with one side open in order to expose the complexes in its interior (the "running" arrow indicates the field of view (FOV) (image adapted from Refs. [69, 70]).

cellular tomograms makes a direct correlation of the different maps impossible. This in turn poses a major problem in deriving from them molecular interaction patterns. An alternative strategy is to combine fluorescent light microscopy and cryo-electron tomography. Specific fluorescent labels can be engineered into proteins of interest or fluorescent labels can be applied and taken up by the cells. Progress has been made in the development of cryogenic light microscopes [57, 58] which can visualize fluorescence in vitrified specimens (Figure 3.14). This makes it possible to record a fluorescence image of the frozen hydrated specimen and identify targets of interest for a subsequent investigation by cryo-electron tomography. After imaging in the light/fluorescence microscope, the specimen is transferred directly to the EM where a tomographic tilt series can be recorded at the identified areas. Software-based methods

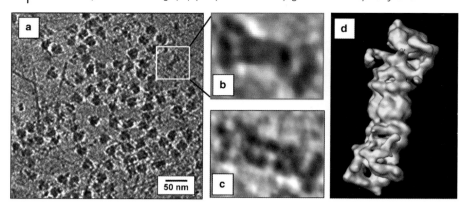

Figure 3.15 Visualization and identification of a 26S Proteasome in *Dictyostelium discoideum* grown directly on an EM grid and embedded in vitreous ice. (a) x–y slice from a tomogram. Dominant features are ribosomes and actin filaments. (b) The magnified "particle" is a 26S proteasome (unaveraged projection of a stack of slices from tomogram; for details, see Refs. [14, 20]). Although in this case the detection and identification was facilitated by the large size (~2.5 MDa) and the peculiar shape of this complex, it indicates that a molecular signature-based approach to mapping cellular proteomes should become feasible. (c) x–y slice of a tomogram of purified 26S proteasomes for comparison and (d) averaged structure of the 26S proteasome based on a 2D cryo-EM single particle analysis (for details see Refs. [71, 72]).

can be used to align the fluorescence image with a low-magnification image from the EM, locating the desired areas in the EM overview image.

The identification of a single 26S proteasome in the cytoplasm of a *Dictyostelium* cell suggested that a template matching approach could be used for mapping cellular proteomes [14] (Figure 3.15). Given that one can see "everything" in a tomogram, it makes sense to probe the tomogram with a library of templates derived from known structures using intelligent pattern recognition algorithms. High-resolution structures of numerous macromolecules are available from X-ray crystallography or NMR. The intent is to locate these known structures and determine the molecular context in which complexes are organized in organelles or cells, with less emphasis on the discovery of novel molecular features and more on determining whether there is some correlation in the spatial arrangement of complexes, relative to one another. To achieve this, a "template matching" strategy is being pursued. Simulations and experiments with "phantom cells," that is liposomes encapsulating macromolecules, indicated that such an approach is feasible [59–61].

However it is computationally intensive, because not only must the positions matching a given template be determined but also their spatial orientations have to be identified. The tomogram must be analyzed with every template and the best match determined for each complex in the tomogram. Ideally such a multi-template search would result in a three-dimensional map with each low resolution complex identified replaced by the higher resolution template it matched in the orientation which was determined (Figure 3.16). At the moment only large complexes, such as the ribosome, have been identified with an acceptable accuracy (>95%) at the routinely attained resolution in tomograms of 4–5 nm [62]. An improvement in resolution to

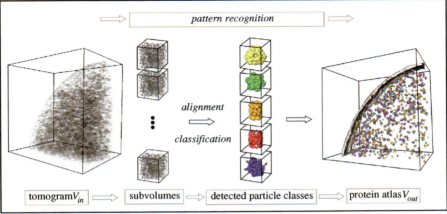

Figure 3.16 Detection and identification of individual macromolecules in cellular tomograms is based on their structural signature. Because of the crowded nature of the cytoplasm and "contamination" with noise, an interactive segmentation and feature extraction is not feasible. It requires sophisticated pattern recognition techniques to exploit the information contained in the tomograms. A volume rendered presentation of a 3D image is presented on the left (V_{in}). Even though some high-density features may be visible, an unambiguous identification of individual structures would be difficult if not impossible given the residual noise. An approach, which has proven to work, is based on template matching: Templates of the macromolecules under scrutiny are obtained by a high- or medium resolution technique (X-ray crystallography, NMR, electron crystallography or single particle analysis). These templates (magnified in this figure (from top to bottom): tricorn, VAT, 20S proteasome thermosome, ribosome) are then used to search the entire volume of the tomogram systematically for matching structures by 3D cross-correlation and the result is refined by multivariate statistical analysis. In principle the 3D image has to be scanned for all possible Eulerian angles φ, ψ and θ around three different axes, with templates of all different protein structures one is interested. The search procedure is computationally very demanding but can be parallelized with respect to the different angular combinations in a highly efficient manner. Finally, the position and orientation of the different complexes can be mapped directly in the 3D image (protein atlas – V_{out}) [73, 74].

2–3 nm would allow the accurate identification of even smaller complexes. The information addressing the spatial relationship of different complexes fosters and complements other proteomic methods and will be indispensable for structural proteomic approaches [63, 64].

3.4
Outlook: The Future is Bright

The achievements in cryo-electron tomography over the past decade have proven the possibilities and the feasibility of this technique for quasi *in vivo* studies of the ultrastructure and larger supramolecular assemblies within whole cells. However, based on the initial developments in cryo-ET, major improvements in instrumentation and sample preparation have to be made in the future to exploit its full potential.

The ultimate goal in structural biology is to investigate the structure-function relationship of molecular complexes and supramolecular assemblies in their native environment, for example in large cells or even tissue, across several dimensions, from the micron level to the sub-nanometer level and if possible within one single experiment, to realize the complete view into the inner space of cells and its constituents. Even at the present practical level, cryo-electron tomograms of organelles and cells contain an imposing amount of information. They are essentially 3D images of entire proteomes and they should ultimately enable us to map the spatial relationships of the full complement of macromolecules in an unperturbed cellular context. However it is obvious that new strategies in sample preparation, advancements in instrumentation and innovative image analysis techniques are needed to make this dream come true, at least to some extent.

In principle there are three major routes to increase the image quality and especially the 3D image quality, which are intrinsically linked to each other: (1) technological developments comprising advanced electron optics, highly resolved high-sensitive detection, improved tilting devices and the geometry of the sample holder, (2) sample preparation and thus sample quality and (3) the design of new software algorithms and acquisition schemes to guarantee a quantitative objective and furthermore exhaustive analysis of the recorded data.

TEM or EM is still a very young method if compared to light microscopy. While the latter was developed and improved over centuries, EM has a history of less than 80 years. Today's microscopes can easily reach a 2D resolution of $\sim$1 Å, clearly orders of magnitude better than what light microscopy can offer. However the electron optical system, determined mainly by the objective lens system and characterized by the spherical and chromatic aberration coefficients, is far from being ideal and still inferior to light optical devices. Simplified one could say: the performance of today's objective lens systems in EM can be described as the attempt to focus through the bottom of a champagne bottle. In this way, the quality of the recorded micrograph in EM suffers greatly from the severe shortcomings of the objective lens system; for example the higher the spherical aberration coefficient C_s, the better the resulting image contrast, but the lower the resolution. However, the spherical aberration coefficient is linked directly to the "spacing" inside the objective lens system and thus determines the maximum achievable tilt range for an angular acquisition. Recent developments in C_s correction technology for TEM (and even for SEM) already showed the possibilities and the final image improvement, especially in material science studies. However, EM investigations of biological samples at cryo-temperatures are characterized by a very low image contrast, due to the very weak scatterers, mainly low Z elements, like carbon, oxygen, hydrogen and so on. Thus high defocus values are more or less mandatory, to increase the image contrast. Unfortunately, by tuning the objective lens to very low defocus values of a couple of microns, low-frequency information is enhanced, while the high frequency information is almost completely obscured if not lost totally. This way C_s correction without any additional equipment is not an option in any cryo-EM study, because the actual improvement in image quality is restricted to a defocus regime in the range of a couple of tenth nanometers, thus orders of magnitude higher than what would be needed for a high-contrast (low-dose) cryo-image. Nevertheless, it might be possible to increase the contrast without defocusing the objective lens, by using so-called phase plates in combination with C_s correctors. The benefit of C_s correction would then become clearly visible, because even the high-frequency information would be accessible. Clearly the design and the application of phase plates is not a new idea, but the technological problems in former times could not be addressed to enable their routine use. For example two major problems were contamination and stability. While placed in the back focal plane, very close to the sample, contamination can cause charging and drift problems. Moreover the stability was low, thus it was mandatory to readjust the phase plate in respect to the central beam, which in the required accuracy is nowadays only possible with piezo-controlled stages. However, today's achievements and possibilities in nano-structuring and nanotechnology make it possible to utilize the already existing designs of phase plates, addressing the problems aforementioned and place them in a normal EM environment. The prospects are good and in the near future experiments will be done to show not only the feasibility of this approach, but also the final image improvement if for example used in combination with C_s correction. However the total sum of elastically scattered electrons will be reduced and thus the large fraction of inelastically scattered electrons has to be removed or even utilized for the final image formation. "Removal" can be done with imaging energy filters, which are nowadays routinely incorporated into the electron microscope setup. They are especially helpful in studies where thick cells or sections are used, because the thicker the material layer to be penetrated by the electron beam the larger is the fraction of inelastically scattered and multiple-scatter electrons. They will have a different energy and thus they will be slightly out of focus, blurring the recorded image. However if the chromatic electrons are removed, the image contrast can be definitely improved, but the number of elastically scattered electrons (forming the image) will be just a fraction of the total sum of electrons leaving the sample. In most cases the amount is only half or one-third of the total applied electron dose to the sample, which contributes to the final image contrast. In this way, the detection has to be very sensitive, so that literally every electron will count. Today CCD cameras are typically used, which detect the signal indirectly and which add noise to the recorded image. Moreover the signal is spread over several pixels, decreasing the lateral resolution and thus the transfer of the recorded image. Direct detection would be beneficial and favorable, because every electron would contribute to the final signal, with no readout noise and the best possible lateral resolution. Developments are on the way to utilize detectors based on CMOS technology, which will guarantee noise-free detection and, if backthinned, a high-resolution recording. However the radiation hardness of these devices, for 300 keV electrons, is at the moment insufficient for a routine use in EM, but improvements are being made to increase the lifetime of these detection devices so that they can be used almost eternally.

Since we cannot increase the total applied dose to the sample, not even with cooling to liquid helium temperature, the only option we have is to increase the performance of the detection device and the electron optics. If for example one utilizes an almost ideal detector, the dose in a recorded image can be as low as possible and thus the number of acquired projections in an angular regime can be as large as possible. In this way the final resolution would be definitely improved. Moreover if high-end electron optical systems are used, for example, C_s or even C_c correctors (which would even utilize the fraction of inelastically scattered electrons for image formation) in combination with phase plates and energy filters, we could harvest not only the low-frequency information responsible for the final image contrast, but also the high-frequency information containing the fine structural details. Moreover with just one single 2D image, one would obtain phase and amplitude information separately.

Besides the improved quality of the single 2D projection, in-phase plate C_s corrected imaging, electron tomography is still hampered by uneven sampling in Fourier space and thus by the fact that information in the direction of the tilt axis cannot be accessed. This missing information (missing wedge), due to the limited tilt range can be decreased by utilizing dual-axis acquisition schemes to a missing pyramid. Clearly with the 90° in-plane rotation of the sample and the acquisition of a two-tilt series of the same sample area, one increases the amount of information by more than 20% and thus reduces the resulting artefacts related to single-axis acquisition schemes. The resolution is not improved, however it is definitely more isotropic. Especially for a subsequent automatic detection of macromolecules and proteins within the tomographic volume, for example by template matching, errors (false positives) can be minimized and thus the quality of the detection can be vastly improved.

At the present resolution of 4–5 nm only very large complexes (ribosomes, 26S proteasomes) can be detected reliably within cryo-electron tomograms of whole cells. However, an improvement in resolution to 2 nm will allow us to detect medium-sized complexes in the range of 200–400 kDa. Some of the problems in the interpretation of tomograms will disappear once a resolution of 2 nm is obtained. At the moment, with a resolution of 4–6 nm, the docking of high-resolution structures to yield pseudo-atomic maps of molecular complexes is computationally demanding and time-consuming. However this procedure of template matching and docking will become straightforward if a resolution of 2 nm can be obtained routinely. One could even think of an "unsupervised" (so to speak "template-free") search of the tomogram. The computational methods described will aid and complement the information provided by other approaches and information, such as localization, labeling studies and the binding properties of molecules. While tomograms with a resolution of 2 nm are a realistic prospect, major technical innovations (see above) will be required if we want to go beyond.

Cryo-ET of whole cells allows us to investigate the structure–function relationship of molecular complexes and supramolecular assemblies in their native environment. It thus makes a fundamental change in the way we approach biochemical processes that underlie and orchestrate higher cellular functions. In the past, molecular interactions were studied mostly in a collective manner, whereas now we have the tools to visualize the interactions between individual molecules in their unperturbed functional environments. Although they share common underlying principles, no two cells or organelles are identical, owing to the inherent stochasticity of biochemical processes in cells as well as their functional diversity. Therefore, it will be a major challenge to extract generic features from the maps, such as the modes of interaction between molecular species. The ultimate goal, the discovery of general rules that underlie cellular processes, has to go beyond observing qualitative features and has to be based on stringent analytical criteria combining the information gathered from different methods for a complete integrative analysis [63–66].

References

1 Knoll, M. and und Ruska, E. (1932) Das Elektronenmikroskop. *Zeitschrift fur Physik*, **78**, 318–339.
2 Sjostrand, F.S. (1988) Ernst Ruska (1906–1988), a genius and a fine person. *Journal of Ultrastructural and Molecular Structural Research*, **101**, 1–3.
3 Glaeser, R.M. (2008) Retrospective: Radiation damage and its associated "information limitations". *Journal of Structural Biology*, **163** (3), 271–276.
4 Dubochet, J. and McDowall, A.W. (1981) Vitrification of pure water for electron microscopy. *Journal of Microscopy*, **124**, RP3–RP4.
5 Adrian, M., Dubochet, J., Lepault, J. and McDowall, A.W. (1984) Cryo-electron microscopy of viruses. *Nature*, **308**, 32–36
6 Taylor, K.A. and Glaeser, R.M. (2008) Retrospective on the early developments of cryoelectron microscopy of macromolecules and a prospective on opportunities for the future. *Journal of Structural Biology*, **163** (3), 214–223.
7 Frank, J. (2006) *Electron Tomography: Methods for Three-Dimensional Visualization of Structures in the Cell*, Springer, New York.
8 Frank, J. (2006) *Three-Dimensional Electron Microscopy of Macromolecular Assemblies: Visualization of Biological Molecules in their Native State*, Oxford University Press, New York.
9 Cormack, A.M. (1963) Representation of a function by its line integrals, with some radiological applications. *Journal of Applied Physics*, **34** (1530), 2722–2727.
10 Hounsfield, G.N. (1972) *A method and apparatus for examination of a body by radiation such as X- or gamma radiation*, The Patent Office, London.
11 Hart, R.G. (1968) Electron microscopy of unstained biological material: the polytropic montage. *Science*, **159**, 1464–1467.
12 Hoppe, W., Langer, R., Knesch, G. and Poppe, C. (1968) Proteinkristallstruktiranalyse mit Elektronenstrahlen. *Zeitschrift fuer Naturwissenschaften*, **55**, 333–336.
13 DeRosier, D.J. and Klug, A. (1968) Reconstruction of three dimensional structures from electron micrographs. *Nature*, **217**, 130–134.
14 Medalia, O., Weber, I., Frangakis, A.S., Gerisch, G. and Baumeister, W. (2002) Macromolecular architecture in eukaryotic cells visualized by cryoelectron tomography. *Science*, **298**, 1209–1213.
15 Hoffmann, C., Leis, A., Niederweis, M., Plitzko, J.M. and Engelhardt, H. (2008) Disclosure of the mycobacterial outer membrane: Cryo-electron tomography and vitreous sections reveal the lipid bilayer structure. *Proceedings of the National Academy of Sciences of the United States of America*, **105** (10), 3963–3967.
16 Beck, M., Lucic, V., Foerster, F., Baumeister, W. and And Medalia, O. (2007) Snapshots of nuclear pore complexes in action captured by cryo-electron tomography. *Nature*, **449** (7162), 611–615.
17 Scheffel, A., Gruska, M., Faivre, D., Linaroudis, A., Plitzko, J.M. and Schueler, D. (2005) An acidic protein aligns magnetosomes along a filamentous structure in magnetotactic bacteria. *Nature*, **440** (7080), 110–114.
18 Kurner, J., Frangakis, A.S. and Baumeister, W. (2004) Cryo-electron tomography reveals the cytoskeletal structure of Spiroplasma melliferum. *Science*, **307**, 436–438.

19 Gruenewald, K., Desai, P., Winkler, D.C., Heymann, J.B., Belnap, D.M., Baumeister, W. and Steven, A.C. (2003) Three-dimensional structure of Herpes Simples virus from cryo-electron tomography. *Science*, **302**, 1396–1398.

20 Baumeister, W. (2005) A voyage to the inner space of cells. *Protein Science*, **14**, 257–269.

21 Grunewald, K., Medalia, O., Gross, A., Steven, A.C. and Baumeister, W. (2003) Prospects of electron cryotomography to visualize macromolecular complexes inside cellular compartments: implications of crowding. *Biophysical Chemistry*, **100** (1–3), 577–591.

22 Ellis, R.J. and Minton, A.P. (2003) Join the crowd. *Nature*, **425**, 27–28.

23 Alberts, B. (1998) The cell as a collection of protein machines: preparing the next generation of molecular biologists. *Cell*, **92**, 291–294.

24 Dubochet, J., Adrian, M., Chang, J.J., Homo, J.-C., Lepault, J., McDowall, A.W. and Schultz, P. (1988) Cryo-electron microscopy of vitrified sections. *Quarterly Review of Biophysics*, **21** (2), 129–228.

25 Al-Amoudi, A., Studer, D. and Dubochet, J. (2005) Cutting artefacts and cutting process in vitreous sections for cryo-electron microscopy. *Journal of Structural Biology*, **150** (1), 109–121.

26 Sartori, N., Richter, K. and Dubochet, J. (1993) Vitrification depth can be increased more than 10-fold by high-pressure freezing. *Journal of Microscopy*, **172**, 55–61.

27 Grimm, R., Baermann, M., Haeckl, W., Typke, D., Sackmann, E. and Baumeister, W. (1996) Energy-filtered electron tomography of ice-embedded actin and vesicles. *Biophysical Journal*, **72**, 482–489.

28 Grimm, R., Singh, H., Rachel, R., Typke, D., Zillig, W. and Baumeister, W. (1998) Electron tomography of ice-embedded prokaryotic cells. *Biophysical Journal*, **74**, 1031–1042.

29 Barnard, D.P., Turner, J.N., Frank, J. and McEwen, B.F. (1992) A 360° single-axis tilt stage for the high-voltage electron microscope. *Journal of Microscopy*, **167**, 39–48.

30 Nickell, S., Hegerl, R., Baumeister, W. and Rachel, R. (2003) Pyrodictium cannulae enter the periplasmic space but do not enter the cytoplasm, as revealed by cryo-electron tomography. *Journal of Structural Biology*, **141**, 34–42.

31 Plitzko, J.M., Frangakis, A.S., Nickell, S., Förster, F., Gross, A. and Baumeister, W. (2002) In vivo veritas: electron cryotomography of cells. *Trends in Biotechnology*, **20** (8 Suppl), S40–S44.

32 Mochel, M.E. and Mochel, J.M. (1986) A CCD imaging and analysis system for the VG HB5 STEM. in Proceedings of the 51st Annual Meeting of the Microscopy Society of America, San Francisco Press, pp. 262–263.

33 Dierksen, K., Typke, D., Hegerl, R., Koster, A.J. and Baumeister, W. (1992) Towards automated electron tomography. *Ultramicroscopy*, **40**, 71–87.

34 Rath, B.K., Marko, M., Radermacher, M. and Frank, J. (1997) Low-dose automated electron tomography: a recent implementation. *Journal of Structural Biology*, **120**, 210–218.

35 Hegerl, R. and Hoppe, W. (1976) Influence of electron noise on three-dimensional image reconstruction. *Zeitschrift für Naturforschung*, **31a**, 1717–1721.

36 Dierksen, K., Typke, D., Hegerl, R. and Baumeister, W. (1993) Towards automatic electron tomography. II. Implementation of autofocus and low-dose procedures. *Ultramicroscopy*, **49** (1–4), 109–120.

37 Koster, A.J., Chen, H., Sedat, J.W. and Agard, D. (1992) Automated microscopy for electron tomography. *Ultramicroscopy*, **46**, 207–227.

38 Braunfeld, M.B., Koster, A.J., Sedat, J.W. and Agard, D.A. (1994) Cryo automated electron tomography—towards high resolution reconstructions of plastic embedded structures. *Journal of Microscopy*, **172** (2), 75–84.

39 Nickell, S., Forster, F., Linaroudis, A., DelNet, W., Beck, F., Hegerl, R.,

Baumeister, W. and Plitzko, J.M. (2005) TOM software toolbox: acquisition and analysis for electron tomography. *Journal of Structural Biology*, **149** (3), 227–234.

40 Saxton, W.O., Baumeister, W. and Hahn, M. (1984) Three-dimensional reconstruction of imperfect two-dimensional crystals. *Ultramicroscopy*, **13** (1–2), 57–70.

41 Zheng, Q.S., Braunfeld, M.B., Sedat, J. and Agard, D.A. (2004) An improved strategy for automated electron microscopic tomography. *Journal of Structural Biology*, **147** (2), 91–101.

42 Ziese, U., Janssen, A.H., Murk, J.-L., Geerts, W.J.C., van der Krift, T., Verkleij, A.J. and Koster, A.J. (2002) Automated high-throughput electron tomography by pre-calibration of image shifts. *Journal of Microscopy*, **205** (2), 187–200.

43 Lawrence, M.C. (1992) Least-squares method of alignment using markers, in *Electron Tomography:: Three-dimensional imaging with the Transmission Electron Microscope* (ed. J. Frank), Plenum Press, New York.

44 Luther, P.K., Lawrence, M.C. and Crowther, R.A. (1988) A method for monitoring the collapse of plastic sections as a function of electron dose. *Ultramicroscopy*, **24**, 7–18.

45 Bracewell, R.N. (1956) Strip integration in radio astronomy. *Australian Journal of Physics*, **9**, 198–217.

46 Radon, J. (1917) Über die Bestimmung von Funktionen durch ihre Integralwerte längs gewisser Mannigfaltigkeiten. *Berichte über die Verhandlungen der Königlich Sächsischen Gesellschaft der Wissenschaften zu Leipzig. Mathematisch-Physische Klasse*, **69**, 262–277.

47 Deans, S.R. (1983) *The Radon Transform and Some of Its Applications*, John Wiley Sons, New York, p. 289.

48 Hoppe, W. and Hegerl, R. (1980) Three-dimensional structure determination by electron microscopy (nonperiodic specimens), in *Computer Processing of Electron Microscopic Images* (ed. P.W. Hawkes), Springer Verlag, Berlin, Heidelberg, New York, pp. 127–185.

49 Radermacher, M. (1992) Weighted backprojection methods, in *Electron Tomography: Three-dimensional imaging with the Transmission Electron Microscope* (ed. J. Frank), Plenum Press, New York, pp. 91–115.

50 Crowther, R.A. and Klug, A. (1971) ART and science or conditions for three-dimensonal reconstruction from electron microscope images. *Journal of Theoretical Biology*, **32**, 199–203.

51 Herman, G.T. (1980) On the noise in images produced by computed tomography. *Computer Graphics & Image Processing*, **12** (3), 271–285.

52 Ladinsky, M.S., Mastronarde, D.N., McIntosh, J.R., Howell, K.E. and Staehelin, L.A. (1999) Golgi structure in three dimensions: Functional insights from the normal rat kidney cell. *Journal of Cell Biology*, **144** (6), 1135–1149.

53 Frangakis, A.S. and Foerster, F. (2004) Computational exploration of structural information from cryo-electron tomograms. *Current Opinion in Structural Biology*, **14**, 1–7.

54 Beck, M., Forster, F., Ecke, M., Plitzko, J.M., Melchior, F., Gerisch, G., Baumeister, W. and Medalia, O. (2004) Nuclear pore complex structure and dynamics revealed by cryoelectron tomography. *Science*, **306** (5700), 1387–1390.

55 Mercogliano, C.P. and DeRosier, D.J. (2007) Concatenated metallothionein as a clonable gold label for electron microscopy. *Journal of Structural Biology*, **160** (1), 70–82.

56 Gaietta, G., Deerinck, T.J., Adams, S.r., Bouwer, J., Tour, O., Laird, D.W., Sosinsky, G.E., Tsien, R.Y. and Ellisman, M.H. (2002) Multicolor and electron microscopic imaging of connexin trafficking. *Science*, **296** (5567), 503–507.

57 Sartori, A., Gatz, R., Beck, F., Rigoet, A., Baumeister, W. and Plitzko, J.M. (2007) Correlative microscopy: bridging the gap between fluorescence light microscopy and cryo-electron tomography. *Journal of Structural Biology*, **160** (2), 135–145.

58 Schwartz, C.L., Sarbash, V.I., Ataullakhanov, F.I., McIntosh, J.R. and

Nicastro, D. (2007) Cryo-fluorescence microscopy facilitates correlations between light and cryo-electron microscopy and reduces the rate of photobleaching. *Journal of Microscopy*, **227** (Pt 2), 98–109.

59 Boehm, J., Frangakis, A.S., Hegerl, R., Nickell, S., Typke, D. and Baumeister, W. (2000) Toward detecting and identifying macromolecules in a cellular context: template matching applied to electron tomograms. *Proceedings of the National Academy of Sciences of the United States of America*, **97**, 14245–14250.

60 Frangakis, A.S., Böhm, J., Förster, F., Nickell, S., Nicastro, D., Typke, D., Hegerl, R. and Baumeister, W. (2002) Identification of macromolecular complexes in cryoelectron tomograms of phantom cells. *Proceedings of the National Academy of Sciences of the United States of America*, **99** (22), 14153–14158.

61 Rath, B.K., Hegerl, R., Leith, A., Shaikh, T.R., Wagenknecht, T. and Frank, J. (2003) Fast 3D motif search of EM density maps using a locally normalized cross-correlation function. *Journal of Structural Biology*, **144** (1–2), 95–103.

62 Ortiz, J.O., Foerster, F., Kuerner, J., Linaroudis, A. and Baumeister, W. (2006) Mapping 70S ribosomes in intact cells by cryo-electron tomography and pattern recognition. *Journal of Structural Biology*, **156**, 334–341.

63 Sali, A., Glaeser, R., Earnest, T. and Baumeister, W. (2003) From words to literature in structural proteomics. *Nature*, **422**, 216–225.

64 Robinson, C.V., Sali, A. and Baumeister, W. (2007) The molecular sociology of the cell. *Nature*, **450** (7172), 973–982.

65 Lucic, V., Leis, A. and Baumeister, W. (2008) Cryo-electron tomography of cells: connection structure and function *Histochemistry and Cell Biology*, **130**, 185–196.

66 Lucic, V., Foerster, F. and Baumeister, W. (2005) Structural studies by electron tomography: From cells to molecules. *Annual Review of Biochemistry*, **74**, 833–865.

67 Gruska, M., Medalia, O., Baumeister, W. and Leis, A. (2008) Electron tomography of vitreous sections from cultured mammalian cells. *Journal of Structural Biology*, **161** (3), 384–392.

68 Koster, A.J., Grimm, R., Typke, D., Hegerl, R., Stoschek, A., Walz, J. and Baumeister, W. (1997) Perspectives of molecular and cellular electron tomography. *Journal of Structural Biology*, **120** (3), 276–308.

69 Lucic, V., Kossel, A.H., Yang, T., Bonhoeffer, T., Baumeister, W. and Sartori, A. (2007) Multiscale imaging of neurons grown in culture: from light microscopy to cryo electron tomography. *Journal of Structural Biology*, **160**, 146–156.

70 Steven, A.C. and Baumeister, W., (2008) The future is hybrid. *Journal of Structural Biology*, **163** (3), 186–195.

71 Nickell, S., Mihalache, O., Beck, F., Hegerl, R., Korinek, A. and Baumeister, W. (2007) Structural analysis of the 26S proteasome by cryoelectron tomography. *BBRC*, **353**, 115–120.

72 Nickell, S., Beck, F., Korinek, A., Mihalache, O., Baumeister, W. and Plitzko, J.M. (2007) Automated cryoelectron microscopy of "Single particles" applied to the 26S proteasome. *FEBS letters*, **581** (15), 2751–2756.

73 Best, C., Nickell, S. and Baumeister, W. (2007) Localization of Protein Complexes by Pattern Recognition, in *Cellular Electron Microscopy* (ed. J. Richard, McIntosh), Methods in Cell Biology, Academic Press, p. 79.

74 Frangakis, A.S. and Rath, B.K. (2006) in *Electron Tomography* (ed. J. Frank), Springer, New York.

75 Al-Amoudi, A., Diez, D.C., Betts, M.J. and Frangakis, A.S. (2007) The molecular architecture of cadherins in native epidermal desmosomes. *Nature*, **450** (7171), 832–837.

Part II
Single Cell Analysis: Technologies

4
Single Cell Proteomics[1]

Norman J. Dovichi, Shen Hu, David Michels, Danqian Mao, and Amy Dambrowitz

4.1
Introduction

Any attempt to explain the function of a protein must account for its spatial and temporal location. Not all proteins are present in all cells and many proteins have a fleeting existence as synthesis, modification and degradation occur while the cell responds to its environment, passes through the cell cycle and recycles its contents. Each cell behaves differently and ultimately proteomics must descend to the level of the single cell to provide an accurate and complete description of protein function and expression. Ideally that study should be correlated with other characteristics of the cell, such as phase of the cell in the cell cycle for proliferating cells.

We cite three examples where single cell proteomics will be valuable. The first is in development. The cells of an embryo undergo profound changes in protein expression as the organism grows from a zygote to a fully developed individual. Similar changes in protein expression occur as pluripotent stem cells progress first into precursor cells and then into terminally differentiated progeny cells. A fundamental understanding of the steps involved in development will be aided greatly by monitoring the proteome on a cell by cell basis.

The second example is in oncology. We hypothesize that the cell to cell heterogeneity in protein expression of a tumor is correlated with the prognosis of that cancer [1]. Just as aneuploidy is correlated with poor prognosis in some cancers, it may prove that the cell to cell heterogeneity in protein expression is correlated with prognosis. If that hypothesis is correct, then large-scale studies of the protein content of single cells will be of great value in the clinic.

The third example is in neuroscience, where individual neurons in the central nervous system are extremely heterogeneous. That heterogeneity reflects differences in the content and architecture of the cells. The creation of a vocabulary to describe

1) This chapter "Single Cell Proteomics" has previously been published in: *Proteomics for Biological Discovery,* edited by Timothy D. Veenstra and John R. Yates; John Wiley & Sons, Inc.

Single Cell Analysis: Technologies and Applications. Edited by Dario Anselmetti
Copyright © 2009 WILEY-VCH Verlag GmbH & Co. KGaA, Weinheim
ISBN: 978-3-527-31864-3

cells and cell types will benefit greatly from the analysis of the protein content of the cell.

Single cell proteomics is of obvious value in basic and clinical studies. Nevertheless, single cell proteomics presents significant challenges because of the minute amount of protein present in the cell. This chapter describes those challenges and several approaches to the characterization of proteins in single eukaryotic cells.

4.2
The Challenge

Most proteins are present at minuscule levels within a cell as shown in Table 4.1. We consider four examples. A 100 μm diameter giant neuron contains perhaps 50 ng of protein, assuming that the cell is 10% protein by weight. The average molecular weight for a protein is about 30 000 g/mol, so that a giant neuron contains perhaps 2 pmol, or 10^{12} copies, of protein.

A typical mammalian cell has a diameter of roughly 10 μm, with a volume of 0.5 pL and a total protein content of 50 pg. Again, assuming an average molecular mass of 30 kDa, the cell contains about 2 fmol total protein, or about one billion copies of protein molecules.

A 5 μm diameter yeast cell contains about 5 pg or 0.2 fmol of protein. This primitive eucaryote functions with about 100 million copies of protein.

A 1 μm diameter bacterium contains only 50 fg or 2 amol of protein. Those organisms survive and replicate using only one million copies of protein.

The number of proteins expressed in a single cell is unknown. An animal might express 10 000 different proteins per cell. If we use this quite arbitrary number, the average protein is expected to be present at the 200 zmol level in a 10 μm diameter mammalian cell. Yeast expresses fewer proteins, perhaps 2000 per cell; the typical protein would be present at the 100 amol level.

However, the concept of the average protein is inappropriate because the distribution of protein expression in a cell is expected to be highly heterogeneous. As an analogy, if Bill Gates were added to the list of authors for this chapter, then the average

Table 4.1 Protein content of a single cell. Prefixes:
n = nano = 10^{-9}; p = pico = 10^{-12}; f = femto = 10^{-15};
a = atto = 10^{-18}; z = zepto = 10^{-21}; y = yocto = 10^{-24}.

	Giant neuron	Mammalian cell	Yeast cell	Bacterium
Volume	0.5 nL	0.5 pL	50 fL	0.5 fL
Total mass	0.5 μg	0.5 ng	50 pg	0.5 pg
Mass protein[a]	50 ng	50 pg	5 pg	50 fg
Moles protein[b]	2 pmol	2 fmol	0.2 fmol	2 amol
Copies protein	1×10^{12}	1×10^{9}	1×10^{8}	1×10^{6}

[a]Assumes the cell is 10% protein by mass.
[b]Assumes an average molecular mass of 20 kDa for proteins.

wealth of the group would be about 10 billion dollars. Clearly this average does not provide insight into the lifestyle of the authors! Similarly the average protein level of a cell does not provide insight into the challenges of the study of a single cell's proteome.

There are limited data available on protein expression levels. In one of the best studies, Gygi *et al.* [2] treated *Saccharomyces cerevisiae* with a radiotracer followed by two-dimensional gel electrophoretic separation of the proteins from the yeast homogenate. Spots were excised and quantitated using scintillation counting. The proteins were identified by mass spectrometry.

The authors obtained expression data for 156 spots corresponding to 124 different gene products. Of these, 11 proteins had expression levels that were estimated to be greater than 100 000 copies (0.2 amol) per cell. This set of 124 gene products accounts for a total of $\sim$8 million copies of protein per yeast cell, which is $\sim$10% of the total protein content of a mid-log yeast cell. It appears that the vast majority of the protein content of the cell was either lost during sample preparation or is found in the many proteins expressed at low levels.

If we extrapolate this data from yeast to a mammalian system, we expect the total protein concentration to be an order of magnitude higher. The number of genes in mammals is also roughly an order of magnitude larger than the number of genes in yeast, so that the average protein level per cell is likely to be similar. If there are $\sim$10 proteins present in a single yeast cell that are expressed at greater than 100 000 copies (0.2 amol or $\sim$5 fg) per cell, then it is reasonable to expect at least 100 proteins would be expressed at that level in a single mammalian cell.

4.3
Single Cell Proteomics: Mass Spectrometry

Mass spectrometry is undoubtedly the tool of choice in protein analysis. Technologies such as multidimensional protein identification technology (MudPIT) and isotope-coded affinity tag (ICAT) provide powerful means of identifying proteins and determining changes in their expression [3, 4]. Unfortunately these methods require relatively large amounts of proteins, certainly much more than contained in a single cell. Smith [5] estimated detection limits of 100 zmol for tryptic peptides by the use of Fourier transform–ion cyclotron resonance (FT-ICR) instrumentation. However, that detection limit is for the amount of peptide introduced into the spectrometer. Significant losses accompany protein isolation, digestion, peptide extraction and chromatographic separation; and it is not clear that analysis of tryptic digests from single cells will be practical.

Instead it appears necessary to perform mass spectrometry on intact proteins to realize single cell analysis. Matrix-assisted laser desorption/ionization (MALDI)-based methods have been used with some success in detecting peptides and low molecular weight proteins from single cells and tissue slices. Whittal *et al.* [6] reported the detection of hemoglobin from a single erythrocyte by MALDI-time of flight (TOF) mass spectrometry. In their approach, the cell was lysed, mixed with protease and then microspotted on a MALDI plate. Each cell contained about 450 amol of hemoglobin and only extremely highly expressed proteins could be detected and identified in a single cell.

Xu et al. [7] reported the use of MALDI-TOF to analyze tissue samples. As few as 10 cells have been used for analysis. That sample generated roughly 10 peaks that exceeded the noise in the baseline. This TOF mass spectrometer has relatively poor mass resolution and does not employ MS/MS analysis, which prevents direct identification of proteins.

Page et al. [8] reported the use of MALDI-TOF to analyze neuropeptides in single giant neurons. This method is not applicable to peptides greater than a few kilodaltons in molecular mass.

Pasa-Tolic et al. [9] considered the use of accurate mass tags for protein analysis with high resolution and accuracy FTICR instruments. That group reported the use of an 11 tesla FTICR allowed determination of a number of proteins obtained from a 5 pg protein sample from a *D. radiodurans* protein homogenate that was injected onto a 15 μm inner diameter capillary liquid chromatography column. The proteome of this prokaryote is certainly much simpler than that of eukaryotes and the introduction of the protein lysate from a single cell onto the chromatography column is not demonstrated; nevertheless these results are encouraging that the protein content of a single mammalian cell will likely be performed using mass spectrometry within the decade.

4.4
Single Cell Separations

The use of separation methods to characterize the composition of a single cell has a fifty year history. The first study in 1953 considered ribosomal ribonucleic acid (rRNA) analysis in single cells based on electrophoresis on a silk fiber [10]. The earliest single cell protein analysis was a study of hemoglobin in single erythrocytes by electrophoresis through an acrylamide fiber, published in 1965 [11]. Repin et al. [12] reported the characterization of lactate dehydrogenase isoenzymes in single mammalian oocytes by electrophoresis in 1975. Ruchel [13] reported the first protein analysis from a single giant neuron from a sea snail the following year.

Kennedy et al. [14, 15] inaugurated the modern era of single cell analysis by using open tubular capillary chromatography for the amino acid analysis of a single giant neuron from a snail. At the same time, Wallingford and Ewing [16] reported the use of a capillary to sample the internal contents of a single giant neuron; they used the same capillary for electrophoresis of biogenic amines. In 1992, Hogan and Yeung [17] reported the use of a specific label to derivatize thiols in individual erythrocytes; once the derivatization reaction was completed, a single cell was injected into a capillary and lysed and the contents were separated by capillary electrophoresis (CE). In 1995, Gilman and Ewing [18] reported the on-column labeling of amines from a cell that had been injected into and lysed within a capillary; the capillary was used for electrophoretic separation of amines. Meredith *et al.* [19] reported the use of a pulsed laser to lyse a cell before analysis of kinase activities by CE. In 2002, Han and Lillard [20] reported the use of cell synchronization based on the shake-off method for the characterization of RNA synthesis in single cells as a function of cell cycle.

Our group's activity in single cell analysis began in collaboration with Monica Palcic and Ole Hindsgaul at the University of Alberta, in a study of carbohydrate metabolism in single yeast and cancer cells [21, 22]. In that project, cells were treated with a fluorescent enzymatic substrate. Cells took up this substrate and metabolized it to create biosynthetic and biodegradation products. To study this behavior in single cells, we developed techniques to manipulate and inject single cells into a fused silica capillary, to lyse the cells within the capillary, to separate the cellular lysate by capillary electrophoresis and to detect the separated lysate by laser-induced fluorescence [21, 23].

In our instrument (Figure 4.1) an inverted microscope is equipped with a capillary manipulator manufactured from Lexan (GE Plastics, Pittsfield, MA) and held by a set of micromanipulators. The capillary is filled with separation buffer that contains a surfactant, such as sodium dodecyl sulfate (SDS). The capillary tip is then centered over the cell of interest; to inject the cell, a brief pulse of vacuum is applied to the distal end of the capillary through a computer-controlled valve, aspirating the cell about 200 μm into the capillary. Mammalian cells are lysed within 30 s by the action of SDS and the osmotic shock from the buffer [23].

Electrophoresis is effected by application of high voltage to the injection end of the capillary, which is held within the interlock-equipped Lexan manipulation block. The distal end of the capillary is placed in a sheath-flow cuvette and fluorescence is excited by an argon ion laser beam and collected with a microscope objective, which images fluorescence onto a pinhole and through a spectral filter to block scattered laser light. The fluorescence is finally detected by a photomultiplier tube and recorded by a computer.

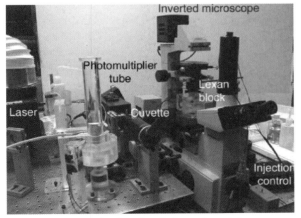

Figure 4.1 Photograph of single cell electrophoresis instrument.

4.5
Ultrasensitive Protein Analysis: Capillary Electrophoresis with Laser-Induced Fluorescence Detection

The single cell methods have been developed for injection, lysis and analysis of single cells by capillary electrophoresis. CE provides exquisite separation of complex mixtures. Laser-induced fluorescence (LIF) provides exquisite sensitivity for highly fluorescent molecules. The combination of the two tools provides an extremely powerful method for the analysis of biological molecules. To illustrate the analysis of a complex mixture, capillary array instruments have been the workhorse of large-scale genomic DNA sequencing efforts [24]. As an example of ultrasensitive analysis, CE-LIF has been used to detect and count single molecules of b-phycoerythrin, resolving isoforms of this protein [25].

The study of b-phycoerythrin was based on that molecule's native fluorescence. That molecule has unusual spectroscopic properties. Most proteins exhibit native fluorescence only when excited in the ultraviolet portion of the spectrum. The molar absorptivity and fluorescent quantum yields of the aromatic amino acids that generate native fluorescence are modest and the background signal from fluorescent impurities tends to be high in this portion of the spectrum, leading to relatively poor detection performance. More practically, lasers that operate in the ultraviolet tend to be expensive and temperamental, which discourages their widespread use.

Instead our approach to ultrasensitive protein analysis relies on the use of derivatization technology to introduce a fluorescent tag that is excited with more robust lasers that operate in the visible portion of the spectrum. These tags have relatively high molar absorptivity and fluorescent quantum yields, which assists sensitive detection. However, tagging chemistry is not without its challenges. We highlight several issues here.

First, it is necessary to perform the labeling reaction on dilute proteins, which may be present at a concentration in the picomole per liter range or lower. The reaction is usually governed by second-order kinetics, so that it is necessary to keep the concentration of the derivatizing reagent in the millimolar range so that the reaction proceeds at a reasonable rate. The vast excess of derivatizing reagent can result in a huge background signal from unreacted reagent. While the reagent itself can be separated from proteins during the electrophoretic step, the sea of fluorescent impurities that accompany the reagent at the part per million level will swamp the signal from the labeled proteins. Fortunately this fluorescent background issue can be eliminated by use of fluorogenic reagents; these reagents are nonfluorescent until they react with the protein, creating a fluorescent product [26].

Second, the use of fluorogenic reagents inevitably results in the production of a complex mixture of fluorescent products. All fluorogenic reagents of which we are aware react with the ε-amine of lysine residues, which is one of the most common amino acids. We have analyzed the yeast genome to estimate the relative abundance of lysine residues. The average open reading frame codes for ∼26 lysine residues, so that most proteins will incorporate more than one fluorescent label [27]. Unfortunately extreme denaturation conditions are required to drive the labeling reaction to completion [28]. These conditions appear impractical when studying the proteome of single cells. If the reaction does not go to completion, then a complex reaction mixture will inevitably be produced. The number of possible fluorescent reaction products is $2^N - 1$, where N is the number of lysine residues (and other primary amines) present in the molecule [29]. The average protein in yeast, with 26 lysine residues, can produce 67 108 863 different reaction products. Unfortunately these fluorescent products from a single protein can have different electrophoretic mobility and multiple labeling results in extremely complex electropherograms that are essentially worthless for protein analysis.

Fortunately we have discovered one solution to the multiple labeling problem. The reaction of proteins with the fluorogenic reagent 5-furoylquinoline-3-carboxaldehyde (FQ), followed by use of appropriate buffers, results in remarkably efficient electrophoretic separations [30, 31]. FQ, unlike other fluorogenic reagents, produces a neutral reaction product; cationic lysine residues are converted to neutral products. In the absence of buffer additives, the heterogeneity in the number of lysine residues that are labeled results in multiple electrophoretic peaks from a single protein. However, we discovered that the addition of an anionic surfactant, such as SDS, to the buffer results in the collapse of that complex envelope into a single, sharp peak for each protein. We believe that the surfactant ion pairs with unreacted lysine residues, producing a neutral complex with the same mobility as the FQ-labeled molecule.

The interactions between SDS and proteins are complicated. At low surfactant concentration, SDS binds specifically to high-energy sites of the protein through electrostatic interactions; the anionic surfactant ion pairs with easily accessible lysine and arginine residues [32–34]. At intermediate concentrations, SDS binds through hydrophobic interactions with the surface of the protein. At higher surfactant concentration, there is a massive increase in the binding of SDS as the protein unfolds and its interior becomes accessible. The surfactant is thought to form a random set of micelles along the protein's backbone. Under saturation conditions, 1.0 g of many proteins will bind $\sim$1.4 g of SDS [32]. This constant binding ratio is important in SDS-PAGE. At high SDS concentration, the complex of protein and SDS is assumed to generate a constant size to charge ratio. Like DNA electrophoresis in polymeric medium, SDS–protein complexes are separated based on size during polyacrylamide gel electrophoresis (PAGE).

4.6
Capillary Sieving Electrophoresis of Proteins from a Single Cancer Cell

We have developed a number of forms of electrophoresis to analyze the protein content of single cells. In the analysis of the protein content of a single cell, we sandwich the proteins from the single cell between two plugs of the FQ derivatizing reagent. A plug of reagent, followed by the single cell and then another plug of reagent are sequentially injected into the capillary. The reagent solution contains SDS to lyse the cell and the capillary tip is heated to $\sim$90 °C to speed the labeling reaction, which typically is performed for 4 min.

We have developed methods to perform the capillary equivalent of SDS-PAGE on single mammalian cells [1, 35, 36]. We find that acrylamide is a difficult polymer to work with; its free radical polymerization is difficult to control in the laboratory. Instead we have investigated a number of commercially available polymers for the separation, including polyethylene oxide and the polysaccharide pullulan, for the separation of proteins from a single cell. These polymers are not crosslinked and have relatively low viscosity, so that a cell can easily be aspirated into the capillary. Separation with buffers containing polymers is called sieving electrophoresis to reflect the size-based separation mechanism.

The capillary sieving separation of the proteins from a single HT-29 human adenocarcinoma cell is shown in Figure 4.2. Perhaps 25 components are partially resolved from this cell. The earliest migrating peaks correspond to low molecular weight proteins, peptides and biogenic amines. The peak migrating at 24 min has a molecular mass of about 100 kDa. The number of components resolved from this single cell is similar to the number of bands resolved on a 10-cm SDS-PAGE separation of the proteins extracted from a few million cells, but less than that separated by a high-resolution gel.

The dynamic range of the separation is exquisite. The noise in the baseline is typically a part per thousand or less of the maximum signal. As discussed later, we are modifying our instrument to incorporate higher sensitivity fluorescence detectors based on avalanche photodiodes. This improved instrument has a fivefold higher sensitivity, which will further extend the dynamic range of the instrument.

We worried about contamination of the single cell electropherogram by residual culture medium. To minimize the amount of culture medium present, we wash the cells and resuspend them in phosphate-buffered saline. Injection of the cellular supernatant generates a low amplitude, featureless background signal that does not interfere with the single cell data.

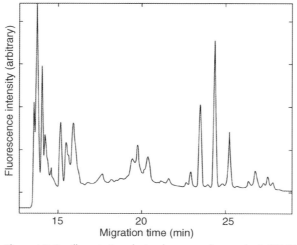

Figure 4.2 Capillary sieving electropherogram from a single HT-29 cell.

4.7
Cell Cycle-dependent Single Cell Capillary Sieving Electrophoresis

We generated a number of capillary sieving electropherograms from single HT-29 cells. Those data showed reasonable reproducibility in the migration time for individual peaks [1]. However, there was a large cell to cell variation in the peak amplitude, averaging 40% in relative standard deviation. Some components varied in amplitude by more than an order of magnitude.

We were confident that the cell to cell variation in peak amplitude was not due to variation in the performance of the electrophoresis or detection system because replicate injections of the same cellular homogenate generated quite reproducible peaks. Instead it became clear that the variation in amplitude was associated either with the cell lysis or was inherent in the cells themselves.

These cells proliferate and a likely cause of cell to cell variation in protein expression is the phase of the cell cycle. The fluorescence microscope of Figure 4.1 can be used to monitor the presence of classic cytometry stains used to treat the cells before analysis. We treated the cells with Hoechst 33 342, which is a vital nuclear stain. This stain is taken up by living cells and transported to the nucleus, where it intercalates within double-stranded oligonucleotides. The intercalated dye becomes highly fluorescent and the resulting fluorescence intensity can be used to estimate the DNA content of the cell. Recently divided cells are diploid, with two pairs of chromosomes, while cells about to undergo mitosis are tetraploid, with four pairs of chromosomes. We use a photomultiplier tube to monitor the fluorescence signal generated by the Hoechst stain before injection into the capillary for electrophoretic analysis and we classify cells into G1 and G2/M phase based on that signal.

We discovered that the cell to cell variation in peak amplitude was indeed dominated by the phase of the cell in the cell cycle [1]. On average, cells in the G2/M phase of the cell cycle generated electrophoresis peaks that were twice the amplitude of cells in the G1 phase; this result is expected because two G1 phase daughter cells are produced by division of one G2/M parent cell. Cells in the G1 phase of the cell cycle generated electropherograms whose peaks had a 27% relative standard deviation while cells in the G2/M phase of the cell cycle generated electropherograms with a 20% relative standard deviation. The majority of the cell to cell variation in protein expression is associated with differences in the phase of the cell in the cell cycle. Residual cell to cell variation in protein electropherograms likely reflects differences of the cells *within* the phase of the cell cycle.

We also performed electrophoretic analysis of an anomalous cell that appeared to contain six pairs of chromosomes. This hexaploid cell was identified based on its Hoechst staining characteristics. This cell line has been karyotyped and some cells in this cell line reveal extra copies of chromosomes. The protein electropherogram of this single unusual cell was more intense than that of other cells and contained one 45 kDa component that was dramatically upregulated compared to G1 and G2/M phase cells. Analysis of this unusual cell reveals a strength of single cell protein analysis. Cells with peculiar properties can be plucked for analysis, whereas preparation of a cellular homogenate for classic analysis will dilute the signal from unusual cells to an undetectable extent.

4.8
Tentative Identification of Proteins in Single Cell Electropherograms

We normalized the single cell electropherograms to their DNA content. The resulting electropherograms did not differ to a large extent with cell cycle. Only one component differed at the 99% confidence limit between the G1 and G2/M phase cells [1]. That component migrated with an apparent molecular mass of 45 kDa, which was the same as the highly expressed protein in the hexaploid cell.

We worked in collaboration with Ruedi Aebersold and Rick Newitt of the Institute for Systems Biology to tentatively identify this protein [37]. First, we prepared a large-scale homogenate of this cell line. Capillary sieving electrophoresis analysis of the homogenate was similar to the average single cell electropherogram but much different from electropherograms generated from the cytosolic, membrane/organelle, nuclear and cytoskeletal/nuclear matrix fractions of this cell line. The single cell analysis appears to sample a homogeneous portion of the single cell's content.

We next used classic SDS-PAGE to separate the cellular homogenate. Fortunately the banding pattern was similar to the capillary electrophoresis peaks and we were able to isolate a 45 kDa band from the gel. We took a portion of the purified proteins contained within this band and spiked the cellular homogenate before capillary electrophoresis analysis. The proteins isolated from the 45 kDa SDS-PAGE gel comigrated with the target 45 kDa peak in the capillary sieving electropherogram. Whichever cellular component demonstrated the cell cycle dependent change in expression was present in the 45 kDa band isolated from the gel and in the 45 kDa peak observed in the single cell electropherograms.

In-gel digestion, liquid chromatographic separation of the extracted peptides, MS/MS analysis of the peptides and database searching were performed to identify the proteins present within the 45 kDa band. Five proteins were identified in the band. However, only one protein was identified from more than one peptide. That protein, cytokeratin 18, is the product of one of the most highly expressed genes in this cell line. It is known to undergo a massive change in phosphorylation state in the G2/M phase of the cell cycle. Careful inspection of the single cell electropherograms revealed that the peak corresponding to this component underwent a mobility shift to faster migration time in the G2/M phase of the cell cycle, consistent with addition of negatively charged phosphate groups to the protein. It was this mobility shift that caused the apparent change in amplitude of the 45 kDa component.

This general approach provides a tedious, but ultimately successful, method for the tentative identification of proteins present in single cells. In the general case, a library of proteins is prepared using classic one- (1D) or two-dimensional (2D) electrophoretic separation of proteins prepared from a large-scale cellular homogenate. The library is split into two parts. One part is used to identify the components using classic mass spectrometric methods and the other is archived in a freezer and used to spike

CE samples, where comigration is taken as evidence for the identity of the component from the single cell. While tedious, this procedure needs to be performed only once for each cell type. Migration patterns appear to be sufficiently robust in CE that a single comigration study should suffice for most applications.

The similarity of the capillary sieving electropherograms and the SDS-PAGE electropherograms is important. The identification of an interesting peak in the single cell data can be simplified by analyzing the corresponding band in the SDS-PAGE electropherogram. However, this study points out a limitation of one-dimensional electrophoresis for the analysis of complex proteomes: only a limited number of components are resolved. A capillary analog of 2D electrophoresis is required to characterize more fully the proteome of a single cell.

4.9
Capillary Micellar and Submicellar Separation of Proteins from a Single Cell

In the ideal case, the second dimension for protein characterization would be isoelectric focusing, to provide the capillary equivalent of classic 2D gels. Unfortunately the multiple labeling issue discussed earlier is devastating in isoelectric focusing, leading to complex and essentially worthless electrophoresis data from fluorescently labeled proteins [38]. We are not aware of a method to correct for this effect.

To generate an alternative form of electrophoresis of single cells, we turned in desperation to a micellar and submicellar electrophoresis [16, 17]. As noted earlier, the addition of SDS to labeled proteins results in the collapse of the multiple labeling envelope to a single sharp peak. However, this phenomenon is only of value if the peaks from different proteins can be resolved. We were concerned that this separation would not be possible. As reported by Oakes [32] nearly 30 years ago, proteins take up a saturating amount of 1.4 g of SDS per gram of protein. If that ratio were accurate, then the charge to size ratio of the SDS-treated protein should be constant. Electrophoresis in the absence of a sieving medium tends to separate proteins based on their size to charge ratio, so that we expected poor resolution of proteins in the presence of SDS and the absence of polymer.

Fortunately this fear was unfounded and we applied the technology to the analysis of the proteins from single cancer cells and from a single cell embryo [39, 40]. The CE separation of proteins from a single *Caenorhabditis elegans* zygote is shown in Figure 4.3. Sample handling was a challenge in this experiment. It was necessary to dissect a single worm to isolate eggs. The egg shell was removed with treatment with chitinase and chymotrypsin before the cell was injected into the capillary and lysed. Roughly 25 components were partially resolved over a quite wide separation window.

This wide separation window is consistently observed with different cell types. We observe that the separation efficiency tends to maximize with surfactant near, but below, its critical micelle concentration. Although the separation in Figure 4.3 was performed with a simple sodium phosphate buffer, we have more recently observed

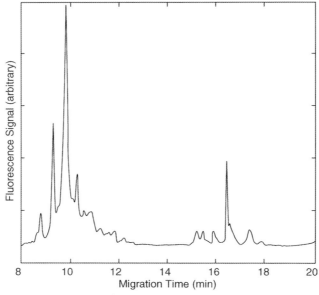

Figure 4.3 Separation of the proteins from a single *Caenorhabditis elegans* embryo.

improved resolution with the addition of alcohol to the separation buffer. We believe that the alcohol decreases the partitioning of hydrophobic proteins from the surfactant, leading to improved resolution.

4.10
Two-Dimensional Capillary Electrophoresis of Proteins in a Single Cell

Giddings [41] realized that multidimensional separations provide resolution that can be as high as the product of the resolution of the individual dimensions. For example, if 32 components can be resolved in one-dimensional separation and 32 components resolved in another type of separation, the combination of those two methods would be able, in principle, to resolve $32 \times 32 = 1024$ components.

Use of conventional two-dimensional gel electrophoresis technology is inappropriate for single cell analysis. The minute amount of protein would be severely diluted. Instead we have developed an automated 2D CE instrument, as shown in Figure 4.4 [42]. In this instrument, our injection system is coupled to a capillary sieving electrophoresis column. The cell is injected and lysed and the proteins are labeled and separated, as in the single capillary system. However, in this instrument, a second capillary is coupled to the exit of the first. When analyte reaches the exit of the first capillary, power supply 1 is turned off and power supply 2 is turned on. Analyte present in the interface is injected onto the second capillary, separated by micellar electrophoresis and detected with our ultrasensitive LIF detector. Once analyte has been separated in the second capillary, power is briefly applied to the first capillary,

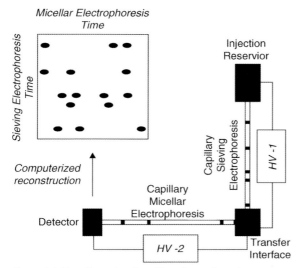

Figure 4.4 Two-dimensional capillary electrophoresis instrument. HV-1 and HV-2 are high-voltage power supplies.

migrating another fraction into the interface for subsequent injection and separation. This process is repeated for 100–200 iterations, performing a comprehensive separation of the proteins in a single cell.

This instrument is based on the coupled liquid chromatography/capillary electrophoresis systems developed by Jorgenson's group in the 1990s [43]. We employ a simple coaxial transfer interface (Figure 4.5), where the tips of the two capillaries are separated by about 50 µm in a buffer-filled chamber. Electrical connection is provided to this chamber so that an electric field can be applied across either capillary as needed. In the figure, a plug of fluorescein is being transferred from the first to the second capillary.

We have analyzed a number of samples and cell types using this instrument. A 2D electropherogram of the separation of the proteins from a single MC3T3 osteoprecursor cell is presented in Figure 4.6. These cells are associated with bone growth and repair; and they are intermediate between primitive stem cells and fully differentiated bone.

The data exist in the computer and may be presented in several different formats. This figure shows the data as an intensity plot, where the density is proportional to the logarithm of the fluorescence intensity. The image has been overexposed to highlight some of the lower intensity components present in the sample. However, this presentation format does not reproduce well the dynamic range of the fluorescence signal, which extends for nearly four orders of magnitude.

There are two streaks that are present in each sieving electrophoresis fraction, one at 10 s and the other at 85 s in the micellar electrophoresis dimension. These streaks are system peaks generated by the difference in buffer composition in the first- and second-dimension capillaries. We believe that the first component is generated by

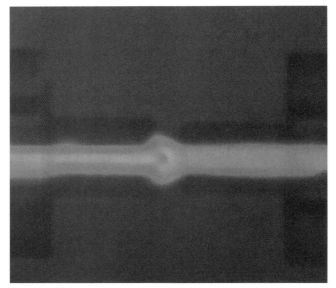

Figure 4.5 Coaxial sample interface. A fluorescent microscope is used to image the transfer of a plug of fluorescein between two capillaries (see color insert).

sieving material that is transferred to the second capillary with each sample. This material has a different refractive index than the sheath buffer. The laser beam can be scattered by this refractive index inhomogeneity, which results in the observed signal. The second streak is of much lower amplitude and represents a minute amount of scattered light, likely due to the presence of small ions used to prepare the buffers.

Figure 4.6 Intensity image of the two-dimensional electropherogram generated from a single MC3T3 cell.

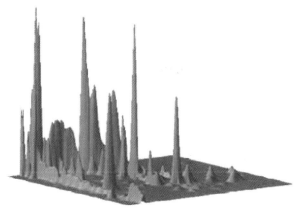

Figure 4.7 Landscape view of the data presented in Figure 4.6.

The intensity plot of Figure 4.6 does a poor job of displaying relative abundances. We instead prefer to use a landscape format to present the data as shown in Figure 4.7. The streaks that extend across the image are resolved into a set of ridges. The highest amplitude component that migrates at low molecular mass generated a fluorescence signal that saturated the detector. Unfortunately the low molecular mass components are poorly resolved by this micellar electrophoresis buffer system.

We are investigating different micellar and submicellar buffer systems to better resolve the complex protein content of a single cell. Once a satisfactory buffer system is available, we will begin the process of tentatively identifying components in the electropherogram. We will perform classic 2D gel electrophoresis on a cellular homogenate and we will excise spots for mass spectrometric identification. As in our 1D capillary sieving electrophoresis experiment, we will spike the sample before 2D electrophoresis to perform comigration analysis. We will employ a combinatorial spiking protocol to speed the identification of peaks.

4.11
Single Copy Detection of Specific Proteins in Single Cells

The multidimensional CE system has detection limits of a few thousand copies of many proteins. We are modifying the instrument by replacing the photomultiplier tube with a single-photon counting avalanche photodiode. This photodetector has a fivefold higher quantum yield, which will result in a modest improvement in detection limit. This improved instrument will certainly have many applications in both basic and clinical studies.

However, many researchers are interested in the study of proteins that are expressed at very low levels. The chemical labeling approaches that are currently available do not provide that capability. We are developing genetic engineering and improved instrumentation to detect and count single copies of a specific protein in a single cell and to monitor the post-translational modifications of that protein. We believe that this technology will have wide application, particularly in the basic sciences.

One serious challenge in the study of a single, specific protein at low copy levels is the signal generated by all of the other proteins present in the cell. Clearly the isolation of a specific protein at very low levels from the complex mixture produced by chemical labeling is unacceptable. In principle, it may prove possible to use a highly fluorescent antibody to tag the protein of interest. Instead we rely on genetic engineering to fuse green fluorescent protein (GFP) to the protein of interest. This genetic engineering step has the advantage of being highly specific to the target protein, albeit with the disadvantage of requiring biochemical steps not available in all laboratories.

An electropherogram of the Gal4/GFP fusion expressed in yeast is presented in Figure 4.8. The bottom trace was generated from a wild-type yeast lysate, while the top was generated from the genetically modified organism, which expresses the Gal4/GFP fusion. The wild-type yeast shows several autofluorescent components that migrate around 4 min; these components are likely endogenous flavins. A part per trillion contamination of the protease inhibitor generates a peak at 2.5 min. The Gal4/GFP fusion creates a set of three peaks that migrate between 3.0 min and 3.5 min after injection. The first peak comigrates with GFP itself and is likely the proteolytic product, where the Gal4 protein has been fully digested, leaving only the tag. The second and third peaks are different phosphorylation states, presumably of the intact fusion protein. Treatment with alkaline phosphatase increases the amplitude of the last peak, which is consistent with removal of a negative charge from the protein.

We have recently begun to investigate detection of GFP in single yeast cells. The walls of these cells are quite robust and resist lysis with simple SDS treatment. We employ a series of enzymatic steps to degrade the carbohydrate and proteinaceous

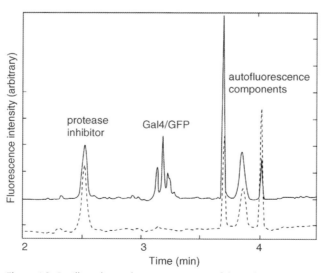

Figure 4.8 Capillary electrophoretic separation of the Gal4/GFP fusion protein from a yeast homogenate.

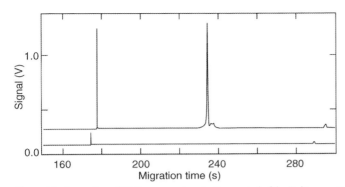

Figure 4.9 Separation of GFP expressed under the control of the Gal1 promoter in a single yeast cell.

components of the wall before lysis. Figure 4.9 presents an electropherogram generated from a single control yeast cell (bottom) and a single engineered cell (top). In this case, the cell had been genetically engineered to express GFP under control of the Gal1 promoter. The cell was grown in the presence of galactose, which resulted in the expression GFP in the single cell.

Single molecule detection has been reported for GFP that is immobilized in a thin gel film and illuminated for long periods under a microscope [44]. Unfortunately simple microscopic detection provides no information on the post-translational modifications of the molecule; in the worst case, the fusion protein may be proteolytically digested, leaving only the GFP tag.

The photophysics of GFP is rather complicated, which makes its detection more challenging in a flowing system as in CE. We use a number of conventional modifications to the instrument to optimize detection of GFP. We minimize the detection volume by tightly focusing the laser beam and by use of a small diameter separation capillary. We optimize the laser power, which is near the photobleaching level for the molecule. We use a high-sensitivity photodiode for detection. We sample the photodetector output at a rate that matches the transit time of a molecule through the laser beam. The result is detection of single molecules of GFP with a signal to noise ratio of about 3, as shown in Figure 4.10.

The molecule generates a burst of fluorescence as it passes through the laser beam. This burst is a millisecond in duration and is detected above the fluctuations in the background signal. We are in the process of modifying our instrument to improve the signal to noise ratio for single molecule detection and we are investigating improvements to the cell lysis conditions. This combination should make routine the characterization of GFP–fusion proteins expressed at extremely low levels.

4.12
Conclusion

Analysis of the protein content of a single mammalian cell is a formidable challenge. Most proteins are present at the zeptomole level and their separation and

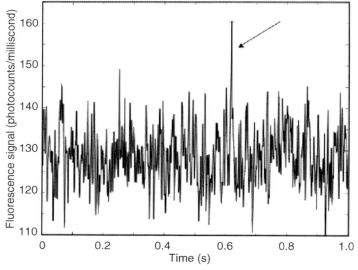

Figure 4.10 The photon burst generated by a single molecule of GFP migrating from a capillary electrophoresis column through an ultrasensitive laser-induced fluorescence detector.

identification push analytical capabilities to the limit. We have developed instruments that allow us to generate 2D electropherograms of the protein content from single mammalian cells and to detect and characterize extremely low levels of specific proteins in yeast and other systems that may be genetically engineered.

These systems are in their infancy and many years of work will be required to optimize their performance and evaluate their utility. We certainly need to improve the resolution of the electrophoretic separation. We need to improve the sensitivity of the 2D electrophoresis instrument to detect proteins expressed at lower levels. We need to demonstrate the tentative identification of a significant fraction of the components in our 2D single cell electropherograms.

Nevertheless, these preliminary results hint at the power of single cell proteomics. We envision applications where we monitor the evolution of protein expression at the single cell level during development and differentiation and in response to specific stresses and hormones. We have begun a number of collaborations to monitor the changes in protein expression associated with cancer progression and viral infection. We are interested in the behavior of extremely rare regulatory proteins associated with regulation of the cell cycle.

As should be obvious from the success of the Human Genome Project, new science flows from the development of new technology. Unfortunately a number of federal funding agencies, particularly those programs with a mandate to support innovative molecular analysis technologies, have been unable to provide continued support for this project. New technology can be brought to maturity only with sustained support.

Acknowledgements

This group is currently supported by the National Human Genome Research Institute, the National Institute of Drug Abuse, the Department of Energy Genomes to Life Program, the Department of Defense Congressionally Directed Medical Research Program and MDS-Sciex. DQM gratefully acknowledges a fellowship from the University of Washington Nanotechnology. We are grateful that these visionary agencies are able to support the development of innovative bioanalytical technology.

References

1 Hu, S., Zhang, L., Krylov, S.N. and Dovichi, N.J. (2003) Cell-cycle dependent protein fingerprint from a single cancer cell: image cytometry coupled with single-cell capillary sieving electrophoresis. *Analytical Chemistry*, **75**, 3495–3501.

2 Gygi, S.P., Rochon, Y., Franza, B.R. and Aebersold, R. (1999) Correlation between Protein and mRNA Abundance in Yeast. *Molecular and Cellular Biology*, **19**, 1720–1230.

3 Gygi, S.P., Rist, B., Gerber, S.A., Turecek, F., Gelb, M.H. and Aebersold, R. (1999) Quantitative analysis of complex protein mixtures using isotope-coded affinity tags. *Nature Biotechnology*, **17**, 994–999.

4 Washburn, M.P., Wolters, D. and Yates, J.R. (2001) Large-scale analysis of the yeast proteome by multidimensional protein identification technology. *Nature Biotechnology*, **19**, 242–247.

5 Smith, R.D. (2002) Trends in mass spectrometry instrumentation for proteomics. *Trends in Biotechnology*, **20** (12 Suppl), S3–S7.

6 Whittal, R.M., Keller, B.O. and Li, L. (1998) Nanoliter chemistry combined with mass spectrometry for peptide mapping of proteins from single mammalian cell lysates. *Analytical Chemistry*, **70**, 5344–5347.

7 Xu, B.J., Caprioli, R.M., Sanders, M.E. and Jensen, R.A. (2002) Direct analysis of laser capture microdissected cells by MALDI mass spectrometry. *Journal of the American Society for Mass Spectrometry*, **13**, 1292–1297.

8 Page, J.S., Rubakhin, S.S. and Sweedler, J.V. (2002) Single-neuron analysis using CE combined with MALDI MS and radionuclide detection. *Analytical Chemistry*, **74**, 497–503.

9 Pasa-Tolic, L., Lipton, M.S., Masselon, C.D., Anderson, G.A., Shen, Y., Tolic, N. and Smith, R.D. (2002) Gene expression profiling using advanced mass spectrometric approaches. *Journal of Mass Spectrometry*, **37**, 1185–1198.

10 Edstrom, J.E. (1953) Nucleotide analysis on the cyto-scale. *Nature*, **172**, 908.

11 Matioli, G.T. and Niewisch, H.B. (1965) Electrophoresis of hemoglobin in single erythrocytes. *Science*, **150**, 1824–1826.

12 Repin, V.S., Akimova, I.M. and Terovskii, V.B. (1975) Detection of lactate dehydrogenase isoenzymes in single mammalian oocytes during cleavage by a micromodification of disc electrophoresis. *Bulletin of Experimental Biology and Medicine*, **77**, 767–769.

13 Ruchel, R. (1976) Sequential protein analysis from single identified neurons of *Aplysia californica*. A microelectrophoretic technique involving polyacrylamide gradient gels and isoelectric focusing. *The Journal of Histochemistry and Cytochemistry*, **24**, 773–791.

14 Kennedy, R.T., St. Claire, R.L., White, J.G. and Jorgenson, J.W. (1987) Chemical

analysis of single neurons by open tubular liquid chromatography. *Mikrochimica Acta*, **I**, 37–45.

15 Kennedy, R.T., Oates, M.D., Cooper, B.R., Nickerson, B. and Jorgenson, J.W. (1989) Microcolumn separations and the analysis of single cells. *Science*, **246**, 57–63.

16 Wallingford, R.A. and Ewing, A.G. (1988) Capillary zone electrophoresis with electrochemical detection in 12.7 microns diameter columns. *Analytical Chemistry*, **60**, 1972–1975.

17 Hogan, B.L. and Yeung, E.S. (1992) Determination of intracellular species at the level of a single erythrocyte via capillary electrophoresis with direct and indirect fluorescence detection. *Analytical Chemistry*, **64**, 2841–2845.

18 Gilman, S.D. and Ewing, A.G. (1995) Analysis of single cells by capillary electrophoresis with oncolumn derivatization and laser-induced fluorescence detection. *Analytical Chemistry*, **67**, 58–60.

19 Meredith, G.D., Sims, C.E., Soughayer, J.S. and Allbritton, N.L. (2000) Measurement of kinase activation in single mammalian cells. *Nature Biotechnology*, **18**, 309–312.

20 Han, F.T. and Lillard, S.J. (2002) Monitoring differential synthesis of RNA in individual cells by capillary electrophoresis. *Analytical Biochemistry*, **302**, 136–143.

21 Krylov, S.N., Zhang, Z., Chan, N.W., Arriaga, E., Palcic, M.M. and Dovichi, N.J. (1999) Correlating cell cycle with metabolism in single cells: combination of image and metabolic cytometry. *Cytometry*, **37**, 14–20.

22 Le, X.C., Tan, W., Scaman, C.H., Szpacenko, A., Arriaga, E., Zhang, Y., Dovichi, N.J., Hindsgaul, O. and Palcic, M.M. (1999) Single cell studies of enzymatic hydrolysis of a tetramethylrhodamine labeled triglucoside in yeast. *Glycobiology*, **9**, 219–225.

23 Krylov, S.N., Starke, D.A., Arriaga, E.A., Zhang, Z., Chan, N.W., Palcic, M.M. and Dovichi, N.J. (2000) Instrumentation for chemical cytometry. *Analytical Chemistry*, **72**, 872–877.

24 Dovichi, N.J. and Zhang, J. (2000) How capillary electrophoresis sequenced the human genome. *Angewandte Chemie (International Edition in English)*, **39**, 4463–4468.

25 Chen, D.Y. and Dovichi, N.J. (1996) Single-molecule detection in capillary electrophoresis: molecular shot noise as a fundamental limit to chemical analysis. *Analytical Chemistry*, **68**, 690–696.

26 Liu, J.P., Hsieh, Y.Z., Wiesler, D. and Novotny, M. (1991) Design of 3-(4-carboxybenzoyl)-2-quinolinecarboxaldehyde as a reagent for ultrasensitive determination of primary amines by capillary electrophoresis using laser fluorescence detection. *Analytical Chemistry*, **63**, 408–412.

27 Pinto, D., Arriaga, E.A., Schoenherr, R.M., Chou, S.S. and Dovichi, N.J. (2003) Kinetics and apparent activation energy of the reaction of the fluorogenic reagent 5-furoylquinoline-3- carboxaldehyde with ovalbumin. *Journal of Chromatography B*, **793**, 107–114.

28 Liu, H., Cho, B.Y., Krull, I.S. and Cohen, S.A. (2001) Homogeneous fluorescent derivatization of large proteins. *Journal of Chromatography. A*, **927**, 77–89.

29 Zhao, J.Y., Waldron, K.C., Miller, J., Zhang, J.Z., Harke, H. and Dovichi, N.J. (1992) Attachment of a single fluorescent label to peptides for determination by capillary zone electrophoresis. *Journal of Chromatography*, **608**, 239–242.

30 Pinto, D.M., Arriaga, E.A., Craig, D., Angelova, J., Sharma, N., Ahmadzadeh, H., Dovichi, N.J. and Boulet, C.A. (1997) Picomolar assay of native proteins by capillary electrophoresis—precolumn labeling, sub-micellar separation and laser induced fluorescence detection. *Analytical Chemistry*, **69**, 3015–3021.

31 Lee, I.H., Pinto, D., Arriaga, E.A., Zhang, Z. and Dovichi, N.J. (1998) Picomolar analysis of proteins using electrophoretically

mediated microanalysis and capillary electrophoresis with laser-induced fluorescence detection. *Analytical Chemistry*, **70**, 4546–4548.

32. Oakes, J. (1974) Protein–surfactant interactions. Nuclear magnetic resonance and binding isotherm studies of interactions between bovine serum albumin and sodium dodecyl sulfate. *Journal of the Chemical Society-Faraday Transactions I*, **70**, 2200–2209.

33. Turro, N.J. and Lei, X. -G. (1995) Spectroscopic probe analysis of protein–surfactant interactions: the BSA/SDS system. *Langmuir*, **11**, 2525–2533.

34. Santos, S.F., Zanette, D., Fischer, H. and Itri, R. (2003) A systematic study of bovine serum albumin (BSA) and sodium dodecyl sulfate (SDS) interactions by surface tension and small angle X-ray scattering. *Journal of Colloid And Interface Science*, **262**, 400–408.

35. Hu, S., Zhang, L., Cook, L.M. and Dovichi, N.J. (2001) Capillary sodium dodecyl sulfate–DALT electrophoresis of proteins in a single human cancer cell. *Electrophoresis*, **22**, 3677–3682.

36. Hu, S., Jiang, J., Cook, L.M., Richards, D.P., Horlick, L., Wong, B. and Dovichi, N.J. (2002) Capillary sodium dodecyl sulfate–DALT electrophoresis with laser-induced fluorescence detection for size-based analysis of proteins in human colon cancer cells. *Electrophoresis*, **23**, 3136–3142.

37. Hu, S., Zhang, L., Newitt, R., Aebersold, R., Kraly, J.R., Jones, M. and Dovichi, N.J. (2003) Identification of proteins in single-cell capillary electrophoresis fingerprints based on co-migration with standard proteins. *Analytical Chemistry*, **75**, 3502–3505.

38. Richards, D.P., Stathakis, C., Polakowski, R., Ahmadzadeh, H. and Dovichi, N.J. (1999) Labeling effects on the isoelectric point of green fluorescent protein. *Journal of Chromatography. A*, **853**, 21–25.

39. Zhang, Z., Krylov, S., Arriaga, E.A., Polakowski, R. and Dovichi, N.J. (2000) One-dimensional protein analysis of an HT29 human colon adenocarcinoma cell. *Analytical Chemistry*, **72**, 318–322.

40. Hu, S., Lee, R., Zhang, Z., Krylov, S.N. and Dovichi, N.J. (2001) Protein analysis of an individual *Caenorhabditis elegans* single-cell embryo by capillary electrophoresis. *Journal of Chromatography B*, **752**, 307–310.

41. Giddings, J.C. (1991) *Unified Separation Science*, John Wiley & Sons, Hoboken, NJ.

42. Michels, D., Hu, S., Schoenherr, R., Eggertson, M.J. and Dovichi, N.J. (2002) Fully automated twodimensional capillary electrophoresis for high-sensitivity protein analysis. *Molecular & Cellular Proteomics*, **1**, 69–74.

43. Bushey, M.M. and Jorgenson, J.W. (1990) Automated instrumentation for comprehensive twodimensional high-performance liquid chromatography of proteins. *Analytical Chemistry*, **62**, 161–167.

44. Dickson, R.M., Cubitt, A.B., Tsien, R.Y. and Moerner, W.E. (1997) On/off blinking and switching behaviour of single molecules of green fluorescent protein. *Nature*, **388**, 355–358.

5
Protein Analysis of Single Cells in Microfluidic Format
Alexandra Ros and Dominik Greif

5.1
Introduction

Single cell analysis allows individual expression studies which are not limited by ensemble averaging effects from cell cycle-dependent states, the different and inhomogeneous cellular response to external stimuli, or the introduction of genomic and proteomic variability during cell proliferation. In ensemble experiments, those influences are difficult to address, because the measurement of the individual cellular response is lost in the bulk average. Furthermore, experiments show that significant cellular phenotypes are only resolvable when studying single cells, emphasizing the importance of studying individual cells rather than heterogeneous populations [1]. Single cell analysis within the framework of systems biology and proteome research requires separation and detection techniques which are effective, sensitive and quantitative. Microfluidic devices have the potential to fulfil these requirements, which is impressively demonstrated by the transfer of proteome-relevant separation techniques to the microfluidic format, such as gel electrophoresis, isoelectric focusing and two-dimensional separation techniques for proteins [2–5].

Furthermore, sensitive detection techniques such as laser-induced fluorescence (LIF), electrochemical detectors and mass spectrometry (MS) can be coupled to the microfluidic format. MS represents probably the most relevant detection technique for proteomic research. State of the art concepts for microchip–MS coupling rely on electrospray ionization (ESI) and were recently summarized [6]. This high potential of microfluidic systems for proteomic research is intensifying interest in cellular analysis within microfluidic chips. Several studies focus on subjects such as mechanical or dielectrophoretical cell manipulation or flow cytometry [7], as well as cell culture and subsequent analysis of cell compounds [8, 9], thereby handling cell cultures on the level of 10^3 cells. However, microfluidic devices capable of analyzinging single cells were only recently been demonstrated and are the subject of the present chapter.

Single Cell Analysis: Technologies and Applications. Edited by Dario Anselmetti
Copyright © 2009 WILEY-VCH Verlag GmbH & Co. KGaA, Weinheim
ISBN: 978-3-527-31864-3

For single cell analysis, microfluidic devices with characteristic length scales of 10–100 µm are predestined due to typical cell dimensions of several microns. In order to significantly contribute to proteomic research, low-abundance proteins with copy numbers <10^5 molecules in a single cell must be detectable. This would require detectors with a sensitivity in the range less than 100 nM considering 10^5 molecules in a typical eukaryotic cell of 10 µm diameter (corresponding to a volume of 1 pL) [10]. This makes clear that single cell analysis in microfluidic format demands new techniques for efficient and sensitive separation and detection.

Microfluidic devices further provide the advantage of adapting the channel dimension to the particular cellular system that is subject of investigations. Thus, the dilution of cellular components can be minimized, which is an important issue due to the extremely low detection sensitivity required. Combining single cell analysis with rapid prototyping techniques such as molding with elastomers further allows for simplified experimental access to the tailored adjustment of microfluidic dimensions within the framework of the specific analytical problem.

LIF detection is a sensitive detection technique which integrates very successfully into microfluidic or capillary based separation devices. Concentration limits in the picomolar range are frequently achieved, rendering this technique suitable for single cell analysis. Thus single cell analysis was first demonstrated with LIF detection in the visible range. First single cell fingerprinting with capillary sieving electrophoresis in one-dimensional [11, 12] and two-dimensional formats [13] using a fluorescent protein stain was pioneered by the group of N. Dovichi. The first examples for single cell analysis in microfluidic device format were explored for the separation and detection of fluorescent dyes [14, 15] with LIF detection. The latter detection technique was also employed for the determination of reactive oxygen species [16] and glutathione in single erythrocytes [17, 18]. Enzymatic products were also detected after cell lysis in microfluidic devices from single leukaemia cells [19] with LIF detection. Alternatively electrochemical detection represents a label-free detection method for electroactive species. This detection method was used for the analysis of a specific small peptide [20] and a vitamin [21] from single cells in microfluidic devices.

More recently, methods for the analysis of proteins from single cells have been demonstrated combining on-chip cell lysis and electrophoretic separation of proteins with subsequent detection in the visible range by LIF [10, 22, 23]. While LIF in the visible range requires adequate on-chip or off-chip labeling steps for the cell component of interest, LIF detection in the UV spectral range allows direct label-free detection of proteins. However, label-free UV-LIF was only recently demonstrated in microfluidic devices [10, 24] and, based on these results, the first label-free single cell protein electropherograms were reported [25]. This chapter presents the necessary microfluidic operations for the protein analysis of single cells, such as the selection, navigation and deposition of single cells in particular microfluidic environments, but also the cell lysis and subsequent separation and detection of proteins. The sensitive LIF detection of proteins originating from single cells in the visible range together with the label-free LIF detection in the UV range is discussed and future aspects of single cell analysis in microfluidic systems are addressed.

5.2
Microfluidic Single Cell Analysis Concept

In order to perform single cell analysis in microfluidic format, one has to design a device in which a single cell from a given population can be selected and subsequently brought into a position in which the desired analysis can be carried out. This position should create a trap for a single cell, so that further fluid manipulation can be performed, such as delivery of external stimuli or lysis reagents. Adequate valves and pumps need further to be integrated if such additional fluidic operations around the single cells are desired. In a minimal approach, the cells can be trapped in a geometrical manner which allows fluid manipulation without displacement of the cells from their position. A subsequent cell lysis step at a desired time further exposes the cell content to the fluidic environment and allows for analysis of these contents in solution, by employing adequate separation principles and detectors.

5.2.1
Single Cell Selection and Trapping

One approach to achieve microfluidic single cell analysis is demonstrated in Figure 5.1a. The simple microfluidic device consists of four linear channels which create the cell injection position at the intersection. This position is bordered by posts acting as physical traps for the cells due to gap distances which are smaller than the diameter of the cell of interest. These posts extend over the entire depth of the microchannel but allow further fluid access into all channels (see Figure 5.1). Such geometrical trapping can also be performed by a gentle suction of cells at microchannel constrictions which are smaller than their diameter. Suction can be performed via the application of negative pressure or hydrodynamically. These methods were demonstrated to operate well for single cell electroporation [26, 27] or pulsed delivery of reagents to single cells [28]. There are several alternative methods to trap single cells in microfluidic devices which do not necessarily rely on geometrical trapping. Other possibilities arise from dielectrophoretic trapping, incorporation into fluid droplets, hydrodynamic focusing or the use of on-chip valves and pumps. The state of the art of these methods was recently reviewed [29].

A critical step in single cell analysis results from the selection of the cells to be analyzed. Microfluidic single cell analysis is in its origin and the large-scale parallelization of this technique has yet to be demonstrated. Therefore, the selection of individual cells of interest from a cell population remains an important step. The following example illustrates the importance of such a selection step. Given that the expression of a fluorescent fusion protein is being studied on the single cell level, only cells with this particular fusion protein are of interest. A preselection step is highly desired, if one recognizes that gene transfection efficiencies resulting in cells which express the fluorescent fusion protein can vary significantly (between 20–50%). It is clear that randomly selecting single cells would lead to irrelevant data acquisition of up to 80%. Figure 5.1b demonstrates a bright field and a fluorescence image of a subpopulation of *Spodoptera frugiperda* (Sf9) insect cells. Comparison of the two

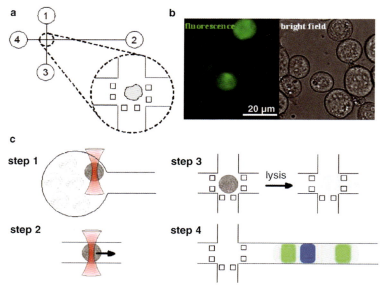

Figure 5.1 (a) Scheme of a microfluidic device for single cell analysis: it is composed of four channels, creating an intersection, which serves as the cell trap. (b) Comparison of fluorescence and bright field microscopy images of a subpopulation of Sf9 cells. Clearly, only three of the seven cells (captured as a whole) express the fluorescent fusion protein. (c) Steps 1 through 4 of the single cell microfluidic analysis concept, which is mainly discussed in this chapter. It consists of a selection step via an OT (step 1), the OT navigation through a microfluidic channel until the geometrical cell trap is reached (step 2). In step 3 the single cell is lysed and subsequently the protein content is analyzed downstream the channel after electrophoretical separation of the protein compounds. The red cone schematically shows the laser beam which is focused via a high numerical aperture objective to create the optical trap. Typically, in single beam optical tweezers Nd:YAG lasers at a wavelength of 1064 nm and a power of 1 W can be employed (see for example Ref. [48]).

images clearly shows that only a subfraction of the population exhibits the fluorescent fusion protein. A selection step prior to analysis thus provides for the minimization of redundant data collection.

A single cell microfluidic analysis approach with a preselection step is given in Figure 5.1c (steps 1 to 4). A cell ensemble is suspended in the sample reservoir of a microfluidic system. In step 1, the cell selection is performed by capturing a single cell of interest by means of optical tweezers (OT). In optical tweezers, a highly focused laser beam creates attractive forces in the piconewton range which are capable of holding spherical dielectric particles, such as biological cells, in so-called optical traps. In step 2, the selected cell is navigated – while being optically trapped by the OT – along the microchannel to an intersection of two channels, which is structured by posts. After release from the OT, the cell is trapped physically at this channel intersection. In step 3, the cell is lysed at the intersection position. Subsequent electrophoretic migration leads to the transport and separation of the protein components of the cell which are detected by an adequate detector downstream from the separation channel in step 4. In this contribution we focus on laser induced fluorescence (LIF) detection of proteins from single cells in a microfluidic device format. Other methods capable of the detection of proteins in solution, such as electrochemical detection, can also be coupled to this single cell analysis approach. However, all experimental results presented within this chapter (see Sections 5.2.2 and 5.3) are based on the above-stated single cell analysis concept with OT preselection.

5.2.2
Single Cell Lysis

Cell lysis can be performed via several procedures, such as mechanical, chemical and electrical methods. In microfluidic systems, however, only the latter two methods are straightforward to integrate (for a schematic of these methods see Figure 5.2a). In microfluidic applications, electroosmotic bulk flow is a versatile method to deliver cells and analyte volumes to microfluidic channels via electrical fields. This is usually performed though the application of (high) voltages to electrodes in the microfluidic reservoirs at the end of the channels. Thus the integration of electrical lysis due to a high voltage pulse is already provided by the experimental setup. Also chemicals can be delivered via hydrodynamic or electroosmotic pumping over trapped and immobilized cells. The literature reports the chemical lysis of cells with ionic

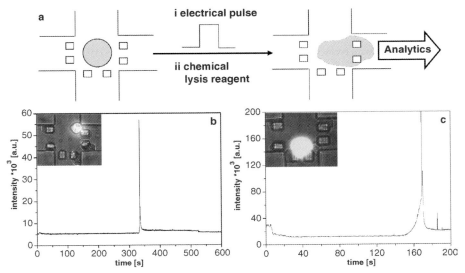

Figure 5.2 (a) Schematic of the cell lysis process at a geometrical cell trap in a microfluidic system. (i) The cell is lysed via a high voltage electrical pulse ($\geq$1000 V/cm, >50 ms). (ii) The cell is lysed with a chemical lysis reagent. (b) Chemical lysis of a single Sf9 cell in a microfluidic device with subsequent downstream visible LIF detection of the fluorescent GFP-fusion protein. Cell lysis was performed with 0.5% SDS in approximately 5 s. (c) Electrical lysis of a single Sf9 cell with an electrical pulse of 50 ms at 1250 V/cm and with subsequent downstream visible LIF detection of the fluorescent GFP-fusion protein. (b) and (c) were adapted from Refs. [22, 10].

detergents, such as soldium dodecyl sulfate (SDS) [10, 30], or nonionic detergents, such as n-dodecyl-β-D-maltidose (DDM) [23]. Insect cells can typically be lysed in a few seconds when exposed to detergent solution. In contrast, electrical lysis employing short, high-voltage pulses can result in much shorter lysis times, in the order of a few milliseconds [22]. For example, applying high-voltage pulses (typically 50 ms, $\geq$ 1000 V/cm) results in the fast lysis of Sf9 cells in a microfluidic injection position. Figure 5.2 demonstrates the detection of the fusion proteine T31N-GFP in single Sf9 cells, both with chemical lysis (Figure 5.2b) and with electrical lysis (Figure 5.2c), followed by electrophoretic transport of the protein component to a fluorescence detection position several millimeters downstream from the lysis position. It is interesting to note that the detected fluorescence signal reflects the different fluorescence intensities of GFP in the two cells (see microscopy image insets in Figure 5.2b, c). Note that those cells were selected via fluorescence inspection from a cell suspension at the entrance of a microchannel via OT and then brought into the lysis position within the microfluidic device (as described in Chapter 2.1).

5.3
Single Cell Electrophoretic Separation and Detection of Proteins

As outlined in the introduction to this chapter, the analysis of the protein content or the proteome of a single cell in comparison to other cells in a population remains a very important and challenging task. In order to obtain a quantitative signal from a single protein, it has to be differentiated from all other protein species. A biologically elegant way to do so is by labeling a specific protein, thus creating a fusion protein via modifying the gene expression of a cell line. Frequently fusions with proteins that exhibit native fluorescence in the visible range are employed, such as GFP and GFP-related proteins which exhibit related fluorophores, such as the yellow fluorescent protein (YFP). This allows the study of the variation in expression levels of a single fusion protein in the cell by means of the fluorescence signal of this specific protein. However this technique is not capable of capturing more complex variations within a group or subgroup of proteins or metabolites in a specific cell, such as the differences in a biological intracellular signal transduction pathway. Moreover in order to capture a fingerprint of a biological cell's protein content, a variety of proteins has to be studied at the same time. Either this would require tagging each protein with a specific label, which is not easy to achieve given hundreds or thousands of different protein species in a single cell. Or the desired information could be provided by labeling all proteins with the same label or function that allows their detection combined with an efficient separation and sensitive detection method. Such a labeling procedure was, for example, recently described by Sun *et al.* with liposome fusion [31], which allows the specific delivery of a nonspecific fluorescent protein label to cells in microfluidic systems.

The most interesting way of analyzing the complexity of proteins in single cells is the label-free detection of those proteins. This can be achieved by analyzing intrinsic protein labels, such as the fluorescence arising from the three amino acids tryptophane, tyrosine and phenylalanine. Thus employing sensitive laser-induced fluorescence detection is one way to address the entirety of the proteins of a single cell. Statistically nearly every protein contains one of these amino acids, such that a sensitive detection of their fluorescence emission in the UV range between 280 nm and 350 nm represents a feasible approach for studying protein composition in single cells.

In this section, two approaches for single cell electropherograms in microfluidic format are presented: First, the separation and detection of fluorescent fusion proteins in the visible range by LIF and, second, the detection of native, nonlabeled proteins from single cells by LIF in the UV range. The experimental approach is realized in microfluidic systems with high UV transparency, fabricated in polymer and quartz-

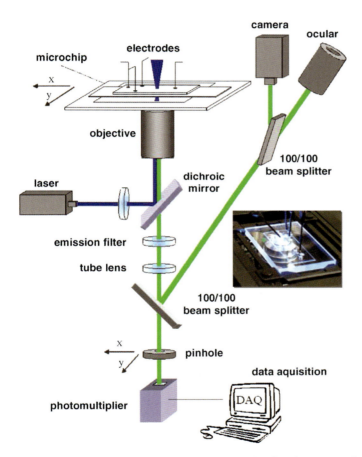

Figure 5.3 Schematic illustration of the LIF detection for single cell analysis realized on an inverted microscope: For excitation in the visible range a laser line of 488 nm from an Ar^+ laser is coupled into the rear port of the microscope. The inset demonstrates a photograph of the completely assembled microfluidic device for single cell analysis mounted on a microscope x/y table. For LIF in the UV range, the optical path was adequately selected for high UV transmission. Filters, mirrors and dichroic were adopted for the excitation wavelength of 266 nm and tryptophan fluorescence characteristics.

based devices. The detection is performed on an inverted microscope equipped with adequate optics for the excitation light and emission wavelengths in the visible or UV range. Figure 5.3 gives an overview scheme of the applied optical setup.

5.3.1
Label-Based Fluorescence Detection

As demonstrated above, signals from fluorescently labeled proteins can be obtained by microfluidic single cell analysis after cell lysis and downstream transport of the analyte to a detection position. Further as recently demonstrated with LIF detection in the visible range in microfluidic systems [32], a detection limit in the femtomolar concentration range can be achieved which is thus adequate for the detection of low-abundance proteins in single cells. Such proteins with copy numbers of $<10^5$ are found in single cells in concentrations smaller than 100 nM (see also Introduction). It is thus very interesting to investigate whether the single cell microfluidic approach can also provide for the separation of fluorescent proteins from single cells. Figure 5.4 provides a first example for the separation of two protein species from a single Sf9 cell [22]. These cells expressed a GFP and a YFP fusion protein, namely the proteins Pep12-YFP ($\sim$50 kDa) and γ-PKC-GFP ($\sim$105 Da). GFP and YFP have slightly different fluorescence excitation and emission maxima. However they can be detected simultaneously with adequate optical filters. In Figure 5.4, one can clearly observe the

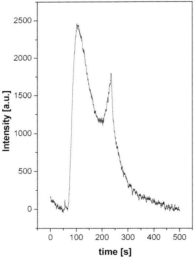

Figure 5.4 Single cell electropherogram of a Sf9 cell expressing both Pep12-YFP and γ-PKC-GFP recorded at 15 mm separation distance (separation buffer: 4% Pullulan, 1% SDS, 0.1% POP-6 (v/v), 100 mM CHES/TRIS). Clearly, the electropherogram demonstrates two peaks for the two fluorescent proteins in the cell (adapted from Ref. [22]). Channels were pretreated with a block copolymer (F108) prior to electrophoretic analysis.

separation of these two protein species from a single Sf9 cell, which was recorded downstream subsequently to a chemical lysis procedure with SDS and the electrophoretical separation of the two species in a sieving buffer. It is clear that the separation and injection parameters further need to be improved. However this result demonstrates the feasibility of this microfluidic single cell approach for protein detection from single cells after cell lysis, thus providing a snapshot of the protein content from single cells at a precise time.

5.3.2
Label-Free Fluorescence Detection

5.3.2.1 UV-LIF in Quartz Microfluidic Devices
In a microfluidic format, the detection of proteins via their intrinsic fluorescence properties of the three fluorescent amino acids can only be carried out in optically suitable materials with low background fluorescence in the UV range. The material of choice thus remains fused silica, which results in significant costs for microfluidic device fabrication. Fused silica is a common material in conventional capillary electrophoretic applications and several native LIF approaches have been described in the past. In 1992, Yeung and coworkers pioneered a LIF detection method with 275 nm excitation light provided by an Ar^+ laser with picomolar detection limits [33]. This method further served for the exocytose monitoring of single mast cells [34] as well as for the separation of hemoglobin variants in red blood cells [35]. Exploiting alternative laser systems, nanomolar detection limits for Trp [36, 37], peptides [38] and proteins [39] were reported.

In microfluidic systems, label-free LIF detection of proteins in the UV range was reported with an excitation wavelength of 266 nm. Belder *et al.* demonstrated an almost micromolar detection sensitivity with standard proteins [24]. Here, the separation of marker proteins in a quartz microfluidic chip designed for single cell trapping, lysis and electrophoretic detection was studied. The quartz microdevices were produced by dry etching, resulting in highly UV-transparent microfluidic channels and thus minimized background fluorescence. Figure 5.5a demonstrates the separation of three proteins with nearly baseline resolution. The injected sample mixture consisted of the three proteins, namely α-chymotrypsinogen A (type II from bovine pancreas, ~25 kDa, with 8 trp), albumin (from egg, ~45 kDa, with 3 trp) and catalase (from bovine liver, ~240 kDa, with 24 trp) at a concentration of 1 mg/mL each. These results demonstrate that quartz microfluidic devices represent a suitable basis for UV-LIF of native proteins. However, users have to consider relatively high production costs.

5.3.2.2 UV-LIF in PDMS Microfluidic Devices
Poly(dimethylsiloxane) (PDMS) is a frequently used polymer for microfluidic investigations. It is particular interesting for cellular microfluidic analysis, as this material is air-permeable and thus allows cell culturing on-chip as well as optimized cell handling due to minimized effects of O_2 depletion. PDMS device fabrication known under the synonym *soft lithography* is further particularly simple to operate. Given a microfabricated master structure with the negative relief of the desired

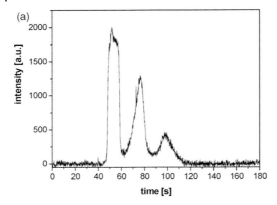

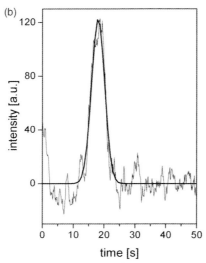

Figure 5.5 (a) Separation of α-chymotrypsinogen A, albumin and catalase (at 0.1 mg/mL each) in the separation buffer (10 mM Tris, pH = 8.2 and 0,1% DDM as dynamic coating reagent) at a separation voltage of 330 V/cm on a whole body quartz chip. (b) Electropherogram of 100 nM injected Trp solution in a PDMS microfluidic chip with carbon black in the PDMS bulk for the reduction of background fluorescence.

microfluidic system, replica molding can be performed under standard dust-free laminar flow conditions. This technique is therefore accessible to most laboratories, including biology laboratories, where single cell analysis is expected to have its most widespread future applications. A typical work flow for PDMS fabrication is given in Figure 5.6, which is adapted from Ref. [40]. Briefly a silicone wafer is spin-coated with a negative tone photoresist and subsequently exposed to UV through a chromium mask with the desired layout. After developing and hard baking, a master wafer with the negative relief structure is created, the surface of which is treated with a silane

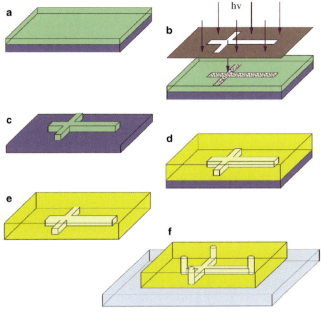

Figure 5.6 Scheme of the workflow for PDMS molding. First, a master wafer from which PDMS microstructures are molded is created (a–c). (a) The negative tone photoresist SU-8 is spin-coated onto a flat wafer. (b) The photoresist is UV exposed through a photomask with the predesigned microfluidic layout. (c) The photoresist is developed and cured to create the negative relief structure on the master wafer. (d) The PDMS mold is subsequently fabricated by pouring the base polymer with its curing agent on the master wafer. (e) The rubber-like elastomeric slab with the microfluidic structures is peeled off the master wafer. (f) After punching reservoir holes for fluid access (not shown) the PDMS microfluidic structure is sealed with a second slab of quartz or PDMS.

reagent to enable repetitive PDMS casting. For the PDMS mold, the base polymer and its curing agent are mixed in a ratio of 10:1 and poured over the master wafer. After curing, the PDMS slab containing the microstructures is peeled off the wafer and reservoir holes are punched for fluid access. To provide an irreversible and tight seal, the PDMS slab and a cover slab are exposed to an oxygen plasma [28]. This treatment also renders the PDMS channel walls hydrophilic, which facilitates filling with aqueous solutions. Once the two exposed surfaces are brought into contact, an irreversible seal is formed. For details on PDMS technology and soft lithography, the reader is referred to comprehensive review articles [41–43].

Two main characteristics of PDMS have to be controlled for single cell electrophoretical separations with UV-LIF detection. First, the surface characteristics of PDMS microfluidic systems significantly influence electroosmotic flow and analyte adsorption in separation experiments. Thus both these properties have to be controlled via adequate surface coatings. Several studies indicated that both dynamic and static surface coatings are suitable to reduce electroomotic flow and analyte adsorption. Among them, poly(ethylene glycols) (PEG) have been studied in detail, demonstrating the successful reduction of electroosmotic flow as well as protein adsorption in static coatings, that is pretreatment methods [44]. Detergent molecules used in dynamic coatings, that is constantly providing the surface treating molecules during the experiment within the separation buffer, also serve as a suitable method to control PDMS surface characteristics within microfluidic systems [45].

Second, the background fluorescence of PDMS has to be considered. Although PDMS is highly transparent in the UV range, particular care has to be taken for excitation wavelengths below 300 nm. It can be demonstrated that PDMS background fluorescence is significantly reduced by the use of adequate optics in microfluidic systems [25]. Another elegant way to reduce undesired background fluorescence was recently demonstrated with the incorporation of carbon black particles in the bulk PDMS. Figure 5.5b demonstrates the successful injection and detection of 100 nM trytophan, indicating a theoretical detection limit of 25 nM [25]. The significant decrease in background fluorescence combined with adequate surface treatments thus approaches the detection limit to the anticipated sensitivity necessary for the detection of even low-abundance proteins in single cells. This demonstrates that such PDMS elastomer-based microfluidic systems are suitable for sensitive LIF detection of proteins in the UV range.

5.3.2.3 Single Cell UV-LIF Electrophoretic Analysis

The above findings demonstrate that detection sensitivities in the nanomolar range can be obtained in both quartz and PDMS microfluidic devices. It is now interesting to study whether protein signals can be detected from single cells. Therefore experiments were conducted with single Sf9 cells in microfluidic layouts, as given in Figure 5.1. According to the optimized fluorescence of tryptophan with the highest quantum efficiency of the three fluorescent amino acids, rather basic conditions should yield highest fluorescence signal. The single cell electropherograms were therefore performed at pH 11. Figure 5.7 exemplarily shows two examples for single

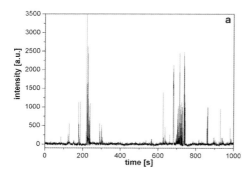
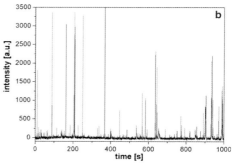

Figure 5.7 (a, b) Baseline corrected electropherograms from single Sf9 cells with native UV-LIF detection ($\lambda_{ex} = 266$ nm). Channels in the PDMS device with a quartz detection window were coated with F108 prior to single cell analysis (separation buffer: 100 mM Tris, 100 mM CHES, 4% Pullulan, pH = 11.0). The quartz window was incorporated in the microfluidic system via processing a 200 μm thick quartz slab onto the master wafer during the PDMS curing process.

cell electropherograms from single Sf9 cells, where ≤ 50 peaks could be resolved. Note that these electropherograms were recorded with a PDMS microfluidic chip with a quartz detection window, providing even for a reduced fluorescence background at the detection position compared to microfluidic channels fabricated in PDMS only. These experiments clearly demonstrate that the microfluidic single cell analysis method combined with an electrophoretic separation of the protein content after cell lysis provides an interesting and powerful tool to study protein expression in single cells. This technique further opens the way for protein fingerprinting from single cells in the future.

5.4
Future Directions in Single Cell Analysis

Single cell analysis of the cell components after decomposing lysis still currently lacks statistically relevant data. To overcome this limitation, it will be necessary in the future to parallelize single cell analysis on microfluidic platforms. For this purpose, single cell selection and trapping must be developed to allow parallel analysis of up to 100 or more single cells on a microchip. There are several techniques which can be further studied in order to fulfil these requirements. One approach relies on OT technology, in which multiple trap OTs could be designed, able to select several cells in one step. Multiple trap OTs have already been developed [46] and have been realized, for example in multifocal microscopy or via holographic optical traps. Their application in microfluidic systems seems straightforward, as the use of OTs has already been proven to be compatible with single cell analysis in microfluidic format. Promising approaches for parallelization also originate from other physical cell trapping principles, such as dielectrophoresis, hydrodynamic trapping or cell suction at microfluidic constrictions. All these approaches have proven their success in single cell trapping; however their use in highly parallelized devices has yet to be demonstrated. This seems mainly be limited by the complexity of microfluidic operations, which however could be solved by more input from electrical and mechanical engineering knowledge into the interdisciplinary field of single cell analysis.

Another interesting approach for single cell analysis results from a recently demonstrated technique, in which the abundance of protein species, that is the number of protein molecules, can be counted in single cells [23]. Future development of this method and combination to parallelized single cell analysis will provide a powerful tool for single cell analysis in proteomic research and systems biology. Furthermore MS represents an important method for the analysis of a variety of molecule classes for cellular analysis, such as metabolites, carbohydrates or proteins. MS not only allows for a label-free identification via precise mass detection of these analytes but can also provide for structural analysis, for example for the identification of biological alterations such as post-translational modifications of proteins or metabolic changes. The sensitivity of mass spectrometers depending on the detection technique and the scanning principle lies between 1000 and 100 000 molecules [47].

Further, MS analysis can be performed on single cells [47] and the coupling of microfluidic separations with MS detection has also been reported [6]. Both online electrospray ionization (ESI) as well as offline matrix assisted laser desorption ionization (MALDI) MS coupled to microfluidic analysis are possible. It thus seems that only a matter of time and thorough experimental skills are needed to achieve single cell analysis in microfluidic format with MS detection. This would make a considerable contribution to the field of single cell analysis for proteomics and systems biology.

Acknowledgements

Prof. Dario Anselmetti is gratefully acknowledged for fruitful discussions, critically reading the manuscript and providing laboratory equipment. Further, the authors thank Dr. Wibke Hellmich, Dr. Katja Tönsing and Dr. Any Sischka for their experimental assistance and enthusiasm. Prof. Karsten Niehaus and Nickels Jensen are acknowledged for providing Sf9 cells and for fruitful discussions. This work was financially supported by the DFG (project An 307/1-2).

References

1 Breslauer, D.N., Lee, P.J. and Lee, L.P. (2006) Microfluidics-based systems biology. *Molecular Biosystems*, **2**, 97–102.
2 Lion, N., Rohner, T.C., Dayon, L., Arnaud, I.L., Damoc, E., Youhnovski, N., Wu, Z.-Y., Roussel, C., Josserand, J., Jensen, H., Rossier, J.S., Przybylski, M. and Girault, H.H. (2003) Microfluidic systems in proteomics. *Electrophoresis*, **24**, 3533–3562.
3 Auroux, P.A., Iossifidis, D., Reyes, D.R. and Manz, A. (2002) Micro total analysis systems. 2. Analytical standard operations and applications. *Analytical Chemistry*, **74**, 2637–2652.
4 Reyes, D.R., Iossifidis, D., Auroux, P.A. and Manz, A. (2002) Micro total analysis systems. 1. Introduction, theory, and technology. *Analytical Chemistry*, **74**, 2623–2636.
5 Vilkner, T., Janasek, D. and Manz, A. (2004) Micro total analysis systems. Recent developments. *Analytical Chemistry*, **76**, 3373–3385.
6 Sung, W.-C., Makamba, H. and Chen, S.-H. (2005) Chip-based microfluidic devices coupled with electrosray ionization-mass spectrometry. *Electrophoresis*, **26**, 1783–1797.
7 Andersson, H. and van den Berg, A. (2003) Microfluidic devices for cellomics: a review. *Sensors and Actuators B*, **92**, 315–325.
8 El-Ali, J., Gaudet, S., Günther, A., Sorger, P.K. and Jensen, K.F. (2005) Cell stimulus and lysis in a microfluidic device with segmented gas-liquid flow. *Analytical Chemistry*, **77** 3629–3636.
9 Lu, H., Gaudet, S., Schmidt, M.A. and Jensen, K.F. (2004) A microfabricated device for subcellular organelle sorting. *Analytical Chemistry*, **76**, 5705–5712.
10 Hellmich, W., Pelargus, C., Leffhalm, K., Ros, A. and Anselmetti, D. (2005) Single cell manipulation, analytics and label-free protein detection in microfluidic devices for systems nanobiology. *Electrophoresis*, **26**, 3689–3696.
11 Hu, S., Zhang, L., Krylow, S. and Dovichi, N.J. (2003) Cell cycle-dependent protein fingerprint form a single cancer cell: image

cytometry coupled with single-cell capillary sieving electrophoresis. *Analytical Chemistry*, **75**, 3495–3501.

12 Hu, S., Zhang, L., Newitt, R., Aebersold, R., Kraly, J.R., Jones, M. and Dovichi, N.J. (2003) Identification of proteins in single-cell capillary electrophoresis fingerprints based on comigration with standard proteins. *Analytical Chemistry*, **75**, 3502–3505.

13 Hu, S., Michels, D.A., Abu Fazal, M., Ratisoontorn, Ch., Cunningham, M.L. and Dovichi, N.J. (2004) Capillary sieving electrophoresis/micellar electrokinetic capillary chromatography for two-dimensional protein fingerprinting of single mammalian cells. *Analytical Chemistry*, **76**, 4044–4049.

14 McClain, M.A., Culbertson, C.T., Jacobson, S.C., Albritton, N.L., Sims, C.E. and Ramsey, J.M. (2003) Microfluidic devices for the high-througput chemical analysis of cells. *Analytical Chemistry*, **75**, 5646–5655.

15 Munce, N.R., Li, J., Herman, P.R. and Lilge, L. (2004) Microfabricated system for parallel single-cell capillary electrophorsis. *Analytical Chemistry*, **76**, 4983–4989.

16 Sun, Y., Yin, X.F., Ling, Y.Y. and Fang, Z.L. (2005) Determination of reactive oxygen species in single human erythrocytes using microfluidic chip electrophoresis. *Analytical and Bioanalytical Chemistry*, **382**, 1472–1476.

17 Ling, Y.Y., Yin, X.F. and Fang, Z.L. (2005) Simultaneous determination of glutathione and reactive oxygen species in individual cells by microchip electrophoresis. *Electrophoresis*, **26**, 4759–4766.

18 Sun, Y. and Yin, X.F. (2006) Novel multi-depth microfluidic chip for single cell analysis. *Journal of Chromatography. A*, **1117**, 228–233.

19 Ocvirk, G., Salimi-Moosavi, H., Szarka, R.J., Arriaga, E.A., Andersson, P.E., Smith, R., Dovichi, N.J. and Harrison, D. (2004) J. beta-galactosidase assays of single-cell lysates on a microchip: A complementary method for enzymatic analysis of single cells. *Proceedings of the IEEE*, **92**, 115–125.

20 Gao, J., Yin, X.F. and Fang, Z.-L. (2004) Integration of single cell injection, cell lysis, separation and detection of intracellular constituents on a microfluidic chip. *Lab on a Chip*, **4**, 47–52.

21 Xia, F., Jin, W., Yin, X. and Fang, Z.-L. (2005) Single-cell analysis by electrochemical detection with a microfluidic device. *Journal of Chromatography. A*, **1063**, 227–233.

22 Ros, A., Hellmich, W., Regtmeier, J., Duong, T.T. and Anselmetti, D. (2006) Bioanalysis in structured microfluidic systems. *Electrophoresis*, **27**, 2651–2658.

23 Huang, B., Wu, H.K., Bhaya, D., Grossman, A., Granier, S., Kobilka, B.K. and Zare, R.N. (2007) Counting low-copy number proteins in a single cell. *Science*, **315**, 81–84.

24 Schulze, P., Ludwig, M., Kohler, F. and Belder, D. (2005) Deep UV laser-induced fluorescence detection of unlabeled drugs and proteins in microchip electrophoresis. *Analytical Chemistry*, **77**, 1325–1329.

25 Hellmich, W., Greif, D., Pelargus, C., Anselmetti, D. and Ros, A. (2006) Improved native laser induced fluorescence detection for single cell analysis in poly(dimethylsiloxane) microfluidic devices. *Journal of Chromatography. A*, **1130**, 195–200.

26 Khine, M., Lau, A., Ionescu-Zanetti, C., Seo, J. and Lee, L.P. (2005) A single cell electroporation chip. *Lab on a Chip*, **5**, 38–43.

27 Valero, A., Post, J.N., van Nieuwkasteele, J.W., Kruijer, W., Andersson, H. and van den Berg, A. (2006) Gene transfer and characterization of protein dynamics in stem cells using single cell electroporation in a microfluidic device. *µTAS Proceedings*, **1**, 22–24.

28 Sabounchi, P., Ionescu-Zanetti, C., Chen, R., Karandikar, M., Seo, J. and Lee, L.P.

28 (2006) Soft-state biomicrofluidic pulse generator for single cell analysis. *Applied Physics Letters*, **88**, 183901-1–183901-4.
29 Roman, G.T., Chen, Y., Viberg, P., Culbertson, A.H. and Culbertson, C.T. (2006) Single-cell manipulation and analysis using microfluidic devices. *Analytical and Bioanalytical Chemistry*, **387**, 9–12.
30 Wu, H.K., Wheeler, A. and Zare, R.N. (2004) Chemical cytometry on a picoliter-scale integrated microfluidic chip. *Proceedings of the National Academy of Sciences of the United States, of America*, **101**, 12809–12813.
31 Sun, Y., Lu, M., Yin, X.F. and Gong, X.G. (2006) Intracellular labeling method for chip-based capillary electrophoresis fluorimetric single cell analysis using liposomes. *Journal of Chromatography. A*, **1135**, 109–114.
32 Ros, A., Hellmich, W., Duong, T. and Anselmetti, D. (2004) Towards single molecule analysis in PDMS microdevices: From the detection of ultra low dye concentrations to single DNA molecule studies. *Journal of Biotechnology*, **122**, 65–72.
33 Lee, T.T. and Yeung, E.S. (1992) High-sensitivity laser-induced fluorescence detection of native proteins in capillary electrophoresis. *Journal of Chromatography*, **595**, 319–325.
34 Lillard, S.J., Yeung, E.S. and McCloskey, M.A. (1996) Monitoring exocytosis and release from individual mast cells by capillary electrophoresis with laser-induced native fluorescence detection. *Analytical Chemistry*, **68**, 2897–2904.
35 Lillard, S.J., Yeung, E.S., Lautamo, R.M.A. and Mao, D.T. (1995) Separation of hemoglobin variants in single human erythrocytes by capillary electrophoresis with laser-induced native fluorescence detection. *Journal of Chromatography. A*, **718**, 397–404.
36 Zhang, X. and Sweedler, J.V. (2001) Ultraviolet native fluorescence detection in capillary electrophoresis using a metal vapor NeCu laser. *Analytical Chemistry*, **73**, 5620–5624.
37 Paquette, D.M., Song, R., Banks, P.R. and Waldron, K.C. (1998) Capillary electrophoresis with laser-induced native fluorescence detection for profiling body fluids. *Journal of Chromatography. A*, **714**, 47–57.
38 Kuijt, J., van Teylingen, R., Nijbacker, T., Ariese, F., Brinkman, U.A.Th. and Gooijer, C. (2001) Detection of nonderivatized peptides in capillary electrophoresis using quenched phosphorescence. *Analytical Chemistry*, **73**, 5026–5029.
39 Chan, K.C., Muschik, G.M. and Issaq, H.J. (2000) Solid-state UV laser-induced fluorescence detection in capillary electrophoresis. *Electrophoresis*, **21**, 2062–2066.
40 Duong, T., Kim, G., Ros, R., Streek, M., Schmid, F., Brugger, J., Ros, A. and Anselmetti, D. (2003) Size dependent free solution DNA electrophoresis in structured microfluidic systems. *Microelectronic Engineering*, **67–68**, 905–912.
41 McDonald, J.C., Duffy, D.C., Anderson, J.R., Chiu, D.T., Wu, H.K., Schueller, O.J.A. and Whitesides, G.M. (2000) Fabrication of microfluidic systems in poly(dimethylsiloxane). *Electrophoresis*, **21**, 27–40.
42 McDonald, J.C. and Whitesides, G.M. (2002) Poly(dimethylsiloxane) as a Material for Fabricating Microfluidic Devices. *Accounts of Chemical Research*, **35**, 491–499.
43 Sia, S.K. and Whitesides, G.M. (2003) Microfluidic devices fabricated in poly (dimethylsiloxane) for biological studies. *Electrophoresis*, **24**, 3563–3576.
44 Hellmich, W., Regtmeier, J., Duong, T., Ros, R., Anselmetti, D. and Ros, A. (2005) Poly(oxyethylene) based surface coatings for poly(dimethylsiloxane) microchannels. *Langmuir*, **21**, 7551–7557.
45 Huang, B., Wu, H., Samuel, K. and Zare, R.N. (2005) Coating of poly

(diemthylsiloxane) with n-dodecyl-β-D-maltoside to minimize nonspecific protein adsorption. *Lab on a Chip*, **5**, 1005–1007.

46 Neuman, K.C. and Block, S.M. (2004) Optical trapping. *Review of Scientific Instruments*, **75**, 2787–2809.

47 Hillenkamp F. and Peter-Katalinic J. (eds) (2007) *MALDI MS*, Wiley-VCH, Weinheim.

48 Sischka, A., Eckel, R., Toensing, K., Ros, R. and Anselmetti, D. (2003) Compact, microscope based optical tweezers system for molecular manipulation. *Review of Scientific Instruments*, **74**, 4827–4831.

6
Single Cell Mass Spectrometry
Elena V. Romanova, Stanislav S. Rubakhin, Eric B. Monroe, and Jonathan V. Sweedler

6.1
Introduction

Single cell mass spectrometry (MS) has become a significant subfield of bioanalytical microanalysis, contributing to multiple discoveries including the characterization of large numbers of cell–cell signaling molecules. This chapter briefly reviews a variety of mass spectrometric approaches to examining the single cell and highlights progress in the analysis of individual cells using this important methodology.

Whereas the microanalysis of cells is normally thought of as a fairly new research area, a general scientific interest in the study of individual cells may have been sparked by the work of Anton van Leeuwenhoek (1632–1723) [1, 2]. A Dutch tradesman and scientist, van Leeuwenhoek substantially improved the quality of primitive microscope lenses, and was the first to observe and describe "animalcules," live unicellular algae and protozoa (1674) and bacteria (1683). In 1838–1839, zoologist Theodor Schwann and botanist Matthias Jakob Schleiden summarized their own findings and observations of others by suggesting that cells constitute the basic units of structure in living organisms [3]. The quest for understanding cellular biology and function at the single cell level had begun.

Early research to examine single cells concentrated on developing and optimizing tools to enable the visualization of cellular morphology, subcellular components and cell division. Perhaps the first chemical information from single cell studies may date as far back as 1878 when young Paul Ehrlich presented his groundbreaking work on the identification of mast cells by applying specific histochemical reactions [4] (reviewed in Ref. [5]). However it was not until the 1950s that the chemical content of individual cells started to be probed by more modern analytical technologies. Methodological developments in the 1970s, characterized by improved staining techniques and better methods for labeling and tracing compounds, permitted exploration of the cell–cell interactions that underlie complex physiological functions. These direct chemical assays evolved using a variety of approaches, including separations and spectroscopic methods, to probe a range of analytes found in single

Single Cell Analysis: Technologies and Applications. Edited by Dario Anselmetti
Copyright © 2009 WILEY-VCH Verlag GmbH & Co. KGaA, Weinheim
ISBN: 978-3-527-31864-3

cells [6–8]. MS played a role in this chemical characterization process, initially with techniques that probed smaller molecules and inorganic ions [9]. The utility of MS greatly increased in the late 1980s following breakthrough developments in soft ionization methods for large biomolecules by Tanaka [10], Fenn [11], Hillenkamp and Karas [12].

Currently a suite of analytical techniques is available to undertake the discovery of biochemical organization and the molecular identity of cells. When combined with gene expression data and physiological activity measurements from cells, MS reveals much about complex signaling processes between and within cells that impacts the activity of an organism as a whole.

6.2
Considerations for Single Cell Chemical Microanalysis using Mass Spectrometry

The cell is often thought of as the most fundamental component of an organism. Recognizing the similarities and differences between distinct cells, each specialized for a specific role within the organism, can lead to unique insights into the functioning of an entire system. Even in the mammalian brain, which contains trillions of cells and demonstrates a high level of redundancy, stimulation of a single neuron can evoke a complete motor action (e.g. whisker movement [13]). Thus, investigating the chemical similarities and differences between the cells in the brain is clearly an important research area.

Due to the inherent complexity of biological tissues, the chemical analyses of proteins, peptides, lipids and metabolites at the system level is challenging. If only a small population of cells in a sample contains a compound of interest, then combining additional cells devoid of that compound will complicate its detection and identification, in part, because the relative concentration of this particular analyte is reduced. Further, chemical species that may potentially interfere with the analyses are often introduced. For example, if 0.1% of the cells in a brain region contain a particular compound, then homogenization of a region dilutes the compound 1000-fold while concentrations of the analytes common to all cells remain the same. Thus, as the heterogeneity of the biological sample increases, the benefits of working with a tissue or its extract decrease. The advantage of MS analysis performed at the level of the individual cell, or even a subcellular compartment, is a reduced anatomical complexity, often leading to improved signal to noise ratio and the detection of analytes unique for the cell under investigation.

There is a tradeoff in single cell analyses – the small amount of sample. Usually, the more material present, the greater the chemical information content resulting from the measurement. Do single cell samples contain analytes in detectable amounts? A typical mammalian cell is 10 μm in diameter, has a volume of 500 fL, and contains approximately 50 pg or 2 fmol of protein [14]. Thus, if only a few compounds are present, these are within the detection range for current MS platforms. Additionally, subcellular compartmentalization of biomolecules can provide for increased local concentrations of analytes present in low mass amounts.

Because many analytical figures of merit scale with concentration rather than mass, sample preparation using low volumes allows low mass samples to be studied at native high local concentrations.

An advantage of MS, relative to many other molecular screening methods used for investigation of single cells, is that the approach does not require a priori information on the identity of the analyte of interest; accordingly MS is well suited to discovery-driven research. Moreover, instead of measuring changes in just a few particular compounds targeted by selective probes, MS detects unexpected changes in peptides, proteins, oligonucleotides and lipids. Unlike immunoaffinity-based methods, MS structurally characterizes analytes without labeling, and complements single cell mRNA investigations [15] by providing information on gene products, including their post-translational modifications (PTMs).

PTMs alter inherent molecular properties such as charge or conformation, which often confer or modify the bioactivity of the modified protein. MS is ideal for discovery and characterization of PTMs because most result in characteristic mass changes to the original molecular mass. The technique also is effective for the study of protein isoforms resulting from differentially expressed genes. Minor changes in protein primary structure associated with alternative splicing or editing of the protein encoding mRNA lead to changes in the molecular mass of the original protein or peptide ion. This is especially true for measurement of signaling peptides because numerous peptides can be produced from a single precursor protein as a result of complex post-translational proteolytic processing or differential gene expression. Peptide diversity can further be increased by a variety of PTMs such as amidation or acetylation, often creating unique peptides. Peptide isoforms with different PTMs frequently are not recognized as distinct using labeling techniques, but can often be deduced by MS due to the specific changes in the peptide molecular mass. It is this ability of MS to measure and identify multiple classes of compounds in single cells that makes it an indispensable tool for elucidation of cell-specific biochemistry.

6.3
Mass Spectrometry as a Discovery Tool for Chemical Analysis of Cells

Mass spectrometry measures the masses of analytes in the gas phase. Briefly, the sample is vaporized and ionized, and the mass to charge (m/z) ratios of the resultant ions are determined directly or indirectly in the mass analyzer [16]. Determination of an accurate molecular mass is often a first step towards molecular structure characterization. The fragmentation of selected molecular ions into smaller ions inside the mass spectrometer can provide information on the original ion structure. This fragmentation/characterization is performed using multiple mass analyzers in a process known as tandem mass spectrometry, or MS/MS.

Sample introduction, vaporization and ionization constitute perhaps the most important steps in the MS analysis of a cell. These processes determine both the analytes that are measured and the quality of the resulting data, and are discussed briefly. The vaporization/ionization processes are often referred to as soft or hard, depending on whether or not they break molecules such as proteins or carbohydrates into smaller fragments. Hard ionization methods are appropriate for inorganic ion and lipid measurements and soft approaches for peptide or protein characterization. Often a "beam" of ions or light is focused to a specific spot and the ablated material is measured. The properties of the beam dictate the type of material ablated and hence the analytes that can be measured. If hard ionization approaches are used for protein analysis, they fragment the protein to the extent that the original proteins cannot be identified. Therefore for the larger molecules the most effective methods of ionization are the soft ionization methods of electrospray ionization (ESI) [11] and matrix-assisted laser desorption/ionization (MALDI) [10, 12]. The ionization of biological samples by both ESI and MALDI imparts low internal energies to the ionized biomolecules, limiting their fragmentation/breakdown so that larger molecules such as proteins, peptides, lipids, oligosaccharides and oligonucleotides can be studied intact.

Prior to the development of MALDI and ESI, hard ionization methods were applied to single cell measurements. The simplicity of sample preparation makes laser desorption/ionization (LDI) techniques suitable for a wide range of biological sample types and a frequent technique for single cell analysis. Hillenkamp was instrumental in creating the laser microprobe mass analysis (LAMMA) technique which allowed a range of lower molecular weight compounds to be detected directly from thin tissue sections. Figure 6.1 shows a section of inner ear with several <5-μm

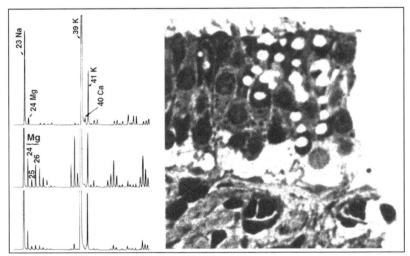

Figure 6.1 Laser desorption ionization using a focused laser allowed samples, a few microns in size, to be probed from a tissue section. In the inner ear section shown here, multiple holes from the laser ablation process can be seen, with the cell outlines clearly visible. Several of the associated mass spectra are shown on the left. The image and spectra are courtesy of F. Hillenkamp and used with permission.

holes ablated from the tissue by the laser, with each spot producing a mass spectrum [17].

A related approach for single cell investigation is bioaerosol mass spectrometry (BAMS) [18], which is a single particle laser desorption/ionization time of flight MS method capable of ionizing analytes localized in single airborne particles. An important feature of this approach is the ability to distinguish particles according to their aerodynamic diameters and therefore eliminate the need to visualize individual cells for positioning on the MS probe. BAMS has been used to examine the composition of various biological particles including the pathogen *Mycobacterium tuberculosis H37Ra* [18]. The simultaneous detection of both negative and positive ions in BAMS mass spectrometers significantly increases the profiling power of this technology [19]. Interestingly, more than two decades before the BAMS work, single *Mycobacterium tuberculosis H37Ra* cells also were studied using LAMMA. Sodium and potassium as well as some metabolites were detected [20]. The MS profiling of pathogenic *Bacillus anthracis* and its harmless cousin *Bacillus cereus* with LAMMA revealed different biochemical fingerprints for the two species [21]. The capability of LAMMA to detect small inorganic molecules in individual cells allowed the quantitative investigation of uranium uptake by the algae *Dunaliella salina* [22].

A large number of LDI approaches have emerged with the development of sampling protocols and approaches for various cell types and even subcellular compartments. For example, desorption/ionization on porous silicon (DIOS) is a matrix-free approach in LDI MS that uses a porous silicon to trap analytes deposited on the surface and laser irradiation to vaporize and ionize them [23]. Matrix-free methods extend the detectable mass range to small molecules, which aids in the elucidation of molecular structure via observation of smaller molecular fragments. Thomas and colleagues [24] have shown that the method enables the detection of analytes at femtomole and attomole levels, induces little or no fragmentation and is compatible with microfluidic technologies. Kruse *et al.* [25] successfully applied DIOS to the analysis of native peptides from neural tissue blots and individual invertebrate neurons cultured directly on porous silicon. More recently individual HEK 293 cells grown on a DIOS substrate were analyzed [26].

Developments in soft ionization methods revolutionized the field of chemical microanalysis. As shown in Figure 6.2, a small volume of tissue extract or solid tissue/cell sample is co-crystallized with an excess of the appropriate ultraviolet (UV)-absorbing chemical matrix for MALDI MS analyses. This matrix is typically an aromatic acid that strongly absorbs the applied laser light. Illumination of the matrix sample mixture by a UV or infrared (IR) laser causes rapid vaporization and ionization of both analyte and matrix. The net effect is the creation of primarily singly charged protonated and/or cationized analyte species into the gas phase. Selection of the most suitable matrix compound is key to determining the optimal ionization of certain classes of analytes and/or target analytes in specific mass ranges [27]. The most important applications of bioanalytical MALDI MS are detection and identification of peptides, proteins, oligonucleotides, oligosaccharides and lipids.

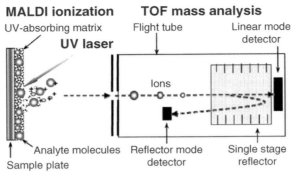

Figure 6.2 Simplified schematic of a time of flight (TOF) mass spectrometer with a MALDI ion source where desorption/ionization of analytes occur. The TOF mass analyzer separates ions according to their mass and charge in both linear and reflectron modes of operation. Two different detectors are shown for two different operating modes. The detected signal is digitized and processed with a variety of software packages.

As an alternative to using a laser, a focused stream of primary ions can be employed for vaporization and ionization of analytes from the surface of tissue and cells. These primary ions generate secondary ions in the material being analyzed; because these secondary ions are the ones that are measured, the approach is called secondary ion mass spectrometry (SIMS). In fact SIMS is one of the older mass spectrometric approaches and has been used for cellular research since the 1970s [28].

Unlike the approaches described so far, ESI is a solution-based ionization method in which a continuous flow of sample, dissolved in a volatile solvent, is delivered through an open-ended capillary that has a high potential applied to it [11]. As a result, an analyte-containing aerosol of charged droplets is emitted at the end of the capillary forming a spray. The spray is directed towards the mass spectrometer inlet by appropriate ion optics. As these droplets travel towards the sampling inlet, the solvent evaporates and "coulombic explosions" generate singly and multiply charged analyte ions. The distribution of charges on the analyte molecules depends on the electrospray conditions and structure of the analyte (i.e., the availability of basic sites). ESI produces fantastic results for many analytes. As one example, Hofstadler *et al.* [29] demonstrated the utility of single cell MS by detecting both the α- and β-chains of hemoglobin from a lysate of a single human erythrocyte using a combination of capillary electrophoresis (CE) and ESI Fourier transform–ion cyclotron resonance MS. However, the performance of ESI MS is degraded by the salty solutions normally associated with direct analysis of cell samples. Although desalting is possible, it is challenging to scale traditional desalting procedures to the nanoliter or smaller volumes of single cells. Recently microminiaturization and hyphenation of analytical approaches began addressing these issues and are showing impressive initial results [30].

While each of the techniques mentioned above have particular strengths, the ability of MALDI MS to probe a wide range of analytes, including peptides and proteins, have made this approach the most common for robust mass spectrometric analysis of single cells. This success is due in part to its ease of use, the availability of high performance instrumentation and the unsurpassed capability of MALDI MS to analyze extremely small quantities of peptides and proteins from chemically heterogeneous samples.

6.4
Single Cell Mass Spectrometric Applications

MALDI MS has sufficient sensitivity, resolving power and throughput for the measurement of a surprising range of compounds in single cells, especially when sample preparation is optimized to achieve precise cell isolation, maximal extraction of analyte from an individual cell and the efficient transfer of the extracted analyte into an optimal volume of MALDI matrix without excessive dilution.

Sample preparation/cell isolation is critical for single cell experiments, especially as many of the fixation protocols designed for optical and electron microscopy are not compatible with MS. One of the techniques available for manipulating single cells is laser capture microdissection (LCM), which utilizes a laser and a thermoplastic film for automated cell dissection from thin tissue sections [31]. More classic methods include manual isolation assisted by either enzymatic dissociation of cells from tissue, application of shear force, or by various stabilization treatments that conserve the morphological and chemical integrity of the resulting cell samples. For example Garden et al. [32] found that brief incubation of freshly dissected molluscan nervous system tissue in aqueous MALDI matrix effectively desalted the sample and simplified the manual isolation of individual identified neurons. Monroe and colleagues [33] have developed an innovative method for massively parallel preparation of single-cell-sized samples from tissue sections. The method is based on the capability of a tissue section to adhere to the surface of a glass bead array formed on a Parafilm M membrane surface. By stretching the membrane, the deposited tissue section is fragmented into small pieces, similar in size to the individual glass beads. As a result, the individual hydrophilic glass beads/tissue pieces are physically and chemically separated by the hydrophobic surface of the Parafilm M membrane, allowing for the simultaneous and individual treatment of each tissue piece with MALDI matrix. This "stretched sample method" enables higher throughput analysis of signaling peptides via the automated acquisition of mass spectra from the separated tissue sections (Figure 6.3).

Small volumes and the decreased mechanical stability of mammalian cells present significant analytical challenges to isolating individual cells for MS investigation. Various approaches have been employed to resolve these challenges. Bergquist [34] cultured PC12 cells directly on stainless steel MALDI sample plates and was able to detect a number of substances with molecular masses above 4000 Da. Fung and Yeung [35] coupled flow cytometry to MS and measured serotonin and histamine in single mast cells. For delicate manual isolation of small endocrine cells of rat pituitary, Rubakhin and colleagues [36] used a stabilization treatment with 30%

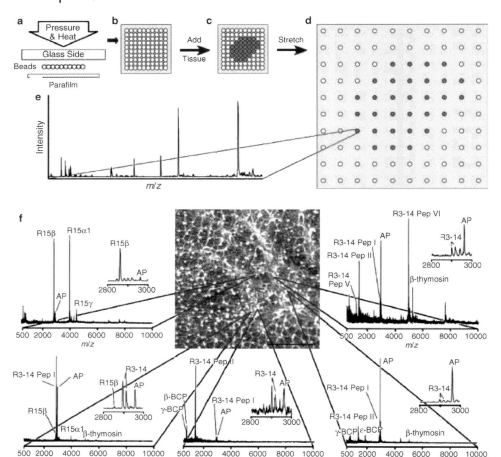

Figure 6.3 Schematic representation of the massively parallel sample preparation process. The stretched sample method is used to generate the sample array of small nearly cell-sized pieces of tissue. (a, b) Formation of bead array on top of Parafilm M. Pressure and heat are used to form the mechanically stable layer. (c) A thin slice of nervous tissue is deposited onto the glass bead layer. (d) The Parafilm M membrane is manually stretched. As a result, the tissue slice is fragmented into thousands of spatially isolated pieces. (e) After MALDI matrix application, individual pieces of tissue may then be investigated with MALDI MS. (f) Representative mass spectra from a 10-μm abdominal ganglion section. They were prepared using the stretched sample method and demonstrate a wide variation in the peptide content of neurons within a ∼750 × 750 μm region of ganglion. Microphotograph of the section after stretching is shown in the top center part of this figure with mass spectra linked to the corresponding bead from which the signal arises. Scale bar, 1 mm. Reproduced with permission from Ref. [33]. Copyright 2006 American Chemical Society.

glycerol in physiological saline. Glycerol treatment greatly improves the mechanical stability of cells and decreases cellular adhesion to the glass surface, which aids in the selection and transfer of specific cells onto a sample holder prior to the application of MALDI matrix [37]. These studies demonstrate that dramatic enhancements in mass sensitivity are achievable with commercial instrumentation using low-volume sampling methods, thus enabling the determination of chemical profiles from even a subcellular biological sample.

The process of MS profiling involves creating a list of compounds that are present in detectable amounts at a specific location (e.g. in a particular cell). Statistical algorithms allow the researcher to analyze the MS profiles for comprehensive qualitative and, at times, quantitative evaluation. Individual cell profiles can be compared against each other to reveal spectral features, so-called markers, that are predictive of class, treatment or condition. MS profiling helps to establish links to normal or pathological health conditions, disease progression, or therapeutic responses, as well as assist in diagnosing conditions that include progressive mutations and unusual PTMs.

As one example, the availability of microbial genomic databases facilitated the development of rapid and accurate microbial identification via specific biomarkers found in unfractionated viral, bacterial and fungal cells [38]. Current methods for the robust taxonomic identification of viral and bacterial cells are based on measuring peptide/protein signature ions in a broad m/z range that are unique and representative of specific species and strains [39, 40]. Measuring masses of individual intact microorganisms is an alternative means for rapid identification of potentially dangerous viruses and bacteria [38]. Both ESI and MALDI platforms have been used for such experiments. The first high-precision mass measurement for intact microorganisms with masses greater than 1×10^{10} Da was reported for *Escherichia coli* [41]. Peng and co-workers achieved an accurate mass measurement approaching 0.1% with a mass distribution of $\pm 3\%$ through repetitive measurements of whole bacterial particles and their clusters using a MALDI quadrupole ion trap mass spectrometer with the elastic light scattering detection.

Perhaps the most successful outcome of single cell MS profiling has been the ability to reveal naturally occurring neuropeptide transmitters, hormones, and modulators. These peptides are released from a cell to trigger an intracellular transduction pathway. Not only are these peptides structurally diverse, but their signaling cascades are variable, such that a tremendous potential of different cellular effects is possible. Because of their important role in cell–cell signaling, peptide mimics are attractive pharmaceutical targets.

The classic approach to assaying neuropeptides has been via the homogenization of select tissues or organs. Measuring peptides in a single cell offers unique insights into prohormone processing and co-transmission. A prohormone is enzymatically processed via a complex series of steps that typically create a suite of peptides, at times, in a cell-specific manner. Single cell MS profiling characterizes such proteolytic processing events for many bioactive (or signaling) peptide gene products and provides insights on cell–cell peptide heterogeneity.

One of the first single cell MALDI MS experiments of this nature was described in 1994 and involved isolated large neurons from the freshwater invertebrate, *Lymnaea stagnalis* [42]. Three types of neurons were examined and known peptides from different prohormones in each cell were detected, in addition to several unknown peptides. Since then, MALDI MS has proven to be a powerful tool for the investigation of the intercellular signaling molecule complement in cells and even subcellular organelles [43–46]. In many cases, new prohormones have been reported, the cells containing them identified, the prohormone processes revealed using single cell MS- and the physiology induced by the resulting peptides have been presented in the same study [44, 47–51]. As a recent example, Romanova and colleagues [52] combined the strategy of using shared peptide gene expression patterns with single cell MALDI MS to identify disparate elements of a neural circuit that contributes to a central command coordinating autonomic activity with escape locomotion in *Aplysia*. MS is also effective in the study of molecular isoforms resulting from differentially expressed genes. For example, Jimenez and colleagues [53] demonstrated that single neurons exhibit a highly complex pattern of peptide gene expression, precursor processing and differential peptide modifications, along with a remarkable degree of convergence in neuromodulatory actions.

The ability to deduce structural information makes single cell MALDI MS helpful in identifying unknown peptides. State of the art time of flight/time of flight (TOF/TOF) and Fourier transform (FT) mass spectrometers have proven capable of *de novo* structure elucidation [54, 55] using sample sizes of less than 100 fmol [56, 57] and typically provide a mass accuracy better than 0.2 Da for fragment ions [58]. This combined sensitivity and mass accuracy is adequate for identifying many peptide PTMs and has a high probability of yielding at least partial sequence information for peptides, as well as identifying the prohormone precursors for unassigned signaling peptides. Several approaches allow identification of unknown signals in individual cells with high confidence. First is the direct fragmentation of a molecular ion to allow sequencing of peptides using MS/MS or post-source decay (PSD) fragmentation methods. Using MALDI tandem double focusing magnetic-orthogonal acceleration TOF MS with collision-induced dissociation (CID) fragmentation, Jimenez and colleagues [59] revealed the primary structures of small cardioactive peptides differentially expressed in the giant identified neuron of the fresh water snail, *Lymnaea*. Based on the structural information provided by MS, they were able to generate selective probes and elucidate the structure of the entire prohormone from the brain-specific cDNA library and map the expression of the prohormone in the central nervous system. Li *et al.* [60] demonstrated that a unique combination of MALDI matrices, employed during sample pre-treatment and embedding, facilitated direct sequencing of a novel decapeptide from a single *Aplysia* neuron by MALDI-TOF MS with PSD/CID. Perry and colleagues confirmed the expression of structurally related cardioactive peptides encoded by a differentially spliced gene in large identified *Lymnaea* neurons using MALDI/PSD-TOF MS [61]. Alternatively unknown peptides can be sequenced by MS/MS in tissue where the cell under investigation originated, which can be an alternative approach to identify peptides of lower abundance that are expressed in numerous adjacent cells within the tissue sample. Accurate mass information obtained in single cell experiments guides the selection of peptides for MS/MS sequencing.

Besides mollusks, the neuropeptide content of individual neurons from other invertebrates, including the cockroach, the corn earworm moth and the crayfish, has been investigated with MS [62–64]. Although invertebrate neurons span a large volume range, most examples of single cell MS have used the larger invertebrate neurons. The majority of mammalian cells are smaller (e.g. low picoliter to femtoliter volumes), containing, at most, attomoles of each analyte. The Sweedler group [36, 65] successfully profiled single mammalian cells, including individual rat intermediate pituitary cells and dorsal root ganglion neurons, using MALDI MS. Remarkably, 20 peaks corresponding by mass to fragments of the pro-opiomelanocortin prohormone and five unidentified peaks were detected in individual pituitary cells, with many of these peptides previously detected in releasate from the stimulated pituitary.

The first successful MALDI MS analysis of vertebrate non-neural single cells was achieved using melanocytes from the amphibian, *Xenopus laevis* [66]. Using LCM and MALDI MS profiling, Xu and coworkers [67] were capable of differentiating between populations of human invasive mammary carcinoma cells and normal breast epithelium. This group has also analyzed proteins in small clusters of mouse colon crypt cells and mouse liver cells [67]. Additionally single red blood cells have been characterized with FTMS, achieving attomole performance sufficient to characterize the most abundant proteins [29, 68].

Miniature sample volumes and the low masses of single cell preparations do not encourage the use of purification and separation methods such as capillary electrophoresis (CE) or chromatography in conjunction with MS. Nevertheless, a number of examples have shown that clever optimization of sampling and a wise selection of analytical instrumentation can overcome this challenge. The pioneering work of Iliffe *et al.* [69] employed gas chromatography MS and stable isotope labeling to measure absolute concentrations of glycine, aspartic acid and glutamic acid in individual identified molluscan neurons. CE has been used quite successfully in enhancing the chemical analysis of single cells due to its requirement for nanoliter volumes of sample and buffer for the analysis. As an example CE/MALDI MS analysis combined with radionuclide detection allowed the quantitation of predicted and unexpected neuropeptides and hormones in a single 40 μm cultured neuron of *Aplysia* [70].

6.5
Subcellular Profiling

MS studies at the single cell level provide unique information on the biochemical organization of heterogeneous tissues, organs, and even unicellular life forms. Although often considered a single entity, the cell includes numerous dynamic processes with significant subcellular heterogeneity. For example the amount of neuropeptides present in different cellular compartments depends on the rate of synthesis in the endoplasmic reticulum, packaging into vesicles in the Golgi apparatus, transport to terminals and release from the presynaptic terminal. These processes are regulated by a plethora of endogenous and exogenous enzymes. Localizing biochemical changes may aid our understanding of mechanisms that control cell–cell signaling. Optical and cellular fractionation techniques are often employed to explore the subcellular biochemical organization of cells. Significant efforts to increase the spatial resolution and sensitivity of MS technologies have led to several breakthroughs that allow measurement of the biochemical composition of cellular regions and even, in some cases, individual organelles.

The subcellular profiling approach is based on probing just one or several discrete locations/components of a cell. Just as with single cell MS, optimizing the sampling process is the key to success. Not surprisingly, the sample preparation steps differ according to the MS technology being used. Perhaps the only universal step is the microscopic visualization of the specimen to locate regions of interest for subsequent isolation and analysis. There are several ways to prepare samples for subcellular MS profiling. One approach is based on the use of a microprobe to volatize analytes from only a small region of a cell; the spatial resolution of MS profiling in this case is, typically, limited by the size of the sampling probe. Here we distinguish methods such as SIMS or MALDI MS imaging that raster a probe on the sample to form a chemical image (as described in the next section) and those that profile selected regions.

Profiling works well with LAMMA technology and is used to measure small molecules at specific cellular locations. For example an increase in aluminum content in the neuromelanin granules of neurons from patients with Parkinson disease was observed by Good *et al.* [71]; they focused a co-axial pulsed high energy neodymium-YAG laser to a $\sim$1 μm spot to vaporize analytes. Likewise, investigation of the elemental content of neurofibrillary tangle-bearing and tangle-free hippocampal neurons from Alzheimer's patients by LAMMA demonstrated significant aluminum deposition in the neurofibrillary tangle region [72]. LAMMA also was employed to study subcellular localization of calcium in crayfish photoreceptor cells [73], sodium and potassium in muscle and stria vascularis specimens [74] and lead in cells of *Phymatodocis nordstedtiana* (Chlorophyta) [75]. Regions of intracellular space can be sampled using guided suction or pressure-driven sampling with a sharp, oil-filled glass microcapillary that penetrates the cell. Employing this approach, 10–100 pL samples can be collected and analyzed using a variety of analytical approaches including single cell MS [76]. A combination of intracellular sampling and biological accelerator mass spectroscopy allows precise quantitative measurement of ^{14}C content in barley (*Hordeum vulgare* L.) leaf tissue [77]. Another interesting approach where individual plant cell content (2–3 pL) was collected is reported by Kajiyama *et al.* [78]. In this work, cell walls of the Torenia plant were locally destroyed with an ArF 193 nm excimer laser pulse. The content of the cell was pushed out by high intracellular pressure appropriate for plant cells, collected by quartz glass micropipette, and analyzed with nano-high-performance liquid chromatography ESI-MS. The set of anthocyanins was identified using MS/MS.

Another sampling strategy involves micromanipulations for selective dissection and isolation of spatially defined cellular regions. The method has an advantage of being independent of the MS probe size. As one example of handling small-volume samples for MS analysis, Rubakhin and colleagues [79] demonstrated the MALDI MS analysis of a single *Aplysia californica* atrial gland secretory organelle, the dense core vesicle (Figure 6.4). Pushing the sample size for this analytic technique down to the single-micron scale and the attoliter to femtoliter volume regime, more than ten peptides from at least four separate genes were detected within each vesicle. A similar sample preparation methodology allows profiling of different regions of neurites and the corresponding cell body of single, isolated neurons from *Aplysia californica* [37].

6.6
Imaging Single Cells with MS

For decades the ability to image the distribution of a compound within a tissue or single cell has been a powerful tool for scientists, yet often requires the development of antibodies or other selective probes. The advancement of MALDI MS imaging (MALDI MSI) has revolutionized the ability to study proteins and peptides in biological materials. MALDI MSI enables the creation of distribution maps for a wide range of analytes in a single experiment without the need for analyte preselection or selective labeling. To create images with MALDI MS, a sample is coated with a thin, homogenous layer of matrix prior to the collection of a series of mass spectra from an ordered array of locations across the sample. Following data collection, the intensity of selected signals are plotted for each mass spectrum (pixel) to produce a series of two-dimensional distribution maps/ion images for each individual signal. Because a complete mass spectrum is collected for each location in the sample, literally hundreds of ion images, each of a different compound, may be created from a single experiment and can include images from both known and uncharacterized molecules. A special issue of the *International Journal of Mass Spectrometry* devoted to MSI highlights the capabilities of this approach [80].

Although the spatial resolution of MALDI MSI, with a few exceptions [36, 81, 82], is insufficient to analyze small individual cells, the technique has produced impressive results from a wide range of samples. Typically, a few hundred signals are observed in any given 25–250 μm tissue region corresponding to a single pixel. Molecular images of a wide array of tissues involving mouse brain [83], pituitary [84] and epididymis [85] have been obtained, as well as a number from other neurological and physiological models, both vertebrate and invertebrate. MALDI MSI also has shown great potential in the study of biomarkers in cancers and other diseases [83, 86, 87].

Recent advancements in high resolution imaging have focused on improving the spatial resolution of analyses to levels that enable investigation at the single cell level. This is being accomplished via modifications to traditional sampling methods [33, 82] and to instruments [81, 88, 89], resulting in a spatial resolution smaller than 4 μm with 500 nm pixel sizes [90]. The imaging of single isolated cells with MALDI MSI is challenging given the small size and quantity of analyte available for analysis. Often, the approaches required to achieve the greatest sensitivity are distinct from those to attain the highest spatial resolution, and so imaging at cellular resolution requires compromises. Even so, individual isolated, cultured neurons from *Aplysia* have been imaged following glycerol stabilization and a quick matrix application [37].

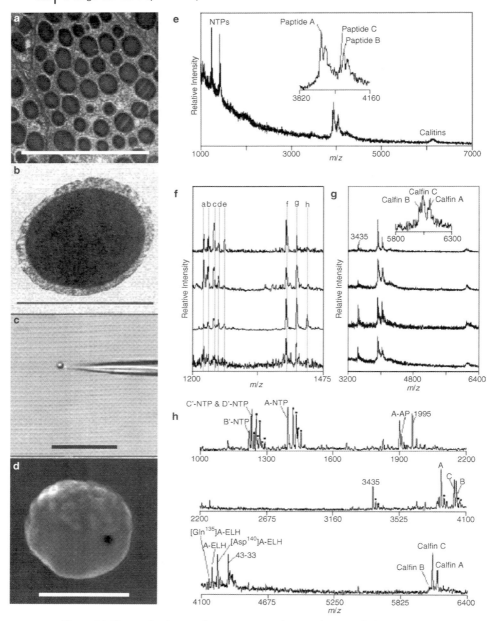

Figure 6.4 The small-volume MALDI MS approach allows isolated individual organelles such as secretory granules from the atrial gland of *Aplysia californica* to be assayed. Transmission electron microscopy images of (a) tissue section (scale bar 10 μm) and (b) an isolated vesicle showing the centrally located electron-dense core surrounded by an amorphous cortex (scale bar 1.5 μm). (c) Video image showing the typical position of a single vesicle on a micropipette tip during the sampling process for MALDI MS (scale bar 10 μm). (d) Environmental SEM image

As another approach, the high spatial resolution of SIMS imaging is highly attractive for the single cell/subcellular study of tissues and isolated cells. As many intracellular processes are spatially localized, the 100 nm to 5 μm spatial resolution commonly obtained with SIMS is nearly ideal for their investigation. Traditionally the SIMS ionization process has been too harsh to enable detection of large organic molecules, although recent advancements in sample preparation and primary ion sources has expanded the mass range and increased the sensitivity of analyses to make molecular imaging of biological specimens attractive. The imaging of atomic ions has been used to diagnose cellular physiological state as well as track labeled compounds. Freeze-fracture preparation methods have been used to study intracellular ionic composition, including precise localization of potassium and calcium, suggesting that little or no analyte redistribution occurs during processing. Importantly, damaged cells may be identified by a reduced potassium signal and the abnormal presence of intracellular calcium [91].

Other imaging studies have found that copper ions accumulate in the cytoplasm of carcinoma PC3 cells and has been linked to metastatic transformation [92]. Metabolism of different compounds may be followed with SIMS using stable ^{13}C and ^{15}N isotopes. Using this approach, the surprisingly specific localization of L-arginine-related compounds in cell nuclei was uncovered [93]. The ability to track and discover the region of integration or effect on subcellular length scales provides valuable insight into the mode of biological action of various compounds such as mutagens [94], antibiotics [95], cholera toxins [96], cocaine [97] and boron neutron capture therapy drugs [98].

Direct tissue SIMS imaging significantly benefits from improved ion sources [99] as well as the development of metal-enhanced [100] and matrix-assisted SIMS [101]. Cholesterol and phospholipid ions have been detected with SIMS imaging and show a differential distribution, with phospholipid signals being elevated in brain regions rich in neuronal cell bodies and cholesterol residing in morphological structures that contain myelinated axons [102, 103]. Application of a MALDI matrix to a tissue slice from *Lymnaea* resulted in detection of molecules with molecular masses >2500 Da, although some loss in the experimental spatial resolution comparatively to direct SIMS imaging of the tissue was observed [101]. By coating a sample with a nanometer of gold, the ion yields from single neuroblastoma cells were found to increase with a

Figure 6.4 (*Continued*)

of a single vesicle (scale bar 1 μm). (e) Analysis of a single vesicle with the inset showing peaks corresponding to peptides A, B, and C. To optimize sensitivity, spectral acquisition parameters are independently optimized for detection of (f) low-mass (1000–3000 Da) and (g) high-mass (3000–7000 Da) atrial gland peptides in different vesicles. The inset shows detection of califins A, B, and C and the putative 3435-Da peptide appears to be encoded by a gene distinct from the ELH family. (h) Representative MALDI mass spectrum acquired from an atrial gland blot. The transmission electron microscopy study was performed at the Center for Electron Microscopy, University of Illinois, Urbana. Scanning electron microscopy was done at the Beckman Institute Imaging Technology Group, University of Illinois, Urbana. The images, with minor modifications, are reproduced with permission from Ref. [79]. Copyright 2000 Nature Publishing Group.

usable mass range extending above 1000 Da. In contrast to the previous approach, no appreciable loss of spatial resolution has been detected, probably due to little or no incorporation of analytes into the metallic coating [100]. Through the use of multiple image acquisitions, megapixel-sized ion images may be created. As an example, the visualization of an entire zebra finch brain tissue section with single cell resolution was demonstrated [104].

SIMS is a surface analysis technique and, as such, the investigation of cellular interiors often requires the use of thin tissue sections or freeze-fracture methodologies. Nonetheless there are examples of lipid membrane analyses of isolated cells that may provide insight into the functional roles of lipids and lipophilic small molecules. Abundant cholesterol is seen in the plasma membrane of blood cells that have been imprinted on a conductive silver substrate, with phosphatidylcholine lipids enriched in the specific membranes [105]. The successful application of SIMS imaging to detect the inclusion of high curvature lipids in the fusion region between two conjugating (mating) protozoan *Tetrahymena* cells demonstrates that this technique helps to develop a better understanding of the mechanisms of a variety of cellular events including endocytosis and exocytosis [106]. Using a C_{60}^+ ion source, Vickerman and coworkers [107] have been able to probe the three-dimensional structure of isolated *Xenopus laevis* oocytes by depth profiling the cellular surface and imaging the exposed subsurfaces.

SIMS imaging has also been used to observe the subcellular localization of vitamin E in single isolated neurons from *Aplysia californica* [108]. Vitamin E was found to be specifically localized in the cellular membrane to the junction region between the neuronal cell body and the extending neurite (Figure 6.5). As expected, on the basis of its antioxidant role, vitamin E was detected in all regions of the neuronal membrane, yet the normalized vitamin E intensity in the junction region was found to be 165% that of the mean value for the entire cell. This specific localization of vitamin E suggests an important biological role that may extend beyond the traditional view on this molecule as the major lipophilic antioxidant.

6.7
Signaling Molecule Release from Single Cells

Following the logic that intercellular signaling molecules are released in an activity-dependent manner, analytes detected specifically in releasates are more likely to function as neurotransmitters, neuromodulators, or hormones. MALDI MS profiling of the activity-dependent release of peptides from individual neurons in whole ganglia and cell cultures was reported using a combined electrophoretic separation and solid-phase extraction (SPE) collection technique [109].

In contrast, releasate detection from an individual cell using MALDI MS is challenging. The issues result from the low amount of peptides released from a single cell and the high inorganic salt concentration in the extracellular matrix that contains the secreted neuropeptides. Hatcher and colleagues [110] developed an improved sample collection paradigm for MALDI MS detection of cellular releasates where single resin particles of SPE material are placed directly onto release sites of

6.7 Signaling Molecule Release from Single Cells

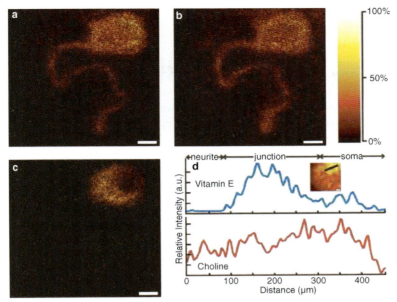

Figure 6.5 Single cell SIMS reveals important subcellular features of individual neurons. Ion-specific images of an isolated single neuron using a relative thermal scale, in which light areas represent a higher ion yield. (a) Cellular lipids are represented by the choline fragment (m/z 86) from sphingosine and the phosphatidylcholine lipid headgroup (0–11 counts). (b) Acyl chain fragment (m/z 69) is also indicative of cellular membranes and serves as an internal standard (0–10 counts). (c) Vitamin E (m/z 430) is localized at the soma-neurite junction as shown by the molecular ion (0–7 counts). (d) Line scans for normalized vitamin E (top) and choline (bottom) begin in the neurite and continue across the cell soma. The line scans have been normalized with the acyl chain fragment signal. Images have a pixel size of 3 μm. Scale bars are 100 μm. Reproduced with permission from Ref. [108]. Copyright 2005 American Chemical Society.

live neurons or tissues, both before and after chemical or electrophysiological stimulation. Exchanging SPE beads from the same location before and after stimulation provides a straightforward comparison to discriminate the complement of secreted peptides from unrelated compounds constitutively present in complex biological environments. But, most notably, precise placement of SPE beads within biological tissues adds a spatial parameter to peptide collection from stimulated regions. Measuring the binding capacity of a single bead, as well as the rate of saturation, was achieved by sensitive and quantitative radionuclide detection of the cytochrome c standard. A notable advantage of single bead SPE extraction is the removal of salts from samples obtained in biological environments. Invertebrate hemolymph, or blood, is essentially isosmotic to seawater (~1000 mOsm), whereas mammalian cerebrospinal fluid and blood has typically one-third this osmolarity. In both cases, without this desalting step, inorganic salts, such as NaCl, are present in quantities that decrease peptide detection with MALDI MS. Recently this bead approach was used to characterize NPY secretion from an *Aplysia* nerve to determine the complement of peptides released with a particular behavior [111].

6.8
Future Developments

Because of the unique capabilities of single cell MS to detect, identify and quantify known and unknown compounds at the cellular level, we predict an exponential growth in the number of applications and in the users of this rapidly advancing technology. The significant improvements in sensitivity and resolution achieved using state of the art mass spectrometers over the past few years has expanded the range of cell types and subcellular structures that can be analyzed. Enhancements to instrumental performance are continuing, so that an even broader range of applications are expected in the future. Of particular importance is the increase in sensitivity of detection for larger and typically less abundant proteins. Improvements in sample collection, signal acquisition and processing undoubtedly will lead to increased throughput of analysis. Above and beyond the identification of new cellular players, quantitation will become even more common. Stable isotope labeling is the mainstream approach in quantitation with MS. There is great promise for single cell MALDI MS in discovering new intercellular signaling molecules that are often present in relatively high quantities in individual cells. Last, improvements in tandem MS will help to identify such compounds without having to isolate them from larger tissue regions. It is exciting to envision the potential that single cell MS may reach in the coming decade.

Acknowledgements

The support of the National Institute on Drug Abuse under Award No. P30 DA 018310 to the UIUC Neuroproteomics Center on Cell to Cell Signaling is gratefully acknowledged. E.B.M. thanks Merck Research Laboratories for fellowship support.

References

1 Leeuwenhoek, A.V. (1999) *The Collected Letters of Antoni Van Leeuwenhoek: 1704-1707 Part XV*, **442** (ed. L.C. Palm), Swets & Zeitlinger, Amsterdam.

2 Leewenhoeck, M. and de Graaf, R. (1673) A Specimen of Some Observations Made by a Microscope, Contrived by M. Leewenhoeck in Holland, Lately Communicated by Dr. Regnerus de Graaf. *Philosophical Transactions (1665–1678)*, **8**, 6037–6038.

3 Harris, H. (1999) *The Birth of the Cell*, Yale University Press, New Haven.

4 Ehrlich, P. (1878) Doctoral Thesis: Beiträge zur theorie und praxis der histologischen färbung, The Medical Faculty, Leipzig University.

5 Crivellato, E., Beltrami, C.A., Mallardi, F. and Ribatti, D. (2003) Paul Ehrlich's doctoral thesis: a milestone in the study of mast cells. *British Journal of Haematology*, **123**, 19–21.

6 Ruchel, R., Loh, Y.P. and Gainer, H. (1977) A technique for the selective extraction of water-soluble polypeptides from identified neurons of *Aplysia*

7 Ruchel, R. (1977) Two-dimensional micro-separation technique for proteins and peptides, combining isoelectric focusing and gel gradient electrophoresis. *Journal of Chromatography*, **132**, 451–468.

8 Matioli, G.T. and Niewisch, H.B. (1965) Electrophoresis of hemoglobin in single erythrocytes. *Science*, **150**, 1824–1826.

9 Kaufmann, R., Hillenkamp, F., Nitsche, R., Schurmann, M. and Wechsung, R. (1978) The laser microprobe mass analyser (LAMMA): biomedical applications. *Microscopia Acta Supplement*, 297–306.

10 Tanaka, K., Waki, H., Ido, Y., Akita, S., Yoshida, Y. and Yoshida, T. (1988) Protein and polymer analysis up to m/z 100.000 by laser ionisation time-of-flight mass spectrometry. *Rapid Communications in Mass Spectrometry*, **2**, 151–153.

11 Fenn, J.B., Mann, M., Meng, C.K., Wong, S.F. and Whitehouse, C.M. (1989) Electrospray ionization for mass spectrometry of large biomolecules. *Science*, **246**, 64–71.

12 Karas, M., Bachmann, D., Bahr, U. and Hillenkamp, F. (1987) Matrix-assisted ultraviolet-laser desorption of nonvolatile compounds. *International Journal of Mass Spectrometry and Ion Processes*, **78**, 53–68.

13 Brecht, M., Schneider, M., Sakmann, B. and Margrie, T.W. (2004) Whisker movements evoked by stimulation of single pyramidal cells in rat motor cortex. *Nature*, **427**, 704–710.

14 Hu, S., Zhang, Z.R., Cook, L.M., Carpenter, E.J. and Dovichi, N.J. (2000) Separation of proteins by sodium dodecylsulfate capillary electrophoresis in hydroxypropylcellulose sieving matrix with laser-induced fluorescence detection. *Journal of Chromatography. A*, **894**, 291–296.

15 Hinkle, D., Glanzer, J., Sarabi, A., Pajunen, T., Zielinski, J., Belt, B., Miyashiro, K., McIntosh, T. and Eberwine, J. (2004) Single neurons as experimental systems in californica. *Hoppe-Seyler's Zeitschrift fur Physiologische Chemie*, **358**, 659–665. molecular biology. *Progress in Neurobiology*, **72**, 129–142.

16 Gross, M.L. and Caprioli, R. (eds) (2003) *Encyclopedia of Mass Spectrometry*, 1st edn, Vol. 1–10, Elsevier, Oxford; San Diego, CA.

17 Meyer zum Gottesberge-Orsulakova, A. and Kaufmann, R. (1985) Recent advances in laser microprobe mass analysis (LAMMA) of inner ear tissue. *Scanning Electron Microscopy*, 393–405.

18 Tobias, H.J., Schafer, M.P., Pitesky, M., Fergenson, D.P., Horn, J., Frank, M. and Gard, E.E. (2005) Bioaerosol mass spectrometry for rapid detection of individual airborne *Mycobacterium tuberculosis* H37Ra particles. *Applied and Environmental Microbiology*, **71**, 6086–6095.

19 Tobias, H.J., Pitesky, M.E., Fergenson, D.P., Steele, P.T., Horn, J., Frank, M. and Gard, E.E. (2006) Following the biochemical and morphological changes of *Bacillus atrophaeus* cells during the sporulation process using bioaerosol mass spectrometry. *Journal of Microbiological Methods*, **67**, 56–63.

20 Seydel, U. and Lindner, B. (1981) Qualitative and quantitative investigations on mycobacteria with LAMMA. *Fresenius Journal of Analytical Chemistry*, **308**, 253–257.

21 Böhm, R. (1981) Sample preparation technique for the analysis of vegetative bacteria cells of the genus bacillus with the laser microprobe mass analyzer (LAMMA). *Fresenius Journal of Analytical Chemistry*, **308**, 258–259.

22 Sprey, B. and Bochem, H.P. (1981) Uptake of uranium into the alga Dunaliella detected by EDAX and LAMMA. *Fresenius Journal of Analytical Chemistry*, **308**, 239–245.

23 Wei, J., Buriak, J.M. and Siuzdak, G. (1999) Desorption-ionization mass spectrometry on porous silicon. *Nature*, **399**, 243–246.

24 Thomas, J.J., Shen, Z., Crowell, J.E., Finn, M.G. and Siuzdak, G. (2001) Desorption/

ionization on silicon (DIOS): A diverse mass spectrometry platform for protein characterization. *Proceedings of the National Academy of Sciences of the United States of America*, **98**, 4932–4937.

25 Kruse, R.A., Rubakhin, S.S., Romanova, E.V., Bohn, P.W. and Sweedler, J.V. (2001) Direct assay of *Aplysia* tissues and cells with laser desorption/ionization mass spectrometry on porous silicon. *Journal of Mass Spectrometry*, **36**, 1317–1322.

26 Liu, Q., Guo, Z. and He, L. (2007) Mass spectrometry imaging of small molecules using desorption/ionization on silicon. *Analytical Chemistry*, **79**, 3535–3541.

27 Gross, M.L. and Caprioli, R. (eds) (2003) *Encyclopedia of Mass Spectrometry: Ionization Methods*, 1st edn, Vol. 6, Elsevier, Oxford; San Diego, CA.

28 Bellhorn, M.B. and Lewis, R.K. (1976) Localization of ions in retina by secondary ion mass spectrometry. *Experimental Eye Research*, **22**, 505–518.

29 Hofstadler, S.A., Severs, J.C., Smith, R.D., Swanek, F.D. and Ewing, A.G. (1996) Analysis of single cells with capillary electrophoresis electrospray ionization Fourier transform ion cyclotron resonance mass spectrometry. *Rapid Communications in Mass Spectrometry*, **10**, 919–922.

30 Lazar, I.M., Grym, J. and Foret, F. (2006) Microfabricated devices: A new sample introduction approach to mass spectrometry. *Mass Spectrometry Reviews*, **25**, 573–594.

31 Mouledous, L., Hunt, S., Harcourt, R., Harry, J., Williams, K.L. and Gutstein, H.B. (2003) Navigated laser capture microdissection as an alternative to direct histological staining for proteomic analysis of brain samples. *Proteomics*, **3**, 610–615.

32 Garden, R.W., Moroz, L.L., Moroz, T.P., Shippy, S.A. and Sweedler, J.V. (1996) Excess salt removal with matrix rinsing: direct peptide profiling of neurons from marine invertebrates using matrix-assisted laser desorption/ionization time-of-flight mass spectrometry. *Journal of Mass Spectrometry*, **31**, 1126–1130.

33 Monroe, E.B., Jurchen, J.C., Koszczuk, B.A., Losh, J.L., Rubakhin, S.S. and Sweedler, J.V. (2006) Massively parallel sample preparation for the MALDI MS analyses of tissues. *Analytical Chemistry*, **78**, 6826–6832.

34 Bergquist, J. (1999) Cells on the target. Matrix-assisted laser-desorption/ionization time-of-flight mass-spectrometric analysis of mammalian cells grown on the target. *Chromatographia Supplement I*, **49**, S41–S48.

35 Fung, E.N. and Yeung, E.S. (1998) Direct analysis of single rat peritoneal mast cells with laser vaporization/ionization mass spectrometry. *Analytical Chemistry*, **70**, 3206–3212.

36 Rubakhin, S.S., Churchill, J.D., Greenough, W.T. and Sweedler, J.V. (2006) Profiling signaling peptides in single mammalian cells using mass spectrometry. *Analytical Chemistry*, **78**, 7267–7272.

37 Rubakhin, S.S., Greenough, W.T. and Sweedler, J.V. (2003) Spatial profiling with MALDI MS: distribution of neuropeptides within single neurons. *Analytical Chemistry*, **75**, 5374–5380.

38 Fenselau, C. and Demirev, P.A. (2001) Characterization of intact microorganisms by MALDI mass spectrometry. *Mass Spectrometry Reviews*, **20**, 157–171.

39 Demirev, P.A., Feldman, A.B., Kowalski, P. and Lin, J.S. (2005) Top-down proteomics for rapid identification of intact microorganisms. *Analytical Chemistry*, **77**, 7455–7461.

40 Dworzanski, J.P., Deshpande, S.V., Chen, R., Jabbour, R.E., Snyder, A.P., Wick, C.H. and Li, L. (2006) Mass spectrometry-based proteomics combined with bioinformatic tools for bacterial classification. *Journal of Proteome Research*, **5**, 76–87.

41 Peng, W.P., Yang, Y.C., Kang, M.W., Lee, Y.T. and Chang, H.C. (2004) Measuring masses of single bacterial whole cells with

a quadrupole ion trap. *Journal of the American Chemical Society*, **126**, 11766–11767.
42 Jimenez, C.R., van Veelen, P.A., Li, K.W., Wildering, W.C., Geraerts, W.P., Tjaden, U.R. and van der Greef, J. (1994) Neuropeptide expression and processing as revealed by direct matrix-assisted laser desorption ionization mass spectrometry of single neurons. *Journal of Neurochemistry*, **62**, 404–407.
43 Li, L., Garden, R.W. and Sweedler, J.V. (2000) Single-cell MALDI: a new tool for direct peptide profiling. *Trends in Biotechnology*, **18**, 151–160.
44 Proekt, A., Vilim, F.S., Alexeeva, V., Brezina, V., Friedman, A., Jing, J., Li, L., Zhurov, Y., Sweedler, J.V. and Weiss, K.R. (2005) Identification of a new neuropeptide precursor reveals a novel source of extrinsic modulation in the feeding system of *Aplysia*. *The Journal of Neuroscience*, **25**, 9637–9648.
45 Hummon, A.B., Amare, A. and Sweedler, J.V. (2006) Discovering new invertebrate neuropeptides using mass spectrometry. *Mass Spectrometry Reviews*, **25**, 77–98.
46 Neupert, S., Johard, H.A.D., Nassel, D.R. and Predel, R. (2007) Single-cell peptidomics of *Drosophila melanogaster* neurons identified by Gal4-driven fluorescence. *Analytical Chemistry*, **79**, 3690–3694.
47 Floyd, P.D., Li, L., Rubakhin, S.S., Sweedler, J.V., Horn, C.C., Kupfermann, I., Alexeeva, V.Y., Ellis, T.A., Dembrow, N.C., Weiss, K.R. and Vilim, F.S. (1999) Insulin prohormone processing, distribution, and relation to metabolism in *Aplysia californica*. *The Journal of Neuroscience*, **19**, 7732–7741.
48 Vilim, F.S., Alexeeva, V., Moroz, L.L., Li, L., Moroz, T.P., Sweedler, J.V. and Weiss, K.R. (2001) Cloning, expression and processing of the CP2 neuropeptide precursor of *Aplysia*. *Peptides*, **22**, 2027–2038.
49 Furukawa, Y., Nakamaru, K., Wakayama, H., Fujisawa, Y., Minakata, H., Ohta, S., Morishita, F., Matsushima, O., Li, L., Romanova, E., Sweedler, J.V., Park, J.H., Romero, A., Cropper, E.C., Dembrow, N.C., Jing, J., Weiss, K.R. and Vilim, F.S. (2001) The enterins: a novel family of neuropeptides isolated from the enteric nervous system and CNS of *Aplysia*. *The Journal of Neuroscience*, **21**, 8247–8261.
50 Li, L., Floyd, P.D., Rubakhin, S.S., Romanova, E.V., Jing, J., Alexeeva, V.Y., Dembrow, N.C., Weiss, K.R., Vilim, F.S. and Sweedler, J.V. (2001) Cerebrin prohormone processing, distribution and action in *Aplysia californica*. *Journal of Neurochemistry*, **77**, 1569–1580.
51 Sweedler, J.V., Li, L., Rubakhin, S.S., Alexeeva, V., Dembrow, N.C., Dowling, O., Jing, J., Weiss, K.R. and Vilim, F.S. (2002) Identification and characterization of the feeding circuit-activating peptides, a novel neuropeptide family of *Aplysia*. *The Journal of Neuroscience*, **22**, 7797–7808.
52 Romanova, E.V., McKay, N., Weiss, K.R., Sweedler, J.V. and Koester, J. (2007) Autonomic control network active in *Aplysia* during locomotion includes neurons that express splice variants of R15-neuropeptides. *Journal of Neurophysiology*, **97**, 481–491.
53 Jimenez, C.R., Spijker, S., de Schipper, S., Lodder, J.C., Janse, C.K., Geraerts, W.P., van Minnen, J., Syed, N.I., Burlingame, A.L., Smit, A.B. and Li, K. (2006) Peptidomics of a single identified neuron reveals diversity of multiple neuropeptides with convergent actions on cellular excitability. *The Journal of Neuroscience*, **26**, 518–529.
54 Standing, K.G. (2003) Peptide and protein *de novo* sequencing by mass spectrometry. *Current Opinion in Structural Biology*, **13**, 595–601.
55 Bienvenut, W.V., Deon, C., Pasquarello, C., Campbell, J.M., Sanchez, J.C., Vestal, M.L. and Hochstrasser, D.F. (2002) Matrix-assisted laser desorption/ionization-tandem mass spectrometry with high resolution and sensitivity

for identification and characterization of proteins. *Proteomics*, **2**, 868–876.

56 Medzihradszky, K.F., Campbell, J.M., Baldwin, M.A., Falick, A.M., Juhasz, P., Vestal, M.L. and Burlingame, A.L. (2000) The characteristics of peptide collision-induced dissociation using a high-performance MALDI-TOF/TOF tandem mass spectrometer. *Analytical Chemistry*, **72**, 552–558.

57 Yergey, A.L., Coorssen, J.R., Backlund, P.S., Jr, Blank, P.S., Humphrey, G.A., Zimmerberg, J., Campbell, J.M. and Vestal, M.L. (2002) *De novo* sequencing of peptides using MALDI/TOF-TOF. *Journal of the American Society for Mass Spectrometry*, **13**, 784–791.

58 Suckau, D., Resemann, A., Schuerenberg, M., Hufnagel, P., Franzen, J. and Holle, A. (2003) A novel MALDI LIFT-TOF/TOF mass spectrometer for proteomics. *Analytical and Bioanalytical Chemistry*, **376**, 952–965.

59 Jimenez, C.R., Li, K.W., Dreisewerd, K., Spijker, S., Kingston, R., Bateman, R.H., Burlingame, A.L., Smit, A.B., Van Minnen, J. and Geraerts, W.P.M. (1998) Direct mass spectrometric peptide profiling and sequencing of single neurons reveals differential peptide patterns in a small neuronal network. *Biochemistry*, **37**, 2070–2076.

60 Li, L., Garden, R.W., Romanova, E.V. and Sweedler, J.V. (1999) *In situ* sequencing of peptides from biological tissues and single cells using MALDI-PSD/CID analysis. *Analytical Chemistry*, **71**, 5451–5458.

61 Perry, S.J., Dobbins, A.C., Schofield, M.G., Piper, M.R. and Benjamin, P.R. (1999) Small cardioactive peptide gene: structure, expression and mass spectrometric analysis reveals a complex pattern of co-transmitters in a snail feeding neuron. *The European Journal of Neuroscience*, **11**, 655–662.

62 Redeker, V., Toullec, J.Y., Vinh, J., Rossier, J. and Soyez, D. (1998) Combination of peptide profiling by matrix-assisted laser desorption/ionization time-of-flight mass spectrometry and immunodetection on single glands or cells. *Analytical Chemistry*, **70**, 1805–1811.

63 Ma, P.W., Garden, R.W., Niermann, J.T., O'Connor, M., Sweedler, J.V. and Roelofs, W.L. (2000) Characterizing the Hez-PBAN gene products in neuronal clusters with immunocytochemistry and MALDI MS. *Journal of Insect Physiology*, **46**, 221–230.

64 Neupert, S. and Predel, R. (2005) Mass spectrometric analysis of single identified neurons of an insect. *Biochemical and Biophysical Research Communications*, **327**, 640–645.

65 Sweedler, J.V., Rubakhin, S.S., Churchill, J.D. and Greenough, W.T. (2003) Assaying the neuropeptides in single mammalian neurons using mass spectrometry. Abstract Viewer/Itinerary Planner. Program No. 326.16. Washington, DC: Society for Neuroscience.

66 Vazquez-Martinez, R.M., Malagon, M.M., van Strien, F.J., Jespersen, S., van der Greef, J., Roubos, E.W. and Gracia-Navarro, F. (1999) Analysis by mass spectrometry of POMC-derived peptides in amphibian melanotrope subpopulations. *Life Sciences*, **64**, 923–930.

67 Xu, B.J., Caprioli, R.M., Sanders, M.E. and Jensen, R.A. (2002) Direct analysis of laser capture microdissected cells by MALDI mass spectrometry. *Journal of the American Society for Mass Spectrometry*, **13**, 1292–1297.

68 Valaskovic, G.A., Kelleher, N.L. and McLafferty, F.W. (1996) Attomole protein characterization by capillary electrophoresis-mass spectrometry. *Science*, **273**, 1199–1202.

69 Iliffe, T.M., McAdoo, D.J., Beyer, C.B. and Haber, B. (1977) Amino acid concentrations in the *Aplysia* nervous system: neurons with high glycine concentrations. *Journal of Neurochemistry*, **28**, 1037–1042.

70 Page, J.S., Rubakhin, S.S. and Sweedler, J.V. (2002) Single-neuron analysis using

CE combined with MALDI MS and radionuclide detection. *Analytical Chemistry*, **74**, 497–503.
71 Good, P.F., Olanow, C.W. and Perl, D.P. (1992) Neuromelanin-containing neurons of the substantia nigra accumulate iron and aluminum in Parkinson's disease: a LAMMA study. *Brain Research*, **593**, 343–346.
72 Good, P.F., Perl, D.P., Bierer, L.M. and Schmeidler, J. (1992) Selective accumulation of aluminum and iron in the neurofibrillary tangles of Alzheimer's disease: a laser microprobe (LAMMA) study. *Annals of Neurology*, **31**, 286–292.
73 Schröder, W.H. (1981) Quantitative LAMMA analysis of biological specimens I. Standards. II. Isotope labeling. *Fresenius Journal of Analytical Chemistry*, **308**, 212–217.
74 Orsulakova, A., Kaufmann, R., Morgenstern, C. and D'Haese, M. (1981) Cation distribution of the cochlea wall (Stria vascularis). *Fresenius Journal of Analytical Chemistry*, **308**, 221–223.
75 Lorch, D.W. and SchÄfer, H. (1981) Localization of lead in cells of *Phymatodocis nordstedtiana* (Chlorophyta) with the laser microprobe analyzer (LAMMA 500). *Fresenius Journal of Analytical Chemistry*, **308**, 246–248.
76 Tomos, A.D. and Sharrock, R.A. (2001) Cell sampling and analysis (SiCSA): metabolites measured at single cell resolution. *Journal of Experimental Botany*, **52**, 623–630.
77 Koroleva, O.A., Tomos, A.D., Farrar, J., Roberts, P. and Pollock, C.J. (2000) Tissue distribution of primary metabolism between epidermal, mesophyll and parenchymatous bundle sheath cells in barley leaves. *Australian Journal of Plant Physiology*, **27**, 747–755.
78 Kajiyama, S., Harada, K., Fukusaki, E. and Kobayashi, A. (2006) Single cell-based analysis of torenia petal pigments by a combination of ArF excimer laser micro sampling and nano-high performance liquid chromatography (HPLC)-mass spectrometry. *Journal of Bioscience and Bioengineering*, **102**, 575–578.
79 Rubakhin, S.S., Garden, R.W., Fuller, R.R. and Sweedler, J.V. (2000) Measuring the peptides in individual organelles with mass spectrometry. *Nature Biotechnology*, **18**, 172–175.
80 Heeren, R.M.A. and Sweedler, J.V. (eds) (2007) International Journal of Mass Spectrometry, Vol. 260.
81 Chaurand, P., Schriver, K.E. and Caprioli, R.M. (2007) Instrument design and characterization for high resolution MALDI-MS imaging of tissue sections. *Journal of Mass Spectrometry*, **42**, 476–489.
82 Jurchen, J.C., Rubakhin, S.S. and Sweedler, J.V. (2005) MALDI-MS imaging of features smaller than the size of the laser beam. *Journal of the American Society for Mass Spectrometry*, **16**, 1654–1659.
83 Stoeckli, M., Chaurand, P., Hallahan, D.E. and Caprioli, R.M. (2001) Imaging mass spectrometry: a new technology for the analysis of protein expression in mammalian tissues. *Nature Medicine*, **7**, 493–496.
84 Caprioli, R.M., Farmer, T.B. and Gile, J. (1997) Molecular imaging of biological samples: localization of peptides and proteins using MALDI-TOF MS. *Analytical Chemistry*, **69**, 4751–4760.
85 Chaurand, P., Fouchecourt, S., DaGue, B.B., Xu, B.J., Reyzer, M.L., Orgebin-Crist, M.C. and Caprioli, R.M. (2003) Profiling and imaging proteins in the mouse epididymis by imaging mass spectrometry. *Proteomics*, **3**, 2221–2239.
86 Chaurand, P., Schwartz, S.A. and Caprioli, R.M. (2004) Assessing protein patterns in disease using imaging mass spectrometry. *Journal of Proteome Research*, **3**, 245–252.
87 Touboul, D., Roy, S., Germain, D.P., Chaminade, P., Brunelle, A. and Laprevote, O. (2007) MALDI-TOF and cluster-TOF-SIMS imaging of Fabry disease biomarkers. *International Journal of Mass Spectrometry*, **260**, 158–165.

88 Luxembourg, S.L., McDonnell, L.A., Mize, T.H. and Heeren, R.M. (2005) Infrared mass spectrometric imaging below the diffraction limit. *Journal of Proteome Research*, **4**, 671–673.

89 Luxembourg, S.L., Vaezaddeh, A.R., Amstalden, E.R., Zimmermann-Ivol, C.G., Hochstrasser, D.F. and Heeren, R.M. (2006) The molecular scanner in microscope mode. *Rapid Communications in Mass Spectrometry*, **20**, 3435–3442.

90 Altelaar, A.F.M., Taban, I.M., McDonnell, L.A., Verhaert, P.D.E.M., de Lange, R.P.J., Adan, R.A.H., Mooi, W.J., Heeren, R.M.A. and Piersma, S.R. (2007) High-resolution MALDI imaging mass spectrometry allows localization of peptide distributions at cellular length scales in pituitary tissue sections. *International Journal of Mass Spectrometry*, **260**, 203–211.

91 Chandra, S. (2001) Studies of cell division (mitosis and cytokinesis) by dynamic secondary ion mass spectrometry ion microscopy: LLC-PK1 epithelial cells as a model for subcellular isotopic imaging. *Journal of Microscopy*, **204**, 150–165.

92 Gazi, E., Lockyer, N.P., Vickerman, J.C., Gardner, P., Dwyer, J., Hart, C.A., Brown, M.D., Clarke, N.W. and Miyan, J. (2004) Imaging ToF-SIMS and synchrotron-based FT-IR micro spectroscopic studies of prostate cancer cell lines. *Applied Surface Science*, **231–232**, 452–456.

93 Chandra, S. (2004) Subcellular SIMS imaging of isotopically labeled amino acids in cryogenically prepared cells. *Applied Surface Science*, **231–232**, 462–466.

94 Quong, J.N., Knize, M.G., Kulp, K.S. and Wu, K.J. (2004) Molecule-specific imaging analysis of carcinogens in breast cancer cells using time-of-flight secondary ion mass spectrometry. *Applied Surface Science*, **231–232**, 424–427.

95 Cliff, B., Lockyer, N., Jungnickel, H., Stephens, G. and Vickerman, J.C. (2003) Probing cell chemistry with time-of-flight secondary ion mass spectrometry: development and exploitation of instrumentation for studies of frozen-hydrated biological material. *Rapid Communications in Mass Spectrometry*, **17**, 2163–2167.

96 Borner, K., Malmberg, P., Mansson, J.-E. and Nygren, H. (2007) Molecular imaging of lipids in cells and tissues. *International Journal of Mass Spectrometry*, **260**, 128–136.

97 Colliver, T.L., Brummel, C.L., Pacholski, M.L., Swanek, F.D., Ewing, A.G. and Winograd, N. (1997) Atomic and molecular imaging at the single-cell Level with TOF-SIMS. *Analytical Chemistry*, **69**, 2225–2231.

98 Oyedepo, A.C., Brooke, S.L., Heard, P.J., Day, J.C., Allen, G.C. and Patel, H. (2004) Analysis of boron-10 in soft tissue by dynamic secondary ion mass spectrometry. *Journal of Microscopy*, **213**, 39–45.

99 Cheng, J., Kozole, J., Hengstebeck, R. and Winograd, N. (2007) Direct comparison of Au(3)(+) and C(60)(+) cluster projectiles in SIMS molecular depth profiling. *Journal of the American Society for Mass Spectrometry*, **18**, 406–412.

100 Altelaar, A.F., Klinkert, I., Jalink, K., de Lange, R.P., Adan, R.A., Heeren, R.M. and Piersma, S.R. (2006) Gold-enhanced biomolecular surface imaging of cells and tissue by SIMS and MALDI mass spectrometry. *Analytical Chemistry*, **78**, 734–742.

101 McDonnell, L.A., Piersma, S.R., MaartenAltelaar, A.F., Mize, T.H., Luxembourg, S.L., Verhaert, P.D., van Minnen, J. and Heeren, R.M. (2005) Subcellular imaging mass spectrometry of brain tissue. *Journal of Mass Spectrometry*, **40**, 160–168.

102 Todd, P.J., Schaaff, T.G., Chaurand, P. and Caprioli, R.M. (2001) Organic ion imaging of biological tissue with secondary ion mass spectrometry and matrix-assisted laser desorption/ionization. *Journal of Mass Spectrometry*, **36**, 355–369.

103 Touboul, D., Halgand, F., Brunelle, A., Kersting, R., Tallarek, E., Hagenhoff, B. and Laprevote, O. (2004) Tissue molecular ion imaging by gold cluster ion bombardment. *Analytical Chemistry*, **76**, 1550–1559.

104 Amaya, K.R., Monroe, E.B., Sweedler, J.V. and Clayton, D.F. (2007) Lipid imaging in the zebra finch brain with secondary ion mass spectrometry. *International Journal of Mass Spectrometry*, **260**, 121–127.

105 Sjovall, P., Lausmaa, J., Nygren, H., Carlsson, L. and Malmberg, P. (2003) Imaging of membrane lipids in single cells by imprint-imaging time-of-flight secondary ion mass spectrometry. *Analytical Chemistry*, **75**, 3429–3434.

106 Ostrowski, S.G., Van Bell, C.T., Winograd, N. and Ewing, A.G. (2004) Mass spectrometric imaging of highly curved membranes during *Tetrahymena* mating. *Science*, **305**, 71–73.

107 Fletcher, J.S., Lockyer, N.P., Vaidyanathan, S. and Vickerman, J.C. (2007) TOF-SIMS 3D biomolecular imaging of *Xenopus laevis* oocytes using buckminsterfullerene (C(60)) primary ions. *Analytical Chemistry*, **79**, 2199–2206.

108 Monroe, E.B., Jurchen, J.C., Lee, J., Rubakhin, S.S. and Sweedler, J.V. (2005) Vitamin E imaging and localization in the neuronal membrane. *Journal of the American Chemical Society*, **127**, 12152–12153.

109 Rubakhin, S.S., Page, J.S., Monroe, B.R. and Sweedler, J.V. (2001) Analysis of cellular release using capillary electrophoresis and matrix assisted laser desorption/ionization-time of flight-mass spectrometry. *Electrophoresis*, **22**, 3752–3758.

110 Hatcher, N.G., Richmond, T.A., Rubakhin, S.S. and Sweedler, J.V. (2005) Monitoring activity-dependent peptide release from the CNS using single-bead solid-phase extraction and MALDI TOF MS detection. *Analytical Chemistry*, **77**, 1580–1587.

111 Jing, J., Vilim, F.S., Horn, C.C., Alexeeva, V., Hatcher, N.G., Sasaki, K., Yashina, I., Zhurov, Y., Kupfermann, I., Sweedler, J.V. and Weiss, K.R. (2007) From hunger to satiety: reconfiguration of a feeding network by *Aplysia* neuropeptide Y. *The Journal of Neuroscience*, **27**, 3490–3502.

7
Single Cell Analysis for Quantitative Systems Biology
Luke P. Lee and Dino Di Carlo

7.1
Introduction

A fundamental goal of cell biology is identifying how cell behavior arises from the dynamic collection of environmental stimuli to which the cell is exposed. From a biosystems science and engineering perspective, there is great interest in how the cell behaves as a system, processing time-dependent input signals into output behaviors (Figure 7.1). The ability to predict this input–output relationship at the level of the individual eukaryotic cell has implications for our understanding of higher level organization of tissues and organisms and more importantly may suggest therapeutic approaches to correct flaws in this organization. Key processes such as stem cell differentiation and self-renewal are intimately connected to the cellular microenvironment or "niche" [1, 2]. Additionally normal morphological design of organisms is dependent on programs of cellular division and apoptosis directed in most cases by temporally and spatially varying extracellular signals [3]. Some of the environmental factors involved include: (1) chemical signals such as cytokines, growth factors, nutrients, and hormones, (2) biological signals such as cell–contact including cell–cell (adherent junctions [4], gap junctions [5], lipid nanotubes [6]) and cell–extracellular matrix (3) contacts (integrin–ECM, proteoglycan–ECM), (4) Physical signals including mechanical (shear stress [7], cytoskeletal tension [8], stretch-activated ion channels [9]), electrical (voltage-gated ion channels [10], galvanotaxis [11]) and thermal (temperature-sensitive ion channels [12], biochemical kinetics [13]) factors.

Ideally with knowledge of the history of this ensemble of environmental stimuli, one could predict the precise behavior that a particular cell would exhibit, thus reaching a deterministic and quantitative description of biological function. Unfortunately cells under seemingly identical environmental conditions often display a distribution of heterogeneous behaviors [14, 15]. This appears to be partly due to probabilistic behavior in the "decision" processes that connect input and output [14, 16–18]. Underlying the links between inputs and outputs are systems of

Single Cell Analysis: Technologies and Applications. Edited by Dario Anselmetti
Copyright © 2009 WILEY-VCH Verlag GmbH & Co. KGaA, Weinheim
ISBN: 978-3-527-31864-3

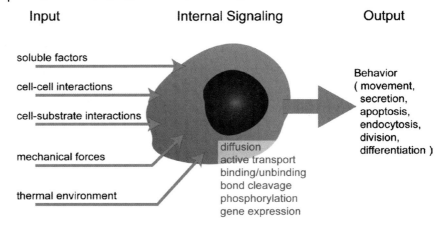

Figure 7.1 Input–output response for the cellular machine. Researchers would like to understand cellular behavior in an analogous way that one understands a complex electrical or mechanical system. Here analogous electrical inputs are instead environmental cues, such as cytokines, ECM, shear stress and temperature. Output is not a voltage wave form but biochemical behavior, such as differentiation, division and apoptosis. Transduction of inputs to outputs requires networks of chemical reactions, localized in space and time. History of chemical reactions is also important (this is not a linear time invariant system). As opposed to an electrical circuit the behavior is not completely deterministic. This is due partly to probabilistic effects occurring with low copy number chemical reactions in the internal signaling pathways that transduce inputs to output behavior. Also, a multitude of environmental inputs that are not always well controlled may lead to further difficulties in predicting behavior.

interconnected molecular interactions (signaling pathways). Cross-talk between pathways, localization of reactions and the small molecule numbers involved in signaling lead to stochastic behavior in these systems [14, 16, 19, 20]. Studies and predictions of the ensemble behavior of these pathways, deemed systems biology, has grown exponentially in the past decade [20]. Due to their ubiquity and importance in cellular function, signaling pathways and the molecules involved have received the most attention amongst researchers interested in the complex behaviors of cells and how these may be controlled, or fixed when gone awry.

Because of the heterogeneity within a population, increased emphasis has been put on analyzing a large quantity of single cells and determining distributions of responses [18, 21, 22]. Thus, allowing development of quasi-deterministic mathematical models of cellular behavior, where well determined environmental inputs leads to a determined distribution of responses amongst a population [18], while the correct behavior for an individual cell still has a probabilistic component.

Although population heterogeneity appears to be partly due to stochastic behavior within signaling pathways, it may be that lack of control of the cellular microenvironment also plays a large role in broadening population distributions. Increased emphasis is now being placed on creating uniform and well controlled *in vitro* environments for cell culture [1, 23, 24], that allow probing of environmental variables in a dynamic and high-throughput fashion. Current standard well plates or Petri dishes lack control of a large amount of environmental factors that may critical for cell behavior including cell–cell, cell–ECM and the soluble environment (Figure 7.2a). Dynamic control of the environment also appears critical for some cellular processes [3] and is lacking in these systems.

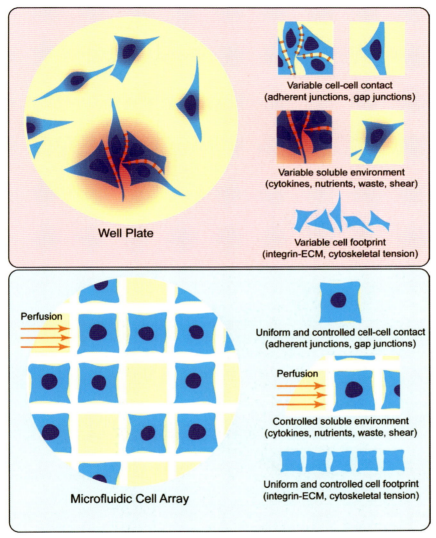

Figure 7.2 The microenvironment in single cell analysis. Most commonly used cell culture techniques, relying on growth of adherent cells in plates or wells, have limited control of cellular environmental factors. Cells are seeded at random for culture and form random distributions after adhering. Soluble nutrients, waste, and paracrine signals are allowed to diffuse and vary depending on location within a dish. Communication by cell contact is not

Bioreactors [25] can allow dynamic control of the soluble environment, but often lack control of the cell–contact environment. New tools, often based on microfabrication and microfluidic technologies, are now allowing improved dynamic control of environmental variables for high-throughput single cell analysis (Figure 7.2b). These experimental technologies combined with systems analysis of signaling pathways are expected to lead to an improved quantitative description of single cellular function.

7.2
Misleading Bulk Experiments

One approach is to create uniform culture environments for individual cells but then analyze cellular response as a blended average parameter. Bulk experiments are often performed to determine average protein or biomolecule content of a large number of cells. Using bulk techniques such as western blots, however, do not provide the correct distribution of a response that is needed to develop mathematical descriptions of cellular behavior, even with multiple experiments [15, 26]. Statistically multiple experiments probing a bulk "blended" parameter such as provided by a western blot yield the standard error of the mean for protein content, since protein content in this context is already an averaged value. This is in contrast to analysis of fluorescence intensity levels of a GFP fusion protein, for example at the single cell level, which can yield the actual distribution of protein content of a population of cells.

Taking this example further, if the distribution of protein levels is not normally distributed but bimodal or some other complex distribution this is only observable at the single cell level (Figure 7.3a). For example when the bulk measured protein level increases linearly during a process, at the single cell level, this could be due to the transfer of cells from the low concentration mode of a bimodal distribution to the higher mode of the distribution [15]. This suggests a bi-stable circuit within the cell

Figure 7.2 (*Continued*)
controlled and variable over space and time. Cell footprint, focal adhesions and cytoskeletal tension are all variable factors that there is no attempt to control. Recent work demonstrates the large effects these parameters have on cellular output behavior. New techniques are being developed to culture cells in more uniform and controlled environments. Presently, cell–cell contacts can be controlled by patterning cells chemically or physically by a variety of methods into groups of single cells and larger. Soluble factors can be controlled by perfusion of the growth media, where the rate of convection of metabolized species, $\sim Cq$, is much larger than the rate of metabolism dN/dt. Here C is the concentration per unit volume of the soluble factor, q is the volumetric flow rate and N is the number of molecules of soluble factor. Finally cell footprint, cytoskeletal tension and focal adhesions can also be controlled by chemical patterning of substrates for cell adhesion using micro-stamping or stenciling approaches. Combined, these techniques may hold promise in reducing the distributions of behavior within a population of genetically identical single cells, thus aiding deterministic-stochastic modeling of single cell behavior.

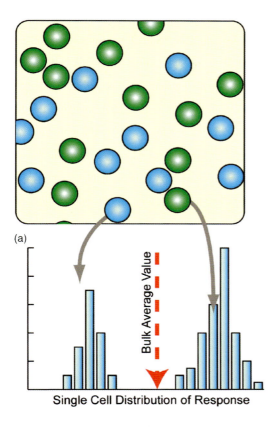

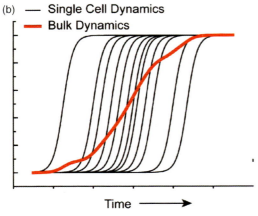

Figure 7.3 Why analyze single cells? Misleading results can be obtained when analyzing bulk blended responses of a population of cells. (a) This is due to population heterogeneity that may manifest itself in non-normal distributions (i.e. bimodal). (b) Bimodal responses may manifest themselves because of fast time-dependent changes in individual cells from one state to another, which vary in time amongst the population. Both the kinetics of the response and the average value can be misinterpreted using bulk techniques.

that controls protein level, not observable using bulk techniques. Furthermore fast time-dependent transitions within individual cells that occur slowly over the population will appear as a gradual change using bulk techniques that could mistakenly be interpreted as a gradual change that occurs at the single cell level (Figure 7.3b).

7.3
Common Techniques for High-Throughput and High-Content Single Cell Analysis

Flow cytometry (FC), laser scanning cytometry (LSC), and automated microscopy (AM) are some of the most widely used techniques for single cell analysis. Well characterized distributions of cellular behavior are often observed using flow cytometry [27]. Briefly, this technique involves the hydrodynamic isolation of individual cells that have been previously labeled using fluorescent dyes that reveal information about the quantity of biomolecules of interest within that cell or on that cell's surface. Additional biological information about cell size and index of refraction can also be gained by observing scattered light. A light source, filtering mechanism and detector are present to observe these signals from individual cells as they pass a particular focal point. The technique is a very high-throughput serial process, which with further feedback is often used to sort populations of cells for further analysis or culture (e.g. fluorescence-activated cell sorting, FACS).

Flow cytometry has been the most successfully used technique for single cell analysis because of the massive throughput, where fluorescent labeling allows quantitative determination of various protein levels amongst a population of cells (GFP fusion proteins, immunofluorescence, fluorogenic substrates to intracellular enzymes) [27, 28]. Time-dependent analysis can also be conducted where different cells are sampled at each time point [29]. This technique was not designed to observe spatial localization of fluorescence within a cell such as is observable using high-content microscopy-based techniques. It should also be noted that the environmental heterogeneity for cells analyzed using this method is dependent on the culture method used prior to analysis and can be reduced by applying new culture techniques to reduce environmental variables such as cell–cell contact and diffusible secretions.

Laser scanning cytometry (LSC), a technique where dyes on surface immobilized cells are excited by a scanning laser, and can be repeatedly interrogated in time is an alternative technique that has been employed [30]. Here, an advantage over FC, time-dependent information can be obtained from the same individual cells, and adherent cells can be maintained in the primary site of culture during analysis. However due to the time required for scanning the sample, LSC has lower throughput when compared to FC. More precisely, time resolution and throughput can be tuned for a particular task, as scanning less cells will lead to an improved temporal resolution for these individual cells. In common implementations of LSC introduction of reagents into a well, dish, or slide are done by pipette where only slow time-dependent changes after solution exchange are desired to be observed.

LSC could also be adapted to novel systems where fast exchange of solutions is possible and (as for FC) controlled culture environments are provided.

Automated microscopy techniques, often termed high-content screening (HCS) or cellomics, have recently provided quantitative insight into cellular behavior and in most cases are applied to observing the response of cells to drug candidate molecules [31, 32]. This technique capitalizes on the recent explosion of computational power by using an automated microscope to take hundreds of images of single cells at different time points and conducting massive image analysis on the collected images to extract useful parameters about cell and nuclear shape as well as various protein distributions within a cell. Usually various cellular components of cells grown in well plates are labeled with fluorescent markers to observe changes over time. In some cases fluorescent labeling necessitates cell fixation, in which temporal measurements are then made on different cell populations. In general, high information content results of cell behavior are averaged using these techniques; however there is the capability of applying AM as a high-throughput technique to determine heterogeneity amongst a population of single cells as well. As with the other conventional techniques, this technique can also be integrated with emerging methods to create uniform culture environments for single cells. This in fact may help reduce noise in the current measurements investigating the mechanism of cellular response to various small molecule candidates.

7.4
Improved Functionality for High-Throughput Single Cell Analysis

7.4.1
Microfluidic Techniques

In order to address the aspects of environmental control, fast timescale measurements, image processing, and secreted biomolecule isolation, several methods of single cell isolation have been developed. A number of microfluidic techniques have been reported to allow optical interrogation of individual cells integrated with fast exchange of reagents. In general, microfluidic techniques employ microfabrication for the miniaturization of fluid channels and conduits. Systems of channels and structures are created that allow dynamic control of reagents and cells through fluid perfusion, and pressure gradients. Wheeler *et al.* [33] demonstrated a semi-permeable physical trap structure that allowed individual cells to be positioned and microfluidic control of applied reagents (Figure 7.4a). This type of setup, although not demonstrated as high throughput, provides advantages over other techniques in that fast optical changes with a "step" introduction of a reagent can be made over time. They employed this advantage to analyze Ca^{2+} waves in response to compound application, not amenable to FC or LSC. Other work using permeable enclosures has also demonstrated analysis of Ca^{2+} signaling in cardiomyocytes [34], and other signaling pathways [35]. Using similar semi-permeable obstacles, we have extended this concept to create uniform arrays of individually trapped cells within a

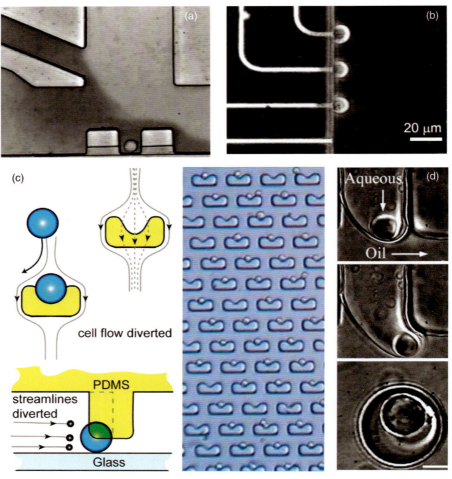

Figure 7.4 Microfluidic single cell analysis. Novel techniques are being developed to control cell environment and conduct dynamic analysis of individual cells in a high-throughput manner. (a) Hydrodynamic trapping of a single cell, with microfluidic ports allowing switching of solutions was demonstrated by Wheeler et al. [33]. Here fast changes in intracellular calcium levels were observable due to application of reagents from an injection microfluidic channel. (b) Work in our laboratory shows single cell trapping on smaller microfluidic pipette-like channels forming a junction with a larger microfluidic channel. In this array, patch-clamp experiments could be performed on individually trapped cells, while compounds are fluidically introduced in the main microchannel [45]. (c) Here a technique developed in our laboratory allows isolation of arrays of individual cells using semi-permeable microstructures that change fluidic resistance after cell loading, sealing the structure to further cells [36]. This is a passive method of creating arrays with control over cell number and contact. Additionally, integration of the arrays in a microfluidic platform allows control of the perfused environment, and fast introduction of reagents. Observation of cellular behavior is done with a standard microscope setup, the device acting as a "functional slide". (d) Microfluidic techniques to isolate cells in a closed volume will be important for secretion analysis of single cells. Here is shown a sequence of images for isolating single cells in droplets within an immiscible fluid [44]. Cells can then be lysed by laser, merged with other droplets for delivery of reagents and analyzed by fluorescence.

microfluidic platform (Figure 7.4c) [36, 37]. In this platform, the microenvironment is well controlled for individual cells, including contact and diffusible stimuli, by isolation and perfusion respectively. Changing trap geometry also allows engineering of the number of cell–cell contacts by trapping groups of cells in proximity. Although throughput is reduced when compared with that of FC, microfluidic integration leads to increased "content", that is fast timescale measurements of tens to hundreds of single cells in parallel.

Other microfluidic techniques hope to replicate the functionality of flow cytometry on a simpler, less expensive platform, with novel functions in some cases. A polymer-based flow cytometer, with integrated pumping and valving, was demonstrated [38]. that allowed both fluorescence interrogation and sorting of individual cells on an inexpensive platform. Fu argues that the level of single cell control in the developed cytometry platform would allow time-dependent kinetic studies, not available with conventional FC. Others have developed chip-based flow cytometry platforms with improved electrical interrogation of cells, not usually seen in traditional FC [39, 40]. Electrical characterization can allow measurement of membrane structure, DNA content and cytoskeletal arrangement that may be complementary to optics-based analyses.

Other researches have used microfluidic techniques to isolate individual cells, but additionally include a small volume of the environment. In these cases, secreted biomolecules can be concentrated and analyzed for individual cells, enabling new types of studies not before available, because secreted elements would diffuse and mix within the environment. Techniques to accomplish this include small microfluidic chambers that can be closed by valves [22, 41, 42]. Using this type of device, single cell studies of gene expression and translation of small copy numbers of β-galactosidase [22] as well as actin [41] and amino acid concentrations [42] were conducted. The study of stochastic protein expression events required a microchamber because the reporter fluorescent product for β-galactosidase diffuses across the plasma membrane. Recently others applied arrays of microfabricated chambers to collecting antibody secretions of individual cells, for selection of antigen-specific antibody production [43]. These types of microchambers may be useful in the future for studies using other membrane permeable fluorogenic probes or investigation of various other cellular secretions at the single cell level.

An alternative approach to creating an enclosed chamber surrounding individual cells is encapsulation of the cell in a droplet present in an immiscible fluid (Figure 7.4d). These methods can be potentially high throughput since many cells can be encapsulated in this manner using a microfluidic droplet generator [44]. Implementation of high-throughput analysis using this technique has been difficult because cell number per droplet is in most cases governed by a Poisson process. Similar secretion and fluorogenic analysis can be performed in this type of platform.

Additional methods hope to bring new functionality to high-throughput single cell analysis not based on optical measurements alone. Ionescu-Zanetti *et al.* [45, 46] have demonstrated a microfluidic platform in which electrophysiological measurements on an array of trapped cells can be performed while optical access is maintained and solutions can be quickly exchanged (Figure 7.4b). The method of single cell isolation is based on producing microfluidic channels to mimic the glass pipette of traditional patch clamp methods. Similar techniques were also used to isolate individual cells for electroporation or gap–junctional communication studies [47, 48].

7.4.2
Array-Based Techniques

Other methods have been developed for isolating single cells in uniform environments without integration in a microfluidic platform. In most cases, these arraying techniques can isolate large quantities of individual cells, but may lack functionality for fast timescale measurements. Environmental factors, like cell–cell contact and cell–ECM contact are usually more controlled in these techniques. Additionally, image processing is simplified for well ordered arrays. In order to form these arrays a multitude of physical interactions have been applied, including mechanical, electrical, optical and chemical means.

Physical trapping structures based on mechanical fit of cells within formed wells are perhaps the simplest and most commonly reported (Figure 7.5a–c) [49–51]. To operate these arrays, cells are seeded on the pre-filled well array and allowed to settle into wells of comparable size. After settling, the remaining cells not constrained by the well are flushed away, leaving a distribution of single cells remaining in individual wells. In some cases, chemical treatments are included to further increase the selectivity of cells in designated trapping areas. These types of structures are quite useful in high-throughput fluorescence-based analysis of individual cells, with simplified image processing and more uniform environment when compared with cells growth on a Petri dish. Fast timescale measurements are difficult to make in these formats because application of a stimulating reagent must be uniform over the array.

Further array-based techniques are based on chemical patterning of surfaces alone with no mechanical confinement. This is usually accomplished by stamping of cell adhesion proteins onto a surface, sometimes followed by attachment of nonadhesive moieties on remaining surfaces (Figure 7.5e) [52–54]. Also a newly developed microfluidic probe [55] shows promise for creation of made to order chemical patterning as one would print a document. Using these methods single cells can be immobilized on adhesive islands of uniform size, and single cells with uniform

Figure 7.5 Single cell arrays. Several methods have been reported for creating arrays of single cells. In most cases the environment is well controlled in these techniques, but they may lack the ability to conduct fast time-dependent observations over the whole array. (a, b) The most common method reported for creating arrays is based on physical trapping. Cells are loaded into microscale wells of similar dimension, only 0–2 cells can fit into the wells, and the remaining cells are rinsed away [49, 50]. *Reproduced by permission of The Royal Society of Chemistry.* (c) When long-term culture is desired chemical treatments may also be added to surfaces to inhibit growth out of the well structure [51]. (d) Holographic optical tweezers have also been used to arrange single cells in arbitrary array patterns. Here red blood cells are

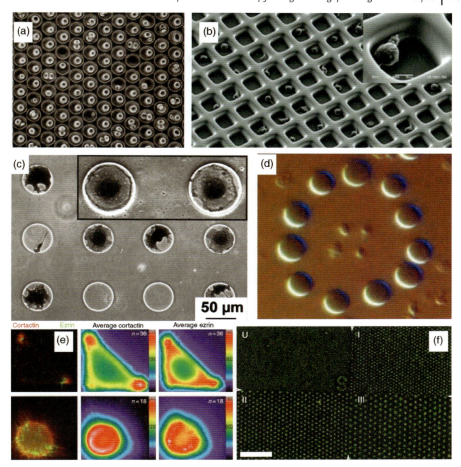

shown in a circular array around platelets. *Image used with permission, copyright 2005 Arryx, Inc., a subsidiary of Haemonetics Corporation.* http://www.arryx.com/PDFdocs/BioRyxApplications.pdf Techniques to reduce heating and cell damage are required before this technique can be more widely adopted. (d) Chemical methods alone have also been utilized to create cell arrays. Control of cell spreading, footprint and morphology will be important for reducing heterogeneity in populations of single cells. Using chemical cell patterning, recent work has demonstrated the effect of cell attachment and morphology on the cell division axis, which appears to be guided by cell morphology and the underlying cytoskeletal organization [52]. *Image reproduced with permission of Nature Publishing Group.* Notably in this work, cell morphology was controlled to such an extent that averages of localization of various proteins could be made for tens of cells at different time points in the cell division process. More general studies on cell behavior will also benefit from employing techniques to normalize cell morphology. Particularly, a reduction in the spread in observed behavior is expected. (f) Dielectrophoretic techniques have also been utilized to create arrays of single cells. Here an image of multicellular arrays of chondrocytes are shown, with cluster size, which encompasses cell–cell and soluble environment, controlled by electrode spacing [63]. *Reproduced with permission of Nature Publishing Group.* The cells were frozen in place by encapsulation in a hydrogel.

shape and environment can be compared. These methods have been used to identify the effect of cell spreading on survival [53] as well as the effect of cell shape and cytoskeletal organization on cell division axis [52]. Demonstrating the uniformity of environment in this technique, the distribution of division axis in tens to hundreds of single cells with uniform extracellular contact environment and attachment were compared. Additionally this method allowed novel analysis where protein spatial location over entire cells could be averaged for tens of single cells (Figure 7.5e). Thus, creation of uniform cell morphologies enabled high-throughput analysis of high-content spatial arrangements of intracellular proteins in adherent cells.

Hybrid chemical techniques to address single cells to known locations have been recently developed where cells are electrically addressed to a location by applying a potential and then relying on an RGD–integrin interaction for binding [56]. Here multiple single cells of different phenotype can be arrayed together in known positions and response to a stimulus under identical experimental conditions can be compared. Along these lines, single cell immobilization based on ssDNA recognition has also been demonstrated. Here cells are bioengineered to contain a known ssDNA coating its surface. Complementary strands are immobilized at known locations in a microfluidic device and cells can be localized to the known locations [57]. Using this technique, although labor-intensive, one could potentially create arbitrary arrangements of single cells, with unprecedented control of cellular environment.

Optical manipulation of single particles and cells may have promise for constructing arrays of single cells for high-throughput experimentation. The challenge has been to create a multitude of potential energy minima to enable trapping and control of large amounts of cells without multiple lasers. Holographic optical trapping has addressed this by creating a three-dimensional field using a single laser and reconfigurable diffractive optical elements in the beam path (Figure 7.5d) [58]. Challenges for this technique include heating the trapped objects. Methods to overcome this include dynamically turning on traps only when cells veer from the defined position, or merging with chemical of mechanical techniques. Some of the promise in single cell analysis using optical manipulation is shown in the study by Wei et al. [59]. Here they investigated contact area dependence of T-lymphocyte signaling upon interaction with antigen presenting cells. Similar studies in well controlled environments with cell–cell signaling could be performed massively in parallel if the promise of optical trapping arrays is fulfilled.

Single cell arrays have also been created by taking advantage of differences in polarizability between cells and the surrounding media with an applied nonuniform (usually AC) electric field. This leads to an electromotive force, analogous to the optical gradient force in optical trapping, on even uncharged particles and is termed dielectrophoresis. By tuning the potential well of the dielectrophoretic trap [60, 61], to encompass the physical size of only a single cell, arrays of single cells and particles can be assembled and individually electronically addressed. An advantage of this technique is cells of particular interest, e.g. with an unusual phenotype, may be isolated and removed for further analysis. This has potential uses for rare cell analysis of outliers in a population. Additionally if cells can be maintained trapped during microfluidic introduction of reagents, fast timescale responses of cells may be investigated. Recently Chiou *et al.* [62] developed a optically controlled electronic substrate that allows on the fly reconfiguration of electrodes for dielectrophoretic movement of particles. They can show isolation of live cells from a population of dead cells using this technique and demonstrate impressive on the fly control using intuitive PowerPoint software as an interface. This technique will have further promise once the challenge of low conductivity solutions required for cell control is overcome. The solutions used are below physiological levels and may lead to aberrant behavior in cell studies. Others have used dielectrophoretic traps along with controlled seeding densities to control cell arrays of chondrocytes (Figure 7.5f) [63]. Here the cell–cell, cell–matrix contact environment was well controlled to study its effects on glycosaminoglycan (GAG) production. It was found that cells clustered in larger groups with more cell–cell and secretion interactions produced lower levels of GAG per cell.

7.4.3
High-Content Separation-Based Techniques

Another class of emerging techniques, labeled chemical cytometry, focuses on collecting a large amount of protein/mRNA content data for single cells, at a given time point. Although these techniques cannot resolve dynamics or localization of cellular processes, there is a high quantity of data on biomolecule levels. In most cases these techniques rely on injection of individual cells into an apparatus that will lyse the cells and perform a separation on the protein or nucleic acids within the lysate [64–66]. Simple 1D separation of proteins have been employed [66] and found to yield different profiles when comparing single cells to a blended population of cells [67]. The authors argue that this is due to longer preparation times with a bulk sample, allowing reactions to occur that were compartmentalized in an intact cell. More recent techniques based on 2D separations [68] provide a more complete high-content protein landscape. Here protein profiles of a native cell can be compared to a cell after exposure to a stimulus.

Implementation of these high-content techniques in microfluidic platforms [42, 69, 70] has been a first step in increasing throughput. McClain *et al.* [70] demonstrated an integrated system for lysis, separation and detection of intracellular enzymatically cleaved fluorescent dyes where $\sim$10 cells/min could be analyzed. For high-throughput operation a significant challenge is adhesion of cellular debris to microchannel walls, which can reduce separation repeatability. The authors successfully addressed this important issue with various surface treatments and additives to the separation buffer, including polyethylene glycol (PEG) and nonionic surfactants. Progress in terms of throughput was recently demonstrated by Wang and Lu [69], who demonstrated a microfluidic system capable of analyzing calcein digestion peaks of up to 80 cells/min. Throughput in these cases where a single separation channel is used is limited by the relative mobility of the various separated species, so future work integrating multiple separation channels may be promising for further increasing throughput for chemical cytometry.

7.5
Example Studies Enabled by Microfluidic Cell Arrays

7.5.1
Pore-Forming Dynamics in Single Cells

A variety of natural toxins act by creating pores in cell membranes. Knowledge of the mechanisms of pore formation is critical in developing molecular therapies to combat infectious disease, permeabilizing cell membranes for the introduction of proteins and genetic material and designing biomimetic toxins to target cancer cells. However for many toxins the method of pore formation is still not well understood. One such case, the bacterial toxin Streptolysin O (SLO), currently has two conflicting models in literature: one hypothesizes that pore formation occurs before the complete assembly of an oligomerized toxin ring, while the other asserts that pore formation takes place only after assembly has occurred [71–73]. We address this question by using a microfluidic single cell-trapping device to analyze the effects of SLO on the membrane permeability dynamics of HeLa cells. These experimental results are compared with a discrete stochastic computational model, providing evidence supporting the hypothesis of simultaneous oligomerization of the toxin and poration of the membrane. Furthermore, using the arrayed single cell technique allows for observation of the variation in toxin response amongst a population of cells and separately is expected to aid in further studies of pore-forming dynamics for other toxins and pore-forming compounds, such as amyloid-β, which is involved in Alzheimer's disease.

7.5.1.1 Microfluidic Single Cell Arrays with Fluorescence Imaging
As shown in Figure 7.6, the method used to investigate the pore-forming mechanism in SLO involves the monitoring of fluorescence intensity loss of dyed cells due to the outward diffusion of calcein through pores formed in the cell membrane by SLO. In general, the calcein-dyed HeLa cells were loaded into a microfluidic cell trapping device (Figure 7.6a) so that the effects of the toxin on individual cells could be observed. Then, a near-step concentration of toxin was introduced into the device while a video recorded cell intensity over time. In order to confirm the timescale of the solution exchange, a fluorescent dye (40 kD fluorescein-dextran) was used in a separate experiment. After toxin introduction the characteristics seen in the resulting intensity loss curves and the stochastic model that is later matched with these results provide some insight into the pore formation mechanism of SLO.

7.5.1.2 Toxin-Induced Permeability
In order to observe the effects of SLO, two concentrations of the toxin were used: 100 U/mL (0.09 µg/mL) and 10 kU/mL (9 µg/mL). Data for both concentrations of SLO experiments that have been corrected for bleaching with DTT are shown in Figure 7.7a, b. Normalized intensity curves are shown for 23 cells with 100 U/mL of SLO and for 28 cells with 10 kU/mL of SLO. The toxin is shown to be introduced at time 0 s for both experiments. All the cells demonstrate a slight increase in

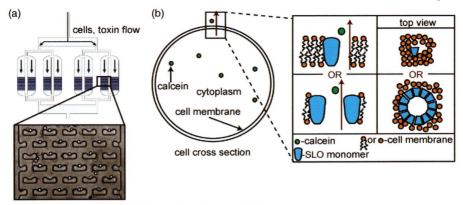

Figure 7.6 Single cell arrays for Streptolysin O quantitative experiments. (a) A schematic and micrograph of the microfluidic device for single cell trapping is shown. (b) The mechanism of intensity measurements and two competing pore formation models are shown.

intensity when the toxin solution is introduced, which is assumed to be an effect from DTT since the control data without toxin showed similar behavior. As expected, it is also evident from the data that the higher concentration of toxin caused a higher rate of decrease in intensity than the lower concentration. More importantly, the experimental data from both monomer concentrations gave similar noteworthy behaviors, in which there was first a delay between toxin introduction and initial membrane permeation and then only smooth transitions in slope while the intensity decreased due to the leakage of calcein (Figure 7.7a, b).

Notice also the large variation in single cell behavior that may be due to a small number of reacting SLO monomers inserting in the membrane, leading to less deterministic behavior. One can calculate the number of molecules at a given time within a region 1000 μm^3 surrounding a cell to be approximately 800 molecules for the lower concentration tested.

7.5.1.3 Stochastic Model of Pore Formation

Next, a stochastic model of the pore formation mechanism of SLO was developed in order to explain more specifically the characteristics seen in the experimental results. The method of pore formation is simulated with this model at each stage, from the attachment of the SLO monomers, to the actual poration of the cell membrane. The simulation is carried out discretely, as the simulated situation evolves at every time step. The time step in each simulation is taken to be one second and the toxin is assumed to be present at time zero seconds. Results for the best fit parameters for this model are shown in Figure 7.7c, d, where because the model is not deterministic, ten simulations with the same parameters yield varying intensity loss curves. As seen in Figure 7.8 the average trends of the best fit simulations also closely resemble those of the actual experimental data at both concentrations.

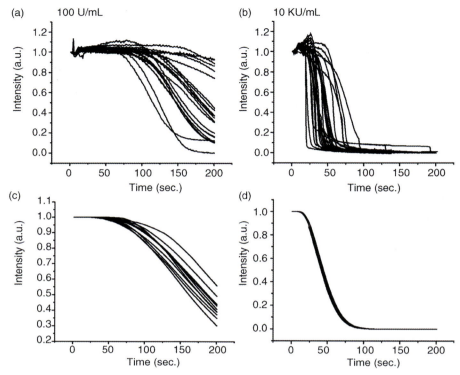

Figure 7.7 Experimental (a, b) and simulated (c, d) results of intensity loss over time. (a) The intensity loss curves for calcein leakage from SLO-porated single cells with 100 U/mL SLO introduced in a stepwise fashion. (b) Intensity curves for 10 kU/mL SLO with a step introduction. For both cases notice a delay time before intensity loss. Each curve represents one cell. (c) Stochastic simulation of intensity loss due to pore-formation with a hybrid model that allows pore insertion before complete assembly, and produces the least error. This simulation is for a concentration of 100 U/mL SLO. (d) Stochastic simulation results are shown for 10 kU/mL SLO. Ten simulations are shown per concentration.

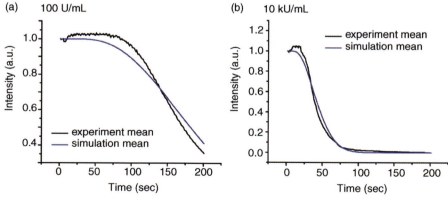

Figure 7.8 Comparison of average trends resulting from experiment and simulation: (a) for a concentration of 100 U/mL of SLO, (b) for a concentration of 10 kU/mL of SLO.

The model spatially represents the cell membrane as a two-dimensional wrapped array, having 2500 entries. Each entry contains fields that indicate whether any toxin monomers or oligomers are attached to that area of the membrane, the number of monomers in the oligomer and whether a transmembrane pore has been formed there. Such specifications allow for the sequential method, simultaneous hypothesis, or something in between to take place.

There are four adjustable parameters involved in the model. Three of them are variables that introduce randomness. This includes the average number of monomers binding to the membrane in the given time step (K_{mem}, the average number of a Poisson distribution), the probability that monomers and oligomers will oligomerize in a time step (K_{bind}), and the probability that monomers and oligomers form a transmembrane pore for a given oligomer size (K_{ins}). Another parameter, D_{mem}, is a measure of particle mobility in the membrane and acts in coordination with K_{bind} to determine which particles have oligomerized. The model assumes that in the short period of a time step only one monomer–monomer, monomer–oligomer, or oligomer–oligomer binding event and one pore-forming event can occur for each particle inserted in the membrane.

7.5.1.4 Amount and Size of Pores for Best Fit Models

After finding the best fit parameters, the simulated data was inspected in more detail to find the average pore sizes and number of pores over time for both concentrations. The mean number of pores was calculated for ten trials for each concentration (Figure 7.10). Initially a linear increase in the number of pores was observed for both concentrations, corresponding to an insertion phase of small oligomers. This was followed by another linear phase with a reduced slope, where most newly attached monomers assemble to pre-inserted pores. After 200 s, the higher concentration ($K_{mem} = 16$) had an average of 64.1 pores in the membrane and the curve was increasing. The lower concentration, however, reached a mean of 13.7 pores, increasing much more slowly than at the higher concentration.

In addition the area of each pore in the modeled membrane was recorded. From this the average radius was found and plotted over time, as shown in Figure 7.9. For the higher concentration the radius of the pores increased rapidly in the beginning and began to stabilize to around 9.3 nm after 183 s. Notice that this is below the radius observed for a completely formed ring (12.5 nm). In contrast the radius for the lower concentration simulation continued to increase and did not saturate within the given 200-s timeframe.

7.5.1.5 Concerning the Pore Formation Mechanism of SLO

Based on the insight gained from the stochastic model in comparison with the experimental results, we support the hypothesis of simultaneous oligomer and pore formation as introduced by Palmer *et al.* [73] and not the sequential model where oligomerization into a complete ring precedes insertion into the membrane [71]. Not only was the experimental data best fit by simulation parameters, which indicated a nonzero probability of pore formation above four monomers, but simulations in

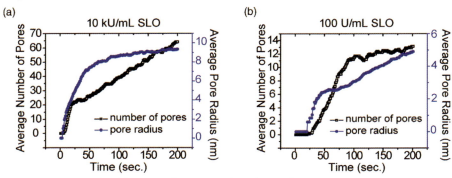

Figure 7.9 Simulated dynamics of pore size and number: (a) for a concentration of 10 kU/mL of SLO, (b) for a concentration of 100 U/mL of SLO.

which 50 monomers were required for insertion yielded incorrect results involving particularly longer delay times than seen in experiments, as well as more punctuated changes in intensity slope. Note that single cell experiments are necessary for this observation but did not yield these punctuated changes in slope. Although the probability of insertion was greater for larger oligomers, incomplete pore complexes still penetrated the membrane, allowing fluorescent dye to diffuse out of the cell. Hence the initial delay times seen in the results most likely correspond to the time that it takes the monomers to diffuse, assemble and oligomerize in the membrane before insertion; however that time is not long enough to allow the formation of a fully oligomerized ($\sim$50 monomer) complex. Furthermore, the smooth transitions in intensity loss shown in the experimental results indicate that the pores formed were not large enough to produce large changes in the leakage of calcein, suggesting that pore sizes increased gradually, consistent with both the insertion of small oligomers (>4 monomers) and the addition of new monomers to small pre-formed pores. Therefore the results from the experimental data, as supported by the best fit probabilities in the stochastic model, suggest that oligomerization and pore formation occur simultaneously.

7.5.1.6 Conclusions on Pore-Forming Dynamics in Single Cells

We have demonstrated that the use of single cell analysis along with stochastic modeling provides insight into the pore formation mechanism of SLO. Without the ability to observe a high density of cells individually, there would be no immediate way to observe the actual response of the cells to the toxin. In other words, only a mean response could be seen and any information-rich variation from cell to cell would have been masked. One example where this was important was in determining that punctuated intensity loss did not occur at the single cell level, supporting the simultaneous insertion and oligomerization model of Palmer and Bhakdi [73]. Also although there are limitations it has been shown that a stochastic model is nonetheless helpful in understanding the physical mechanism of pore formation

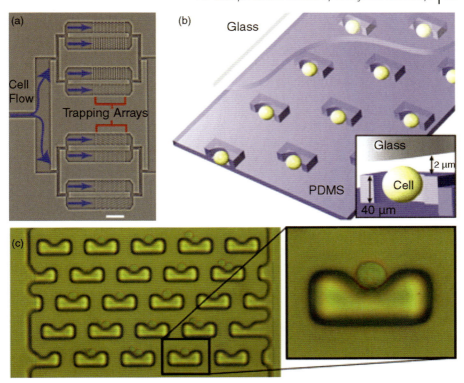

Figure 7.10 Single cell trapping arrays. (a) Photograph of the cell trapping device is shown demonstrating the branching architecture and trapping chambers with arrays of traps. The scale bar is 500 μm. Cell and media flow enters from the left and enters the individual trapping chambers where it is distributed amongst the individual traps. (b) Diagram of the device and mechanism of trapping is presented. Traps are molded in PDMS and bonded to a glass substrate. Trap size biases trapping to predominantly one or two cells. The diagram is flipped from the actual device function for clarity; a functioning device is operated with the glass substrate facing down towards the earth. Inset shows the geometry of an individual trap. The device is not drawn to scale. (c) High-resolution brightfield micrograph of the trapping array with trapped cells is shown. In most cases cells rest at the identical potential minimum of the trap, while in some cases two cells are trapped in an identical manner amongst traps. The magnified insert shows details of the trapped cell. Trapping is a gentle process and no cell deformation is observed for routinely applied pressures.

where the random events occurring for small numbers of reacting monomers provide a certain amount of variation seen from cell to cell.

7.5.2 Single Cell Culture and Analysis

It is important to quantify the distribution of behavior amongst a population of individual cells to reach a more complete quantitative understanding of cellular processes. Improved high-throughput analysis of single cell behavior requires uniform conditions for individual cells with controllable cell–cell interactions, including diffusible and contact elements. Uniform cell arrays for static culture of adherent cells have previously been constructed using protein micropatterning techniques but lack the ability to control diffusible secretions. Here we present a microfluidics-based dynamic single cell culture array that allows both arrayed culture of individual adherent cells and dynamic control of fluid perfusion with uniform environments for individual cells [37]. In our device no surface modification is required and cell loading is completed in less than 30 s. The device consists of arrays of physical U-shaped hydrodynamic trapping structures with geometries that are biased to trap only single cells. HeLa cells were shown to adhere at a similar rate in the trapping array as on a control glass substrate. Additionally rates of cell death and division were comparable to the control experiment. Approximately 100 individual isolated cells were observed growing and adhering in a field of view spanning $\sim 1\,mm^2$, with more than 85% of cells maintained within the primary trapping site after 24 h. Also more than 90% of cells were adherent and only 5% had undergone apoptosis after 24 h of perfusion culture within the trapping array. We anticipate uses in single cell analysis of drug toxicity with physiologically relevant perfused dosages as well as investigation of cell signaling pathways and systems biology.

7.5.2.1 Single Cell Trapping Arrays

Trapping arrays were successfully fabricated and tested. The device consists of branched trapping chambers linked in parallel (Figure 7.10a), while the arrays within the chambers consist of U-shaped PDMS structures that are 40 µm in height and are offset from the substrate by 2 µm (Figure 7.10b, c). Each chamber contained between four and five traps over its width (Figure 7.10c). Also, each row of traps was asymmetrically offset from the previous row (Figure 7.10c). It was qualitatively observed that asymmetric rows of traps were better at filling throughout the chamber when compared to symmetrically offset rows. Several depths of trap were examined for the best isolation of individual cells (10 µm, 15 µm, 30 µm, 60 µm). It was found that traps with a depth of 10 µm most consistently trapped individual HeLa cells (average diameter $\sim$15 µm) [36]. For other cell types with different average diameters, the optimum trap size should vary. Additionally since there is a distribution of cell sizes amongst a population, there may be a bias to trap smaller cells that can more easily occupy the trapping sites.

7.5.2.2 Arrayed Single Cell Culture

We demonstrate culture of ordered arrays of single HeLa cells under constant perfusion of media + 10% FBS. For a flow rate of 25 µL/min time-lapse images were taken every 3 min of a trapped array of HeLa cells on an incubated microscope stage. After 12 h small changes in morphology were observed away from a spherical morphology towards an adherent morphology. Also, cell division was observed in a few cases. After 24 h a majority of cells displayed an adherent morphology. In some cases cells were observed to escape the trapping structures as well. The behavior of several cells in the trapping structure over time is shown for dividing and adhering

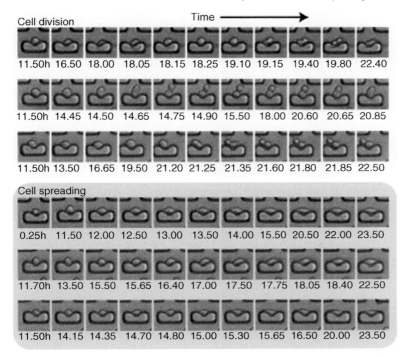

Figure 7.11 Uniform cell behavior. Characteristics of growth for single trapped cells are shown. Frames from a movie of cell growth in the array are shown demonstrating both cell division (first three rows) and morphologies indicative of cell adhesion (rows 4 through 6). Notice the uniformity in morphology observed amongst adherent and amongst dividing cells. The hours after seeding are shown underneath each image. After division daughter cells remained within the trapping region.

cells in Figure 7.11. It should be noted that in most cases after cell division both daughter cells remain isolated in the trapping structure. Another interesting observation is the directionality of adherence in HeLa cells that are trapped. It is observed that a large fraction of growing HeLa cells have a long axis parallel to the long axis of the trapping structure. It also appears that the cells became adherent to the PDMS structure as opposed to the glass substrate in these cases. This may be due to serum containing adhesion-promoting proteins that may adhere to the hydrophobic PDMS surface biasing attachment. Adhesion on the PDMS structures may limit microscopic analysis in some cases, due to diffraction at the interface of the trap. To limit adhesion, future studies could employ treatments with high concentrations of bovine serum albumin (BSA) that will coat the PDMS surface.

Quantitative analysis of the dynamics of cell adhesion, death, division and escape from traps were performed for a 24-h period and are plotted in Figure 7.12a. Here it was observed that 50% of cells displayed adherent morphology after 15 h. After 24 h 6% of cells showed characteristics of apoptosis, while 15% had escaped from the vicinity of the initial trapping site. The high level of maintenance within the trapping

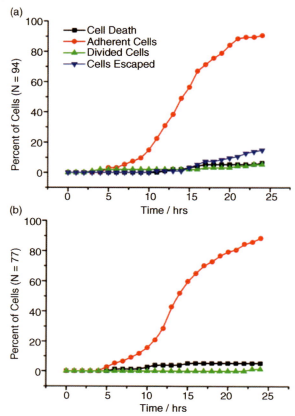

Figure 7.12 Cell behavior in trapping structures and the control substrate. (a) Cell adhesion, division and death are reported every hour for individual cells in the single cell array. (b) The same characteristics are plotted for culture on a control glass slide without perfusion.

structures after 24 h may be due to shear sheltering within the trapping structure. Additionally 5% of cells had undergone cell division after 24 h. These results were compared to cell behavior in a control experiment using the same glass substrate with no traps or perfusion (Figure 7.12b). In this experiment 50% of cells were adherent after a similar 14 h, while 5% of cells were apoptotic after 24 h and only 1% of cells had undergone cell division. The requirement for a cell to be considered adherent was a length $1.3\times$ its width.

7.5.2.3 Conclusions on Arrayed Single Cell Culture

We have demonstrated a microfluidics-based hydrodynamic trapping method for creating arrays of single adherent cells with dynamic control of perfusion possible. HeLa cells were cultured and a high level of maintenance in the original position of trapping was observed after 24 h. Additionally, cell division, adhesion and apoptotic behavior was comparable to static culture on the same substrate, indicating that cells were not stressed above normal culture conditions. After cell division, daughter cells were also observed to be maintained within the original trapping structure. As compared with previous single cell arrays, cell–cell communication by both contact and diffusible elements is a controllable parameter in this device. We anticipate this technique will be useful in single cell studies of metabolism, pharmacokinetics, drug toxicity, shear stress activation and chemical signaling pathway activation and inhibition.

7.6
Conclusions and Future Directions

We have highlighted three important trends in single cell analysis technology. The first is increased control of the fluid environment using microfluidic technology to allow positioning of cells and application of reagents to those cells in a well controlled manner. These methods are allowing parallel analysis of fast timescale changes that occur at the single cell level and are not observable in the bulk, including transient Ca^{2+} signaling, ion channel activity and intracellular transport. A second growing area is improved microscale environmental control and uniformity. These include the soluble environment, as well as cell–cell, cell–substrate and mechanical interactions. Since the cellular microenvironment has an immense impact on cellular behavior, new single cell analysis methods in which these parameters are controlled and uniform will inevitably lead to an improved understanding of the input–output response of individual cells and also how cells interact in higher-order structures. The third is increasing throughput in high-content techniques like chemical cytometry, where single cell variation of several intracellular separated biomolecules can be observed simultaneously.

In most cases these emerging techniques have not yet been adopted by the cell biology community. This is mostly due to two factors that need to be addressed with most techniques. The first is ease of use: most of the techniques are still at the prototype stage where devices require significant training before operation. One important trend and future direction is the development of easy to use systems with robust operation. Arguably it may not be the role of academic scientists to develop prototypes into usable products, but attention to the ability to transfer to a robust device is important. Additionally an increased number of small startup corporations in this sector may optimize previously developed prototypes in the literature and serve the cell biology community. A second stumbling block to adaptation by cell biologists is that the effect of environmental factors on cell behavior, although accepted as important, has not been quantified in most cases. Further systematic experimental biology in which distributions of behavior for cells in controlled environments are compared to distributions observed in current culture environments will be required to further convince wide usage of these emerging techniques. Once accepted, wide usage of these techniques is expected to greatly assist in understanding the dynamics of the cell as an engineering system and applications in cell bioassays molecular diagnostics.

Acknowledgements

The authors would like to thank Liz Wu and Josephine Shaw for their work on presented examples as well as the Whitaker Foundation Graduate Fellowship for funding (D.D.).

References

1 Khademhosseini, A., Langer, R., Borenstein, J. and Vacanti, J.P. (2006) *Proceedings of the National Academy of Sciences of the United States, of America*, **103**, 2480–2487.
2 Moore, K.A. and Lemischka, I.R. (2006) *Science*, **311**, 1880–1885.
3 Hogan, B.L.M. (1999) *Cell*, **96**, 225–233.
4 Gumbiner, B.M. (2005) *Nature Reviews Molecular Cell Biology*, **6**, 622–634.
5 Wei, C.J., Xu, X. and Lo, C.W. (2004) *Annual Review of Cell and Developmental Biology*, **20**, 811–838.
6 Rustom, A., Saffrich, R., Markovic, I., Walther, P. and Gerdes, H.H. (2004) *Science*, **303**, 1007–1010.
7 Helmke, B.P. and Davies, P.F. (2002) *Annals of Biomedical Engineering*, **30**, 284–296.
8 Ingber, D.E. (2003) *Journal of Cell Science*, **116**, 1157–1173.
9 Geiger, B. and Bershadsky, A. (2002) *Cell*, **110**, 139–142.
10 Ahern, C.A. and Horn, R. (2004) *Trends in Neurosciences*, **27**, 303–307.
11 Mycielska, M.E. and Djamgoz, M.B.A. (2004) *Journal of Cell Science*, **117**, 1631–1639.
12 Lee, H. and Caterina, M.J. (2005) *Pflugers Archiv-European Journal of Physiology*, **451**, 160–167.
13 Lucchetta, E.M., Lee, J.H., Fu, L.A., Patel, N.H. and Ismagilov, R.F. (2005) *Nature*, **434**, 1134–1138.
14 Rao, C.V., Wolf, D.M. and Arkin, A.P. (2002) *Nature*, **420**, 231–237.
15 Lidstrom, M.E. and Meldrum, D.R. (2003) *Nature Reviews Microbiology*, **1**, 158–164.
16 Raser, J.M. and O'Shea, E.K. (2005) *Science*, **309**, 2010–2013.
17 Rosenfeld, N., Young, J.W., Alon, U., Swain, P.S. and Elowitz, M.B. (2005) *Science*, **307**, 1962–1965.
18 Mettetal, J.T., Muzzey, D., Pedraza, J.M., Ozbudak, E.M. and van Oudenaarden, A. (2006) *Proceedings of the National Academy of Sciences of the United States of America*, **103**, 7304–7309.
19 Bhalla, U.S. (2004) *Biophysical Journal*, **87**, 733–744.
20 Kholodenko, B.N. (2006) *Nature Reviews Molecular Cell Biology*, **7**, 165–176.
21 Yu, J., Xiao, J., Ren, X.J., Lao, K.Q. and Xie, X.S. (2006) *Science*, **311**, 1600–1603.
22 Cai, L., Friedman, N. and Xie, X.S. (2006) *Nature*, **440**, 358–362.
23 Griffith, L.G. and Swartz, M.A. (2006) *Nature Reviews Molecular Cell Biology*, **7**, 211–224.
24 Yasuda, K.J. (2004) *Journal of Molecular Recognition*, **17**, 186–193.
25 Meier, S.M., Huebner, H. and Buchholz, R. (2005) *Bioprocess and Biosystems Engineering*, **28**, 95–107.
26 Teruel, M.N. and Meyer, T. (2002) *Science*, **295**, 1910–1912.
27 Nolan, J.P. and Sklar, L.A. (1998) *Nature Biotechnology*, **16**, 633–638.
28 Krutzik, P.O. and Nolan, G.P. (2006) *Nature Methods*, **3**, 361–368.
29 Martin, J.C. and Swartzendruber, D.E. (1980) *Science*, **207**, 199–201.
30 Bedner, E., Melamed, M.R. and Darzynkiewicz, Z. (1998) *Cytometry*, **33**, 1–9.
31 Eggert, U.S. and Mitchison, T.J. (2006) *Current Opinion in Chemical Biology*, **10**, 232–237.

32. Perlman, Z.E., Slack, M.D., Feng, Y., Mitchison, T.J., Wu, L.F. and Altschuler, S.J. (2004) *Science*, **306**, 1194–1198.
33. Wheeler, A.R., Throndset, W.R., Whelan, R.J., Leach, A.M., Zare, R.N., Liao, Y.H., Farrell, K., Manger, I.D. and Daridon, A. (2003) *Analytical Chemistry*, **75**, 3581–3586.
34. Li, P.C.H., de Camprieu, L., Cai, J. and Sangar, M. (2004) *Lab on a Chip*, **4**, 174–180.
35. Li, X.J. and Li, P.C.H. (2005) *Analytical Chemistry*, **77**, 4315–4322.
36. Di Carlo, D., Aghdam, N. and Lee, L.P. (2006) *Analytical Chemistry*, **78**, 4925–4930.
37. Di Carlo, D., Wu, L.Y. and Lee, L.P. (2006) *Lab on a Chip*, **6**, 1445–1449.
38. Fu, A.Y., Chou, H.P., Spence, C., Arnold, F.H. and Quake, S.R. (2002) *Analytical Chemistry*, **74**, 2451–2457.
39. Cheung, K., Gawad, S. and Renaud, P. (2005) *Cytometry Part A*, **65**, 124–132.
40. Sohn, L.L., Saleh, O.A., Facer, G.R., Beavis, A.J., Allan, R.S. and Notterman, D.A. (2000) *Proceedings of the National Academy of Sciences of the United States of America*, **97**, 10687–10690.
41. Irimia, D., Tompkins, R.G. and Toner, M. (2004) *Analytical Chemistry*, **76**, 6137–6143.
42. Wu, H.K., Wheeler, A. and Zare, R.N. (2004) *Proceedings of the National Academy of Sciences of the United States of America*, **101**, 12809–12813.
43. Love, J.C., Ronan, J.L., Grotenbreg, G.M., van der Veen, A.G. and Ploegh, H.L. (2006) *Nature Biotechnology*, **24**, 703–707.
44. He, M.Y., Edgar, J.S., Jeffries, G.D.M., Lorenz, R.M., Shelby, J.P. and Chiu, D.T. (2005) *Analytical Chemistry*, **77**, 1539–1544.
45. Ionescu-Zanetti, C., Shaw, R.M., Seo, J.G., Jan, Y.N., Jan, L.Y. and Lee, L.P. (2005) *Proceedings of the National Academy of Sciences of the United States of America*, **102**, 9112–9117.
46. Sabounchi, P., Ionescu-Zanetti, C., Chen, R., Karandikar, M., Seo, J. and Lee, L.P. (2006) *Applied Physics Letters*, **88**, 183901.
47. Lee, P.J., Hung, P.J., Shaw, R., Jan, L. and Lee, L.P. (2005) *Applied Physics Letters*, **86**, 223902.
48. Khine, M., Lau, A., Ionescu-Zanetti, C., Seo, J. and Lee, L.P. (2005) *Lab on a Chip*, **5**, 38–43.
49. Rettig, J.R. and Folch, A. (2005) *Analytical Chemistry*, **77**, 5628–5634.
50. Revzin, A., Sekine, K., Sin, A., Tompkins, R.G. and Toner, M. (2005) *Lab on a Chip*, **5**, 30–37.
51. Ostuni, E., Chen, C.S., Ingber, D.E. and Whitesides, G.M. (2001) *Langmuir*, **17**, 2828–2834.
52. Thery, M., Racine, V., Pepin, A., Piel, M., Chen, Y., Sibarita, J.B. and Bornens, M. (2005) *Nature Cell Biology*, **7**, 947–953.
53. Chen, C.S., Mrksich, M., Huang, S., Whitesides, G.M. and Ingber, D.E. (1997) *Science*, **276**, 1425–1428.
54. Thomas, C.H., Collier, J.H., Sfeir, C.S. and Healy, K.E. (2002) *Proceedings of the National Academy of Sciences of the United States of America*, **99**, 1972–1977.
55. Juncker, D., Schmid, H. and Delamarche, E. (2005) *Nature Materials*, **4**, 622–628.
56. Toriello, N.M., Douglas, E.S. and Mathies, R.A. (2005) *Analytical Chemistry*, **77**, 6935–6941.
57. Chandra, R.A., Douglas, E.S., Mathies, R.A., Bertozzi, C.R. and Francis, M.B. (2006) *Angewandte Chemie International Edition*, **45**, 896–901.
58. Grier, D.G. (2003) *Nature*, **424**, 810–816.
59. Wei, X.B., Tromberg, B.J. and Cahalan, M.D. (1999) *Proceedings of the National Academy of Sciences of the United States of America*, **96**, 8471–8476.
60. Voldman, J., Gray, M.L., Toner, M. and Schmidt, M.A. (2002) *Analytical Chemistry*, **74**, 3984–3990.
61. Taff, B.M. and Voldman, J. (2005) *Analytical Chemistry*, **77**, 7976–7983.
62. Chiou, P.Y., Ohta, A.T. and Wu, M.C. (2005) *Nature*, **436**, 370–372.
63. Albrecht, D.R., Underhill, G.H., Wassermann, T.B., Sah, R.L. and Bhatia, S.N. (2006) *Nature Methods*, **3**, 369–375.

64 Li, H.L. and Yeung, E.S. (2002) *Electrophoresis*, **23**, 3372–3380.
65 Woods, L.A., Roddy, T.P. and Ewing, A.G. (2004) *Electrophoresis*, **25**, 1181–1187.
66 Zhang, Z.R., Krylov, S., Arriaga, E.A., Polakowski, R. and Dovichi, N.J. (2000) *Analytical Chemistry*, **72**, 318–322.
67 Krylov, S.N., Arriaga, E., Zhang, Z.R., Chan, N.W.C., Palcic, M.M. and Dovichi, N.J. (2000) *Journal of Chromatography B*, **741**, 31–35.
68 Hu, S., Michels, D.A., Fazal, M.A., Ratisoontorn, C., Cunningham, M.L. and Dovichi, N.J. (2004) *Analytical Chemistry*, **76**, 4044–4049.
69 Wang, H.-Y. and Lu, C. (2006) *Chemical Communications*, 3528–3530.
70 McClain, M.A., Culbertson, C.T., Jacobson, S.C., Allbritton, N.L., Sims, C.E. and Ramsey, J.M. (2003) *Analytical Chemistry*, **75**, 5646–5655.
71 Heuck, A., Tweten, R.K. and Johnson, A. (2003) *The Journal of Biological Chemistry*, **278**, 31218–31225.
72 Palmer, M., Vulicevic, I., Saweljew, P., Valeva, A., Kehoe, M. and Bhakdi, S. (1998) *Biochemistry*, **37**, 2378–2383.
73 Palmer, M., Harris, R., Freytag, C., Kehoe, M., Tranum-Jensen, J. and Bhakdi, S. (1998) *Embo Journal*, **17**, 1598–1605.

8
Optical Stretcher for Single Cells
Karla Müller, Anatol Fritsch, Tobias Kiessling, Marc Grosserüschkamp, and Josef A. Käs

8.1
Introduction

It is often an essential and vital goal in the fields of biotechnology, medicine and cell biology to effectively assess the composition of heterogeneous populations of cells, often utilizing this information as a means by which to facilitate their division into populations displaying homogeneous properties. These cell samples may be derived from a variety of sources, such as blood, fine needle aspirations and biopsies, from which researchers would like to detect both the existence of certain cell types as well as changes in their properties, such as those brought about by malignant alterations. The principal aims of such studies are the diagnosis of certain diseases, such as cancer or malaria, and the isolation of certain cell types, such as adult mesenchymal stem cells, which can be subsequently used in the study of therapeutic treatments.

There are a number existing techniques for sorting cells; however most of these have serious drawbacks. Current characterization techniques rely on using fluorescent markers for staining either individual cells in suspension or full tissue samples. These molecular markers work by selective binding to specific components of the desired cells, or by staining the entire cell body. It is also possible to bind fluorescent dyes to antibodies against specific cell surface proteins and thus to mark the cells using these tracers. The presence of the desired cell type can then be determined with a suitable microscope. For adhered cells, one common evaluation tool is the laser scanning cytometer (LSC) [1] while, for cell suspensions, a fluorescence activated cell sorter (FACS) is frequently employed [2, 3]. FACS machines have the additional advantage of providing the ability to perform both the evaluation and sorting of samples. Methods for sorting fluorescent cells using microfluidic devices have also been successfully employed [2, 4]. Furthermore, one can mark the desired cells with magnetic nanoparticles bound to selective antibodies, using magnetic fields to subsequently separate and sort the different subpopulations (magnetically activated cell sorter, MACS) [5, 6]. In all of these techniques, one potentially detrimental side effect is the contamination of the investigated cells with foreign molecules, which may render the selected cells

Single Cell Analysis: Technologies and Applications. Edited by Dario Anselmetti
Copyright © 2009 WILEY-VCH Verlag GmbH & Co. KGaA, Weinheim
ISBN: 978-3-527-31864-3

unsuitable for further processing and use. Methods employing noncontaminating density gradient medium cell sorting only allow a coarse separation of different cell types, giving nowhere near the precision necessary for many applications demanding highly homogeneous cell populations [5, 7].

The approach presented here for the selection and detection of certain cell types is stain-free and based on the inherent mechanical properties of different cells, as defined by the cell's cytoskeleton.

The cytoskeleton is an integral part of eukaryotic cells; it serves a variety of important tasks, spanning the entire cell, defining cellular shape, serving as a transport system and acting as a stable yet dynamic scaffold. The main components of the cytoskeleton are actin filaments, microtubules and intermediate filaments. Together with numerous cytoskeletal accessory proteins such as crosslinkers, cappers, nucleators and molecular motors, these form the bulk of the functional cytoskeleton [8].

From the viewpoint of polymer physics, the filaments of the cytoskeleton can be classified into three stiffness categories as defined by their persistence length LP, the length over which the internal correlation of a polymer's orientation under Brownian motion is lost. The bending energy $UBend$ of a polymer chain is directly proportional to the persistence length, therefore allowing this characteristic measure to be used as a clear classification of polymer type based on flexibility [9]. Microtubules are considered rigid rods since their persistence length of approximately 2 mm far exceeds the typical filament length. Actin filaments are semiflexible polymers, having persistence length of approximately 7 µm, which is on the order of the typical filament length. The orientation of intermediate filaments becomes uncorrelated at very short distances of $LP < 1$ µm; thus, they are considered very flexible polymers.

The properties of networks of actin have been extensively studied *in vitro*, both in terms of structure and as elastic properties under different conditions (e.g. varying concentrations, with or without crosslinkers, or in the presence of active motor proteins). The elastic plateau shear modulus G_0 scales in a highly nonlinear fashion with the actin concentration c_A, $G_0 \sim c_A^{2.5}$ [10]. Therefore if a cell changes its functions due to differentiation or to malignant alteration, the corresponding cytoskeletal changes can be sensitively detected by cell elasticity measurements.

Several techniques have been developed in order to measure a cell's biomechanical properties. Using micropipette aspiration, Ward *et al.* showed that a direct correlation exists between an increase in deformability and progression from a nontumorigenic cell line into a tumorigenic, metastatic cell line [11]. In another technique known as the microplate method, a cell is deformed between two planes, while the plate displacement is controlled with a piezo element [12]. The method of magnetic bead rheology displaces a paramagnetic bead located on the cell surface by a magnetic field and the displacement is observed with a microscope [13]. Scanning force microscopy (SFM) can also be used to probe a cell's viscoelastic response. By this method, a cantilever with a spherical shape indents the cell with the force necessary for deformation, giving a measure of the local viscoelastic properties of the cell [14]. SFM elasticity measurements performed on adhered cells in which different cytoskeletal components were selectively depolymerized indicated that F-actin is the main contributor to cell elasticity [15].

While the aforementioned methods for the determination of the elastic properties of cells can serve as an accurate means of measurement, the optical stretcher presented in this chapter allows contact free optical deformation of cells. This provides minimally invasive access to biomechanical properties and sorting without any contaminating or expensive molecular markers. As a result, such a method opens up a wealth of new possibilities in the areas of research, diagnostic medicine and biotechnology.

8.2
Theory, Methods and Experimental Setup

The optical stretcher is composed of an optical trap, a microfluidic chip for cell delivery and a phase-contrast microscope equipped with a CCD camera that enables real-time imaging.

Cell trapping and stretching with laser light is shown in Figure 8.1. First, a cell in suspension is attracted by two counter-propagating divergent laser beams (Figure 8.1a) and then pulled towards and trapped in the center (Figure 8.1b). If the laser power is increased, the cell stretches along the laser axis (Figure 8.1c) [16].

Due to the low opacity of cell biological matter (which allows a laser beam to pass through the cell without being absorbed) phase-contrast microscopy is an excellent choice for monitoring cell deformations. Phase-contrast microscopy uses the differences in refractive index between the cells and their surrounding medium as a basis for imaging nondyed cells.

A CCD camera captures the optically induced whole cell deformation by using self-written edge detection software; and it thus permits the extraction of numerical data and calculation of the cells' mechanical properties.

In contrast to other techniques, such as micropipette aspiration, SFM measurements or magnetic bead rheology that examine local properties, the optical stretcher probes the whole cell elasticity of cells in a well defined spherical shape.

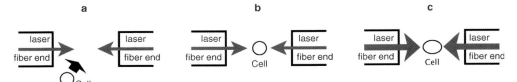

Figure 8.1 Illustration of the optical stretcher. A cell is attracted by two counter-propagating divergent laser beams with Gaussian intensity profile emerging from fiber ends (a) and aligns between the fibers (b). As soon as the power is increased (bold arrows) the cell starts to stretch out along the laser axis (c).

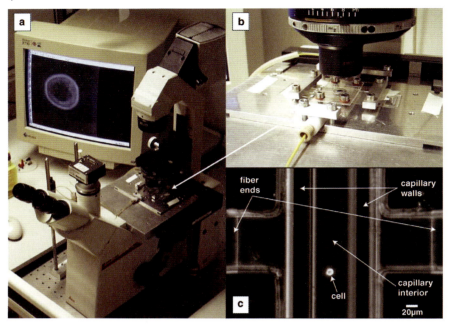

Figure 8.2 (a) Overview of the experimental setup. All parts (laser, camera, fluidics) are computer-controlled. The optical fibers run from the fiber-laser to the microfluidic flow chamber, which is mounted onto an inverted phase-contrast microscope. (b) The microfluidic flow chamber is shown in detail. (c) Phase contrast image of the setup. The optical fibers are arranged perpendicularly to the capillary tube, which transports the cells between the laser beams. The cells are trapped and stretched sequentially. The deformation is recorded by video microscopy and subsequently analyzed.

Combined with a microfluidic chip for cell delivery, the optical stretcher is an ideal tool to measure a statistically relevant amount of cells quickly. Currently a rate of 100 cell measurements per hour can be achieved; however planned advances in design will allow much higher measurement rates in future (Figure 8.2).

8.2.1
Fundamentals of Optical Stretching

Any dielectric and transparent material can be trapped by this method as long as the refractive index of the particle exceeds the refractive index of the surrounding medium, making cells ideal candidates for such trapping. An optical double beam trap is used for the optical stretcher (OS). In contrast to conventional optical traps (optical tweezers) which require a tightly focused laser beam for 3D trapping, the OS performs the trapping and stretching of dielectric materials by using two unfocused laser beams without any additional optics. Therefore this technique is economical, easy to use and reliable at the same time.

Two different regimes for providing a theoretical description can be distinguished, determined by the ratio of the incident light's wavelength λ to the diameter D of the irradiated particle. In the ray optics regime, the particle is very large compared to the wavelength ($D \gg \lambda$), whereas in the Rayleigh regime the opposite is true ($D \ll \lambda$). The calculation of optical forces for particle sizes $D \approx \lambda$ is nontrivial. For a full theoretical description, the proper solution of Maxwell's equations with the appropriate boundary condition is required. Most cells have a typical diameter of 10 µm; the wavelength of the laser light is ~1 µm. Hence it is sufficient to describe the interaction between cell and laser within the limits of the ray optics regime.

8.2.1.1 Ray Optics

In the ray optics regime, the size of the object is much larger than the wavelength of the light and a single beam can be tracked throughout the particle. The incident laser beam can be decomposed into individual rays with appropriate intensity, momentum and direction. These rays propagate in a straight line in uniform, nondispersive media and can be described by geometrical optics.

A light ray propagating along the x direction in a medium with refractive index n_0 hits a sphere with radius r and refractive index n_1 at a height $y_0 < r$, as indicated by the red arrow in Figure 8.3. The angle α_0 between the incident ray and the surface normal is then determined by:

$$\sin\alpha_0 = \frac{y_0}{r} \tag{8.1}$$

The momentum of the light ray is described by the momentum of a single photon:

$$\vec{p}_0 = p_0 \vec{e}_x = \frac{En_0}{c}\vec{e}_x \tag{8.2}$$

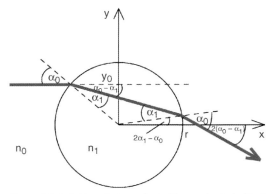

Figure 8.3 A light ray being refracted by a spherical object. Outside the object, the refractive index is n_0, inside it is n_1. The coordinate system is centered at the object's center; the object has a radius r. The incident ray (red) is parallel to the x axis and hits the object at a height y_0 which corresponds to an incident angle α_1. According to Snell's law the ray is refracted at the front and at the back of the object.

where E is the energy of the incoming photon, c is the speed of light in a vacuum, n_0 is the refractive index of the surrounding medium.

According to Snell's Law:

$$n_0 \sin \alpha_0 = n_1 \sin \alpha_1 \tag{8.3}$$

where the angle α_1 between the refracted ray and the surface normal is determined knowing n_1, which is the average refractive index of the cytoplasm. The surrounding medium has a refractive index $n_0 = 1.335$, close to that of water, while the cytoplasm's refractive index is slightly higher (i.e. fibroblasts: $n_1 = 1.370$ [17]). It is absolutely necessary that the refractive index inside the cell is higher than that in the surrounding medium, otherwise the effect would be reversed and cause the trap to be unstable.

The momentum of the light ray inside the cell has an x and a y component:

$$\vec{p}_1 = \frac{En_1}{c}(\cos(\alpha_0-\alpha_1)\vec{e}_x + \sin(\alpha_0-\alpha_1)\vec{e}_y). \tag{8.4}$$

As the ray exits the cell it is refracted once more. Immediately before exiting the cell, the angle between the ray and normal to the cell surface is α_1, while after refraction at the exiting surface the angle between the exiting ray and normal to the surface is α_0 again.

The exiting ray then has a momentum:

$$\vec{p}_2 = \frac{En_0}{c}(\cos(2(\alpha_0-\alpha_1))\vec{e}_x + \sin(2(\alpha_0-\alpha_1))\vec{e}_y) \tag{8.5}$$

Momentum conservation requires that:

$$\vec{p}_0 = \vec{p}_1 + \Delta\vec{p}_1 + \vec{p}_{\text{reflected}} \tag{8.6}$$

as well as:

$$\vec{p}_1 = \vec{p}_2 + \Delta\vec{p}_2 + \vec{p}_{\text{reflected}} \tag{8.7}$$

At the initial refraction, the momentum transferred to the cell boundary (note: the smaller contributions of reflections, in particular multiple reflections, are ignored for simplicity in this explanation of the optical stretching effect, but are necessary for an accurate data analysis) is:

$$\Delta\vec{p}_1 = \vec{p}_0 - \vec{p}_1 = \frac{E}{c}[(n_0 - n_1\cos(\alpha_0-\alpha_1))\vec{e}_x - n_1\sin(\alpha_0-\alpha_1)\vec{e}_y] \tag{8.8}$$

At the second refraction it is:

$$\Delta\vec{p}_2 = \vec{p}_1 - \vec{p}_2 = \frac{E}{c}[(n_1\cos(\alpha_0-\alpha_1) - n_0\cos(2(\alpha_0-\alpha_1)))\vec{e}_x \\ + (n_1\sin(\alpha_0-\alpha_1) - n_0\sin(2(\alpha_0-\alpha_1)))\vec{e}_y] \tag{8.9}$$

Figure 8.4 illustrates the impulse transferred to the cell. The momentums $\Delta\vec{p}_1$ and $\Delta\vec{p}_2$ that are transferred to the cell's boundary are perpendicular to the surface as shown in Figure 8.4(i) and (ii). Overall the momentum vector of the photon changes its direction by an angle of $2(\alpha_0 - \alpha_1)$, as illustrated in Figure 8.4(iii).

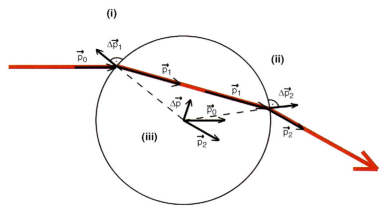

Figure 8.4 Momentum transfers due to refraction. The incident ray carries a momentum p_0, which is changed in direction and size to p_1 by initial refraction (i). Refraction changes again the momentum to p_2; its direction is different from p_0 (ii). The entire momentum change is shown in (iii).

8.2.1.2 Resulting Forces

From the transferred momentums it is possible to construct a force profile by knowing:

$$F = \frac{dp}{dt} = \frac{dE}{dt}\frac{n}{c} = P\frac{n}{c} \tag{8.10}$$

P is the total power of the laser light, n the refractive index, c is the speed of light in a vacuum. The laser light coming from one fiber is a bundle of rays with a radial Gaussian intensity profile $I(\alpha_0)$ given in (8.11) representing the number of photons with energy E per area and time. The radial profile gives the problem a rotational symmetry around the x axis:

$$I(\alpha_0) = \frac{2P}{\pi w^2} \exp\left(-\frac{2(r\sin\alpha_0)^2}{w^2}\right) \tag{8.11}$$

At the end of the fiber the beam's profile has a waist w_0. It increases with distance, as given by (8.12) and shown in Figure 8.5 (dashed lines):

$$w(x) = w_0\sqrt{1 + \left(\frac{\lambda x}{n_0 \pi w_0^2}\right)^2} \tag{8.12}$$

The Gaussian intensity profile causes a gradient force that attracts an incoming cell and draws it along the y–z direction towards the center of the intensity profile. The force in the x direction is called the scattering force which pushes the cell along the x axis. However these two force components stem from just one single physical effect, as discussed.

This process is illustrated in Figure 8.6. If a second fiber is placed opposite to the first fiber and the laser beam has the same Gaussian intensity profile, the total

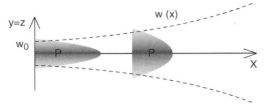

Figure 8.5 Gaussian profile of the laser beam. The laser beam exits the fiber with beam waist w_0. The beam waist increases with distance (dashed lines), the intensity profile (red) flattens in order to maintain the area which corresponds to the total power P. 95% of the total power are distributed within the circle that has the beam waist as its radius.

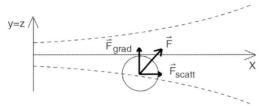

Figure 8.6 Examples for gradient force and scattering force in one laser beam. The cell (circle) is driven by the gradient force towards the place of highest intensity, where the gradient force is balanced. The cell is pushed along the laser axis by the scattering force.

scattering force from both beams is balanced and is zero at the midpoint between the fibers. The cell is, however, attracted by the gradient force and stably trapped in the center between the two fibers.

After a ray of light travels through the particle, its momentum changes in direction and magnitude. This impulse is picked up by the particle. The force due to the directional change of a ray's momentum has components in the forward direction as well as to the side; however there are many rays incident on the particle. The net force has only a forward component due to the rotational symmetry of the system. This symmetry is broken if the particle is not centered exactly between the two laser fibers on the optical axis of the Gaussian beam. In this case the particle feels a restoring force (Figure 8.7).

At closer inspection it becomes obvious that forces are actually applied at discontinuities in refractive indices, that is at the surface. The net force is due to the combination of all surface forces (described by the Maxwell surface tensor). For a rigid object, such as a glass bead, the net force is the only force that matters. For a soft object the forces on the surface become important and lead to a deformation of the object [16].

Figure 8.8 shows a representative time dependent strain curve $\gamma(t) = \frac{\Delta r(t)}{r}$ for a measurement with the optical stretcher. While step stress is applied there is a rapid deformation, followed by a slower increase in strain and finally a plateau

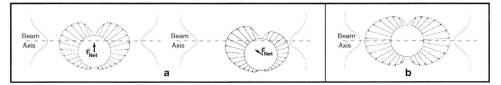

Figure 8.7 Stress maps of optical forces acting on a sphere irradiated by two counter propagating Gaussian laser beams. (a) If the cell is out of the trap's center, the transferred momentum distribution causes a restoring net force. (b) Since the sum over all surface forces (net force) equals zero, the cell is resting at the center of the trap. However, the local optically induced stress stretches the cell. This is the basis for this novel optical tool. Surprisingly the applied radiation pressure leads to an expansion of the cell instead of compression as one would expect.

region is observed. When the stress is reduced the strain relaxes. The observed extension and relaxation behavior is viscoelastic.

8.2.2
Microfluidics – Laminar Flow

With biological variability in mind, there is a need to collect statistics on any population of cells; and an adequate amount of experiments is necessary. Therefore, a technique managing the delivery of cells to the laser beam precisely and

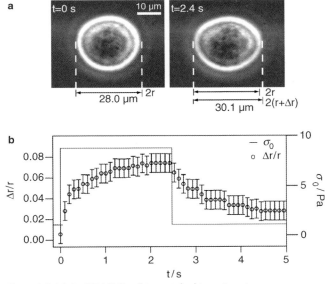

Figure 8.8 (a) An NIH/3T3 cell is stretched in a step stress experiment for 2.5 s. (b) The resulting radial deformation $\Delta r/r$ (i.e. axial strain), reveals a time-dependent, viscoelastic extension and relaxation behavior.

quickly is mandatory. A microfluidic flow chamber fulfills these requirements, taking advantage of the properties of laminar flow on the micron scale.

It is well known that the Navier–Stokes equation (here for incompressible fluids):

$$\rho \left(\frac{\partial}{\partial t} + \vec{u} \cdot \nabla \right) \vec{u} = -\nabla p + \mu \nabla^2 \vec{u} + \vec{F} \quad (8.13)$$

(ρ fluid density, $\vec{u}$ fluid velocity field, p pressure, μ dynamic viscosity) describes various kinds of fluids over a wide range of scales.

As the qualitative behavior of fluids changes with reduction of the system size, microfluidic flows cannot be simply understood as a miniaturized normal "macroflow". Depending on the environmental conditions, the flow can be laminar and time-reversible or chaotic with perturbations. A measure to distinguish these different regimes is the Reynolds number:

$$\text{Re} = \frac{\rho U L}{\mu} \quad (8.14)$$

(U characteristic velocity scale, L characteristic length scale) which can be understood as the ratio between inertial forces and viscous forces. If the Reynolds number is much smaller than 1, inertial forces are negligible and viscous forces dominate. This phenomenon is known as laminar flow. Equation (8.13) then appears in a simpler form:

$$\nabla p = \mu \nabla^2 \vec{u} \quad (8.15)$$

called the Stokes equation. The microfluidic system which is used in the optical stretcher works at a Reynolds number of magnitude 10^{-3}. The flow can thus be considered laminar, which indicates time-reversibility and no mixing except through diffusion due to the lack of turbulence.

These properties allow the accurate transport of cells. As long as the velocity of the cell and its surrounding medium differ, it experiences a drag force, which depends on the difference in flow velocity of the cell and surrounding medium. This drag force appears until cell and local medium velocity are equal. Consequently when the flow is stopped the moving cell experiences a drag force against its moving direction until it stops.

The drag force is very large compared to the cell's momentum (which is expressed by a small Reynolds number); thus cells react instantly to flow velocity changes, enabling the required precession and quickness on a microscale.

8.3
Applications

As the optical stretcher is a relatively young technology (patented 2000; J. Käs and J. Guck), new possibilities and uses steadily arise. To get a feeling for the current applications of this technique, some recent research done with the optical stretcher is presented in the following section.

8.3.1
Cancer Diagnostics

Measuring single cell deformation has been shown to be a sensitive marker to distinguish healthy cells from malignant cells. Malignant cells are deformed consistently more than their healthy counterparts, due to a weakening of the intracellular cytoskeleton. In general there is a tight connection between the cytoskeleton and specific cell function. The cytoskeleton's mechanical properties vary nonlinearly with small changes, such as in the concentration of actin or cross-linker, leading to a sensitive external method to measure internal changes in the cell cytoskeleton [10].

Preliminary studies revealed that the optical stretcher as a technique is capable of discriminating metastatic from non metastatic cells. In one study, cell lines of healthy (MCF-10) and malignant nonmetastatic (MCF-7) breast tissue were cultivated. The cancerous cell line was chemically transformed by 12-O-tetradecanoylphorbol-13-acetate (TPA) into a metastatic cell line (modMCF-7). By analyzing cell deformation due to induced stress, a relationship was found between cell elasticity and cell malignancy. In general, the more malignantly transformed a cell was, the more elastically compliant it was observed to be and the higher was its deformation under stress.

Noncancerous cells (MCF-10) showed the smallest deformations, MCF-7 cells appeared more deformable and modMCF-7 cells had the highest deformations [17].

Figure 8.9 shows that there is a clear separation of the different populations with respect to cell deformability. In order to reach the significance level to separate between the cell populations, only a small number of cells are needed (around 50). The effective measurement capacity of this technique is far higher than the minimal number of cells needed for differentiation of various populations, therefore the optical stretcher is well suited as a precise diagnostic tool, based on objective data.

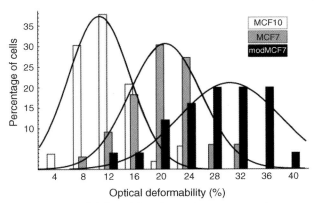

Figure 8.9 Optical deformability of normal, cancerous, and metastatic breast epithelial cells. Three populations of the MCF cell lines are clearly distinguishable in the histograms of the measured optical deformability.

The ability to distinguish malignant from healthy cells has additionally been reproduced in other cell types, for example in mouse fibroblasts (NIH/3T3 or BALB/3T3) and their malignantly transformed cells (SV-T2). The optical deformability of the malignant cells was found to be higher compared to nonmalignant cells, consistent with the earlier presented measurement.

8.3.2
Minimally Invasive Analysis

Since only small numbers of cells are needed for diagnosis, minimally invasive fine needle biopsies and cytobrushes can be used to collect a sufficient tissue sample, avoiding a more harmful and expensive biopsy procedures. In early experiments, healthy and cancerous keratinocytes and their respective deformations were studied (Wottawah Doctoral Dissertation 2006). Keratinocytes are the cells found in the epithelium of human oral mucosa. As observed in other cell types, malignant keratinocytes extend more than healthy cells and are thus more elastically compliant.

In Figure 8.10 distributions of the tensile creep compliance $D(t)$ demonstrate the difference between cells from cancer and healthy patients, where $D(t)$ is the measured relative radial extension divided by the peak stress σ_0 and the geometric factor F_G.

$$D(t) = \frac{\Delta r(t)}{r\sigma_0 F_G}$$

F_G is a geometric correction factor that incorporates the azimuthal distribution of stress and the architecture of the cell being deformed. $D(t)$ can be considered

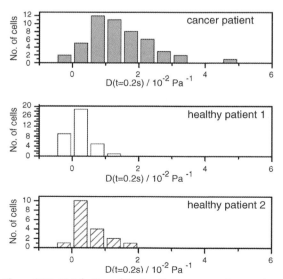

Figure 8.10 Distribution of the tensile creep compliance values (normalized strain) for normal and cancerous cells at $t = 0.2$ s.

as a normalized strain, so that measurements of different peak stress can be compared [18, 19]. The different compliance of normal and cancerous keratinocytes suggests the use of cellular deformability as a quantitative cancer marker.

8.3.3
Stem Cell Characterization

Optical stretching is able to determine not only malignant transformations of a cell, but also normal processes in cells, such as their differentiation. In the course of a cell's progression from a nonspecialized stem cell to a mature fully functioning cell, the appropriate cytoskeleton for the cell role develops. The changes in a cell cytoskeleton are connected to changes in the mechanical properties of the cell, which are measurable with the optical stretcher, as initially demonstrated with leukocytes and their precursors.

Adult mesenchymal stem cells derived from bone marrow are of great interest for therapeutical approaches in tissue regeneration, yet one difficulty is that they are harvested from a heterogeneous population. The ability of the single cells to develop into osteoblasts, chondroblasts, muscle cells, adipocytes and fibroblasts varies within the stem cell population [20]. Differences related to the cytoskeleton can be detected with the optical stretcher and the cells most suitable for differentiation can be selected.

8.4
Outlook

The quick and sensitive determination and separation of cell types renders the optical stretcher potentially useful for a variety of uses in both clinical and research settings. One potential clinical vision for the optical stretcher is the immediate screening for oral cancer in a dentist's office. During the annual dental examination, suspicious and potentially cancerous lesions can be sampled by simple cytobrushes and subsequently analyzed with the optical stretcher. In general, the earlier a cancerous tissue is identified, the better the chances are for successful treatment. The ability of the optical stretcher to sort stem cells without molecular markers by using inherent differences in optical deformability additionally presents an important improvement for stem cell research. The isolated cells are not contaminated, offering a new perspective for cell based regenerative medicine. In combination with a microfluidic system, the preferred stem cells in a heterogeneous population could be sorted out and potentially used to treat diseases such as Parkinson's or diabetes.

The above are only two possible uses for the optical stretcher that have already been initially tested in preclinical studies. In principle, every biological process altering the cytoskeleton and the cell's viscoelastic properties could be investigated and identified. Furthermore the optical stretcher facilitates important fundamental research on cellular components, such as the actin cytoskeleton.

References

1 Tarnok, A. (2002) *Cytometry*, **50** (3), 133–143.
2 Orfao, A. (1996) *Clinical Biochemistry*, **29** (1), 5–9.
3 Tung, J.W. (2004) *Clinical Immunology (Orlando, Fla.)*, **110** (3), 277–283.
4 Sia, S.K. (2003) *Electrophoresis*, **24** (21), 3563–3576.
5 Smits, G. (2000) *Archives of Gynecology and Obstetrics*, **263** (4), 160–163.
6 Miltenyi, S. (1990) *Cytometry*, **11** (2), 231–238.
7 Boyum, A. (1984) *Methods in Enzymology.*, **108**, 88–102.
8 Lodish, H. (2000) *Molecular Cell Biology*, W. H. Freeman and Company, New York.
9 Doi, M. and Edwards, S.F. (1986) *The Theory of Polymer Dynamics*, Clarendon Press.
10 Gardel, M.L. (2004) *Science*, **304**, 1301–1305.
11 Ward, K.A. (1991) *Biorheology*, **28**, 301–313.
12 Thoumine, O. (1997) *Biorheology*, **34**, 309–326.
13 Hoffmann, B. (2006) *Proceedings of the National Academy of Sciences of the United States of America*, **103**, 10259–10264.
14 Mahaffy, R.E. (2000) *Physical Review Letters*, **85**, 880–883.
15 Rotsch, C. (2000) *Proceedings of the National Academy of Sciences of the United States of America*, **96** (3), 921–926.
16 Guck, J. (2000) *Physical Review Letters*, **84**, 5451–5454.
17 Guck, J. (2005) *Biophysical Journal*, **88**, 3689–3698.
18 Wottawah, F. (2005) *Physical Review Letters*, **94**, 098103.
19 Wottawah, F. (2005) *Acta Biomaterialia*, **1** (3), 263–271.
20 Baksh, D. (2004) *Journal of Cellular and Molecular Medicine*, **8** (3), 301–316.

Part III
Single Cell Analysis: Applications

9
Single Cell Immunology
Ulrich Walter and Jan Buer

9.1
Introduction

In the field of immunology, methods to study cells at the single cell level are necessary for several reasons: to monitor gene expression in complex tissues, to define different cell subsets, to get insights into both their functional capacity and their specificity and to track antigen-specific cells. Single cell analysis is also indispensible to investigate cell–cell interactions on the subcellular level.

Substantial progress in the development of hardware, software and especially new reagents such as monoclonal antibodies and fluorochromes has enabled researchers to unravel the fundamental mechanisms of immune defense in unprecedented sophisticated scientific approaches. These techniques include single cell PCR, fluorescence-activated cell sorting, several live cell fluorescence microscopic techniques (confocal laser scanning microscopy, total internal reflection fluorescence microscopy, Förster resonance energy transfer imaging, two-photon laser scanning microscopy), enzyme-linked immunospot assay, *in situ* hybridization and electron microscopy. This chapter aims to briefly introduce these techniques and to give significant examples of how single cell analysis has contributed to our understanding of the immune system.

9.2
Single Cell Gene Expression Profiling

9.2.1
Single Cell (Multiplex) RT-PCR

One of the applications of single cell analysis is to assess gene expression within a complex tissue. In type 1 diabetes (T1D), the insulin-producing β cells are destroyed in a T cell-dependent manner, resulting in lifelong dependency on exogenously administered insulin of affected individuals.

Single Cell Analysis: Technologies and Applications. Edited by Dario Anselmetti
Copyright © 2009 WILEY-VCH Verlag GmbH & Co. KGaA, Weinheim
ISBN: 978-3-527-31864-3

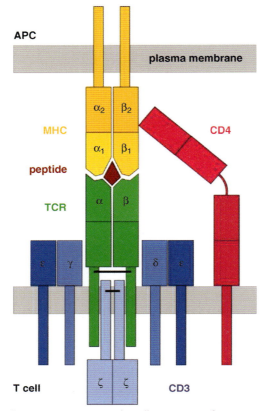

Figure 9.1 MHC-restricted T cell recognition of antigenic peptides.

As T cells are the protagonists of the examples, we have chosen to demonstrate how single cell analysis has contributed to our understanding of the immune system and would like to briefly introduce these major players within the immune system. T cells express the so-called T cell receptor (TCR) in complex with other molecules on their surface (Figure 9.1).

These receptors allow the T cells to recognize antigenic peptides in the binding groove of major histocompatibility (MHC) molecules on the surface of other cells. The vast majority of T cells (those expressing a $\alpha\beta$ T cell receptor) can roughly be subdivided into two classes depending on the expression of either a CD8 or a CD4 co-receptor, restricting the T cell to MHC class I or class II molecules, respectively.

The core of the TCR complex is the TCR, which consists of covalently bound highly variable α and β chains. This heterodimer recognizes a cognate peptide presented in the context of a specific MHC molecule. It is associated with the CD3 complex, which is made up of the invariable CD3 γ, δ, ε and ζ chains. The CD3 complex is crucial for productive antigen recognition, as it mediates signaling upon binding of a cognate pMHC complex. Co-receptors are CD4 or CD8 molecules which bind to invariant parts of MHC class II and class I molecules, respectively, restricting CD4+ T cells to MHC class II and CD8+ T cells to MHC class I. Figure 9.1 shows antigen recognition of a CD4+ T cell. The scenario in the case of a CD8+ T cell is in principle the same, with three differences. Whereas the shown MHC class II heterodimer consists of two transmembrane chains, each of which having two domains, MHC class I molecules consist of a three-domain transmembrane α chain and the noncovalently associated β_2 microglobulin chain, which does not span the membrane. In contrast to the single-chain, four immunoglobulin domain CD4 co-receptor, the CD8 co-receptor consists of two single-domain chains covalently associated by a disulfide bond. Finally, whereas MHC class I molecules bind short peptides of 8–10 amino acids, the length of peptides bound by MHC class II molecules is not constrained.

A prominent duty of MHC class I-restricted CD8+ T cells, once they have differentiated into cytotoxic T lymphocytes (CTL), is to destroy virus-infected cells. As virtually any cell of the body can be infected by viruses, almost all cells express (or can be triggered to) MHC class I on the cell surface, presenting peptides derived from endogenous proteins (including viral proteins in the case of an infection) to CD8+ T cells. In contrast, many CD4+ T cells have rather modulatory/regulatory functions, for example as T helper cells to provide necessary help to B cells to mount an antibody response. Whereas the expression of MHC class I is widespread, the expression of MHC class II, to which CD4+ T cells are restricted, is largely confined (but not strictly limited) to so-called antigen-presenting cells (APC). One task of these cells is to process exogenous, pathogen-derived proteins to present corresponding peptides to CD4+ T cells in the context of MHC class II.

In the reference model of human T1D, the nonobese diabetic (NOD) mouse, progression towards diabetes is dependent on both CD4+ and CD8+ T cells, but the exact contribution of either T cell subset is still under investigation. It was previously generally believed that β cells do not express MHC class II and that CD8+ T cells are responsible for β cell killing. However some models demonstrate that antigen-specific destruction of the β cells occurs in the complete absence of CD8+ T cells. As antigen-specificity is dependent on TCR/MHC interactions, these observations imply that MHC class II molecules are expressed by β cells.

β Cells are located within spherical cell clusters in the pancreas, the so-called islets of Langerhans, together with four additional types of hormone-producing cells. Moreover, progression to overt hyperglycemia is accompanied by infiltration of the islets with different types of immune cells, including MHC class II expressing varieties. Thus, classic methods like staining with antibodies or *in situ* hybridization may not only be too insensitive but also too imprecise to localize the expression of molecules on the surface of β cells in islets that are heavily infiltrated. We therefore chose single cell multiplex reverse transcription (RT)-PCR as the approach of choice. Islets of Langerhans were isolated from mice and dissociated into single cell suspensions from which individual cells were isolated by micromanipulation. After disruption of the cell, the mRNA was reverse transcribed and the resulting cDNA was amplified in two rounds of PCR. In the first round, all the primers for the genes to be monitored were included and the resulting product was split for a second round, in which each individual gene was amplified separately in a nested or semi-nested PCR. Using this protocol, we were able to identify single β cells on the basis of preproinsulin expression. Moreover, we could also determine co-expression of other molecules and were thus able to demonstrate that β cells express MHC class II and that expression is drastically upregulated during progression to overt diabetes [1]. These findings suggest that autoreactive CD4+ T cells, known to be crucial for disease progression, can directly interact with the target β cell. We used this technique also to test β cells for expression of molecules that potentially can cause apoptosis in β cells upon binding of the corresponding ligand [2].

In the above settings, multiplex PCR had to be used to identify a certain cell type derived from a complex tissue. This technique also proved to be useful for studying the co-expression of effector molecules on a per cell basis within a homogenous population of CD8+ memory T cells. Upon an encounter with pathogens, memory T cells are generated enabling the immune system to fight this pathogen faster and more efficiently upon reinfection. To get insights into the molecular mechanisms underlying this enhanced efficiency, we applied single cell multiplex RT-PCR to study gene expression in naïve and memory CD8+ T cells in a TCR transgenic model. We found not only that memory T cells acquired effector functions faster than naïve cells, but also that memory cells simultaneously expressed genes for two to three of the three tested effector molecules. Thus, in contrast to naïve cells, where the expression of these genes was mutually exclusive on a per cell basis, memory T cells were multifunctional [3].

Single cell PCR has also been applied to study TCR rearrangements. As the adaptive immune system is supposed to deal with every potential pathogen, the TCR repertoire has to be highly diverse. One mechanism by which this is achieved is the genomic rearrangement of the variable (V), diversity (D) and joining (J) segments of the β chain and the V and J segments of the α chain. Each of these segments is in general randomly chosen out of a set of up to 80 different variants, resulting in high combinatorial diversity, which is further vastly increased by imprecise joining, by the addition of P- and N-nucleotides and by α/β pairing combinations. This is reflected in a healthy individual by a more or less random selection of the different V, D and J segments to form TCRs. However in the case of an inflammation (e.g. an infection) the few T cells which rearrange a TCR that recognizes an immunogenic pathogen-derived peptide (presented in the context of MHC) with sufficient avidity then undergo clonal expansion, resulting in a substantial overrepresentation of the corresponding rearranged TCR within the whole repertoire. Thus analysis of rearranged TCR genes provides a molecular fingerprint allowing identifying and tracking of disease-associated, *in vivo* expanded T cells.

This is especially true for the hypervariable complementarity-determining region 3 (CDR3) because the CDR3α and CDR3β loops dominate the contact with the peptide/MHC complex (pMHC). Thus, analyzing the TCRs of antigen-specific cells (identified by tetramer-staining, see below) by single cell RT-PCR in combination with sequencing provides a method to closely characterize T cell repertoires [4].

The Doherty group and others used this "spectratyping" approach to analyze in detail the CD8+ T cell response characteristics of influenza-specific CD8+ T cells. Although the CD8+ T cells of C57BL/6 mice proliferate in response to at least six viral epitopes, most studies focused on T cells specific for peptides of nucleoprotein 336–374 (NP336–375) and acid polymerase 224–236 (PA224–236) presented in the context of MHC class I H-2D^b, because CD8+ T cell responses for both epitopes are prominent during primary response and after secondary challenge. Whereas the structurally "flatter" H-2D^b-NP336–375 epitope (recognized by a less diverse TCR repertoire) is shared by most of infected mice ("public" repertoire), TCRs specific for the "protruding" H-2D^b-PA224–236 tend to be more individual, that is they differ among infected mice ("private" repertoire) [5]. Although the more numerous "private" TCRs specific for H-2D^b-PA224–236 [6] are detected earlier than the less prevalent "public" TCRs recognizing H-2D^b-NP336–375, the higher rate of synthesis of NP336–375 as compared to PA224–236 seems to allow for both the emergence of equivalent numbers of memory T cells and the NP336–375 recall response to dominate over the PA224–236 response after challenge. As the size of the response is substantially reduced when the NP336–375 is disrupted in its native configuration but re-expressed by insertion in a less abundant protein, the immunodominance hierarchy seems, in this model, to be a direct consequence of T cell precursor frequency and antigen dose [7]. However although the clonal expansions occurring after primary infection are mirrored in the memory CD8+ T cell pool, clones that dominate the primary response do not necessarily dominate the secondary response as well. Rather, the process of selecting dominating T cell clones in secondary responses seems to be stochastic with minority primary populations often prevailing [6]. Many studies on T cell development/differentiation and "shaping of the T cell repertoire" used single cell PCR [8–14]. Note that some of the mentioned studies performed PCR with genomic DNA as a template [8, 9, 11].

In the above-mentioned studies of CD8+ T cells responses to viral infection, the immunodominant epitopes are known and specific T cells are readily available for further investigation by tetramer staining. However the situation is different in many diseases, like autoimmune disorders, where the target epitopes (or even the whole autoantigenic protein(s)) have yet to be identified. In these instances, like in the cases of multiple sclerosis and polymyositis, spectratyping has been applied to monitor clonal expansion of tissue-infiltrating CD8+ T cells, which is indicative of an antigen-driven response and suggests involvement of the CD8+ T cell subset to the overall progression to the diseases [15–17]. In the latter report, note that laser capture microdissection microscopy was used to analyze single muscle-infiltrating T cells.

9.2.2
Quantitative Single Cell Multiplex RT-PCR

Although single cell RT-PCR has been proven to be very useful to address a variety of questions in the field of immunology, this technique has its limitations. As the amount of mRNA from a single cell is so minute, samples cannot be split and both RT and the first round of PCR have to be performed for the whole sample with all the primers present in the same tube. This can result in nonspecific inhibition of amplification, especially when more than a handful of genes is studied. Moreover, the template switching required for the two-step amplification may introduce potential bias. These effects may result in uncontrollable false-negative results and in many studies, efficiency had not been controlled at all. But single cell multiplex PCR is in general very sensitive, as even genomic DNA (see above) can be successfully amplified from a single cell. Thus it is not clear whether a positive PCR signal is actually translated into a physiologically significant number of molecules synthesized.

Recently Peixoto *et al.* brought single cell RT-PCR to a whole new level, circumventing all these mentioned problems. They described a protocol allowing for the quantification of gene expression for up to 20 genes by maintaining an abundance relationship through all the steps [18]. This was achieved by careful primer design for gene-specific RT and PCR and optimized cycling conditions, which prevented competition between different amplifications, tube to tube variability due to template switching and false positives by nonspecific signaling of SYBR green, enabling them to quantify mRNA over the impressive range of $2.0–1.28 \times 10^9$ molecules.

This study also highlighted the importance of assessing gene expression on a per cell basis rather than at the population level for functional genetic profiles. By analyzing a bulk population of monoclonal CD8+ T cell population *ex vivo* after antigen stimulation by real-time PCR, they found Granzyme B, perforin, TGF-β and IFN-γ were expressed. These results suggested that this CD8+ T cell population differentiates into cells expressing TGF-β and IFN-γ and that these cells are cytotoxic, as they express Granzyme B and perforin, both of which are necessary for the lysis of target cells. Analysis of the very same population on a single cell level, however, revealed a very different scenario. Whereas the vast majority of cells actually expressed TGF-β, only a few scored positive for IFN-γ. Moreover, Granzyme B and perforin were rarely co-expressed, making it unlikely that this population is at all cytolytic. Thus, conventional quantitative assays determining population averages may be highly misleading.

9.3
Fluorescence-Activated Cell Sorting

Fluorescence-activated cell sorting (FACS) is one of the most routinely used and most powerful technologies in immunology. In FACS, individual cells are held in a thin stream of fluid and passed through one or more laser beams, causing light to scatter and fluorescent dyes to emit light. Photomultiplier tubes convert the light emitted to electrical signals. Whereas the forward scatter allows estimation of a cell's size, the side scatter reflects cell complexity/granularity. Invented by the Herzenberg group in 1968 [19], FACS initially allowed the measuring of three parameters: forward and side-scatter characteristics and one fluorescence signal. An additional advantage of this technology is that cells fulfilling given parameters can be sorted, that is separated from negative counterparts. The generation of monoclonal antibodies [20] (substantially boosting sensitivity and selectivity) and the development of two-color and (by the mid-1980s) four-color analysis allowed the multiparametric analysis of blood cells on a per cell basis, to discriminate phenotypically distinct subsets of leukocytes. The number of different surface markers on lymphocytes that can be detected by monoclonal antibodies is continuing to grow. In 1982 a classification was established annotating these markers as "cluster of differentiation" (CD) antigens. The latest entry in the corresponding NIH webpage (http://mpr.nci.nih.gov/prow/) is CD339. Today with the availability of a multitude of different fluorochromes (see below), FACS technology can assess up to 17 fluorescence emissions. The impact of this technology on immunology research cannot be overestimated, as it allows for the identification and characterization of functionally different cell types. For example, T cells can be identified by the expression of the cell surface marker CD3. Within the CD3+ population, CD4+ and CD8+ cells can be distinguished, which have, as mentioned above, very different important functions within the immune system. Additional markers, such as CD25, CD45, CD62L, CD28, CD27 and CCR7, to mention a few, allow us to further divide these T cell populations into naïve, effector and memory populations or regulatory T cells. However we are just at the beginning. It is clear that a given combination of cell surface markers does not, so far, unambiguously identify a functionally homogenous cell population, but rather narrows down the candidate populations in which we may eventually find such a homogenous population. This may be achieved by adding markers to the immunophenotyping scheme.

The CD antigens not only provide a matrix for the classification of phenotypically different cell populations which may then be further investigated for their functions, for example after sorting. As many of these molecules play important roles with regard to virtually any aspect of the immune system, the CD surface markers themselves (given their function has yet been revealed) also provide insights into the efferent or afferent function of a cell population expressing the corresponding marker.

In addition to surface antigens, FACS allows us to also detect intracellular proteins after the staining of permeabilized cells, such as interleukins, which are messenger molecules crucial for orchestrating immune responses, chemokines, main regulators of lymphocyte trafficking and other effector and regulatory molecules [21–24]. It is clear that most aspects of cell function are regulated by intracellular signaling networks involving the phosphorylation or dephosphorylation of intracellular proteins and lipids. With the advent of phospho-specific antibodies that recognize the phosphorylated form of proteins [25], FACS can be applied to study these signaling cascades in primary single cells [26–29]. This likely will be very useful in understanding native-state tissue signaling biology, complex drug actions and dysfunctional signaling in diseased cells. Phospho-specific antibodies have also allowed important insights in TCR signaling within the immunological synapse and microclusters, as discussed below.

A very important advance was the development of what immunologists call "tetramers". As the interaction between TCR and pMHC is in general characterized by low affinity and fast off-rates, pMHC are in general not suited to detect specific T cells. To overcome this problem, multimers (mostly using the streptavidin–biotin interaction) of the pMHC complex have been constructed to increase the avidity of binding of such complexes to the specific TCR [30]. These tetramers allow us to identify T cells that recognize a specific peptide bound by a certain MHC molecule. Thus if the antigenic epitope is known, antigen-specific T cells can be distinguished from their bulk (bystander) counterparts not specific for the antigen in question. This tool thus enables for the close monitoring of antigen-specific responses. In the field of virology, identifying epitopes dominating an inflammatory response was greatly facilitated by the fact that the viral genome encodes only for a small set of proteins. Therefore virologists were among the first to fruitfully adopt this new technology. For most autoimmune diseases, in contrast, the quest for the autoantigen(s) involved in disease progression is in most instances still under way. Recently however, tetramer technology for the first time enabled a detailed study of the CD8+ T cell population targeting the immunodominant CD8+ T cell autoantigen in the NOD mouse model of T1D. In this mouse model, which closely mirrors the spontaneous human disease, a large fraction of islet-associated CD8+ T cells uses highly homogenous TCRα chains (Vα17-Jα42); and that population is a significant component of the earliest islet CD8+ infiltrates (see Ref. [31] for references). Moreover, they are unusually frequent in the periphery [32]. Their contribution to the diabetogenic process is further underscored by the fact that the disease can be transferred to healthy mice by injection of these cells [33]. Although the natural ligand of these T cells was not initially known, the use of tetramers consisting of a mimotope and the H-2K^d MHC class I molecule has allowed us to establish that the progression of insulitis to overt diabetes is invariably accompanied by the cyclic expansion/contraction of this T cell pool. This provides, in the mouse model, a tool to predict the disease by simple quantification of autoreactive T cells in peripheral blood [34]. Meanwhile it has been shown that these CD8+ T cells target a peptide of islet-specific glucose-6-phosphatase catalytic subunit-related protein (IGRP) [32]. This population of $IGRP_{206-214}$/H-2K^d-reactive CD8+ T cells undergoes avidity maturation with diabetes progression [35]. Although mechanisms of central and peripheral tolerance selectively limit the contribution of these high-avidity T cells at the earliest stages of the disease, they do not abrogate their ability to progressively accumulate in inflamed islets and eventually kill the β cells [31].

In addition to antibodies specific for surface or intracellular molecules, an array of dyes is available to monitor characteristics such as DNA and RNA content, proliferation, cell membrane changes and redox state (see Ref. [36] for references). Although a multitude of organic fluorophores and chemically synthesized fluorescent dyes has been generated, their simultaneous use is often limited due to overlapping emission spectra and the need for different excitation sources. To date, the practical limit of approaches relying on organic fluorophores has been 12-color flow cytometry. The recent engineering of quantum dot semiconductor nanocrystals may help to overcome these limitations. Quantum dots are inorganic crystals of cadmium selenide, coated in a zinc sulfide shell [37, 38]. Besides the fact that they are quite photostable, the main advantage of quantum dots is that they have a relatively narrow emission spectra, reducing the need for adjustment of spectral overlap. Chattopadhyay *et al.* recently used these quantum dots to extend the capabilities of polychromatic flow cytometry to resolve 17 fluorescence emissions. They applied this technique to study simultaneously the T cell populations specific for four distinct viral epitopes from a single individual and, on a single cell basis, revealed variations within complex phenotypic patterns that would otherwise remain obscure [39].

9.4
Live Cell Fluorescence Microscopy

The ability to visualize, track and quantify molecules and events in living cells with sufficiently high spatial and temporal resolution is of highest importance for our understanding of dynamic biological systems. Although bright field imaging of living cells with transmitted light has been technically improved substantially (differential interference contrast, reflection contrast microscopy, digitally recorded interference microscopy with automatic phase shifting), we would like to focus on the exciting field of live cell fluorescence microscopy. Clearly, with the development of fluorescent protein technology, light microscopy has been revolutionized. Progress in this field has been triggered by the discovery of green fluorescent protein (GFP) from the jellyfish *Aequorea victoria*, found to fluoresce under excitation without the need of substrates or coenzymes [40]. Mutagenesis studies have yielded variants with improved characteristics (folding kinetics, expression properties, photostability) and different absorbance and emission spectra, including enhanced GFP (eGFP). In addition, other fluorescent proteins have been engineered, like the spectral variants of the unrelated red fluorescent protein (see Ref. [41] for references). These fluorescent proteins can be fused to virtually any protein of interest and used in combination with a multitude of highly sophisticated techniques as minimally invasive markers to track and quantify individual or multiple proteins, to monitor protein–protein interactions, to study biological events and signals and as photo-modulatable proteins to follow the fate of protein populations within a cell.

One of the fields within immunology where live cell fluorescence microscopy has allowed major discoveries is the study of the immunological synapse (IS). The IS is a highly regulated spatio-temporal interface consisting of a specialized large-scale molecular segregation of surface receptors and signaling components between T cells and APC for recognition and activation of T cells.

9.4.1
Confocal Laser Scanning Microscopy

Compared to conventional microscopy, images obtained by confocal imaging are less blurred due to two technical characteristics: first, a single point of excitation light

Figure 9.2 Molecular markers of the different regions of the IS.

(or sometimes a group of points or a slit) is scanned across the specimen; second, a detector aperture obstructs the light that is not coming from the focal point. Thus the focus of excitation and detection are confocal, reducing the background signal from surrounding areas and allowing for high-resolution 3D images. However, the resolution of conventional widefield microscopy has been substantially improved by subsequent deconvolution of the data series, enabling a mathematical reassignment of the out of focus light back to its point of source.

Using confocal microscopy, it was found that there is a spatial segregation of several proteins at the contact interface between T cells and APC [42] (Figure 9.2). While TCR and protein kinase accumulate in the central "bull's eye" region (the central supramolecular activation cluster, cSMAC), the surrounding ring structure (the peripheral (p)SMAC) is rich in leukocyte function-associated antigen 1 (LFA-1) and talin. Outward to the edge of the IS is the distal (d)SMAC which is enriched in CD45 [43]. Meanwhile, an increasing number of molecules have been reported to accumulate in the IS, including CD28 and cytotoxic T lymphocyte-associated antigen 4 (CTLA-4) [44] which are crucial for IS formation.

The IS at the contact zone between T cells and APC can be subdivided into three functionally different regions according to the distribution of certain molecules. The outer dSMAC is enriched in the tyrosine phosphatase CD45 which is important for TCR signaling. It is followed towards the center by the pSMAC which displays the integrins LFA-1 and VLA-4, adhesion molecules which are important for T cell activation, the cytoskeletal integrin linker talin and the transferrin-receptor (Tf-R). The central SMAC contains TCR complexes, the co-stimulatory molecule CD28 and protein kinase PKCθ, among others. The cSMAC seems to be important for endocytic (e.g. TCR degradation) and exocytic processes (e.g. directed secretion of cytokines).

Initially it was believed that the IS formed as a stable structure and that TCR signaling was both initiated and sustained by the cSMAC. This perception has evolved into a more dynamic concept. The cSMAC has been demonstrated to form by the convergence of TCR microclusters (MC) emerging in the periphery of the developing IS [45]. Moreover TCR signaling has been shown not only to occur before the cSMAC forms [45, 46] (note that the latter report uses the above-mentioned phospho-specific antibodies) but also to be dispensable for T cell activation [47] and even CD8+ T cell killing [48]. Two reports applying total internal reflection fluorescence microscopy (TIRFM) for single cell analysis highlight the role of MC as the site for the induction of initial activation signals upon TCR engagement.

9.4.2
Total Internal Reflection Fluorescence Microscopy

TIRFM provides a means to image processes within very close proximity to a glass coverslip, which enables a selective visualization of surface regions of adhered cells. It uses evanescent waves (which are generated only when the incident light is totally reflected at the glass–water interface) to selectively illuminate and excite fluorophores in a restricted region of the specimen immediately adjacent to the glass–water interface. As the evanescent electromagnetic field decays exponentially from the interface, it penetrates to a depth of only approximately 100 nm into the sample, allowing for high-resolution imaging.

This technique has been employed to analyze the role of MCs in IS formation and T cell activation in more detail [49, 50]. Whereas early TCR microclusters contain a number of TCRs (up to ~150) sufficient for detection by wide-field and confocal microscopy, TCRs in microclusters generated after cSMAC formation become scarce (<20) for detection by these techniques. Using TIRFM, the Dustin group has shown that TCR MC form continuously in the IS periphery and move towards the cSMAC. Whereas in the periphery, these MC concentrate activated kinases and adapter proteins emblematic of early TCR signaling, as demonstrated using phospho-specific antibodies, the majority of kinases and adaptors dissociate from the MC before translocation to the cSMAC. Moreover confocal microscopy was employed to demonstrate that while the elimination of MCs is correlated with the loss of Ca^{2+} signaling, loss of the cSMAC does not [51]. Taken together these findings suggest that MCs, rather than cSMAC, are the site for the induction of T cell activation and impose the re-evaluation of the function of cSMAC. Confocal microscopy has been applied to address this issue. Although initial T cell activation signals are induced within a few minutes, continuous stimulation over several hours is required for the final induction of T cell activation, such as cytokine secretion and proliferation [52]. The framework of the IS including cSMAC may be essential for maintaining a prolonged T cell/APC interaction and thus late signal transduction. cSMAC also seem to be the site of active endocytosis of TCRs and their reorganization through intracellular trafficking [53], a mechanism probably important for balancing T cell activation. cSMACs may also be important for lineage commitment by regulating direction of cytokine secretion and by sequestering of its receptors [54].

T cells are very sensitive in recognizing cognate pMHC complexes. CD4+ as well as CD8+ T cells can respond to even a single agonist pMHC ligand, while IS formation requires about ten agonists [48, 55]. Interestingly when agonistic peptides are present at low frequency, endogenous peptide–MHC complexes contribute to T cell activation in the form of heterodimers of agonist peptide–MHC and endogenous peptide–MHC complexes, stabilized by CD4 [55, 56]. Similar results were obtained for CD8+ T cells using confocal microscopy and Förster resonance energy transfer (FRET) analysis.

9.4.3
Förster Resonance Energy Transfer Imaging

One of the few techniques that is able to detect molecular interactions at the subcellular level relies on fluorescence energy transfer from one fluorophore (the donor) to an acceptor. As this resonant coupling is a very short range effect (within a few nanometers), observation of energy transfer suggests actual physical interaction between donor and acceptor molecules. FRET can be measured by different fluorescence microscopic methods (see Ref. [57] for review), for example by quantitation of donor fluorescence recovery after acceptor photobleaching or fluorescence lifetime imaging microscopy (FLIM). The Gascoigne group was the first to apply this technique to visualize molecular interactions in live immune synapses [58]. By applying FRET analysis, they showed that nonstimulatory peptides contribute to antigen-induced CD8/TCR interaction at the IS [59]. Other studies measuring FRET gave further information as to how molecular interactions are involved in T cell activation. Some examples include insights into how the CD4 monomer/dimer ratio tunes the activation threshold during initial engagement [60], how antigen recognition is translated into T cell responses by differential recruitment of CD8 to the TCR [61] and on implications of the oligomeric state of B7-1 and -2 in signaling [62].

9.4.4
Two-Photon Laser Scanning Microscopy

In two-photon laser scanning microscopy (TPLSM), a chromophore is excited not by a single photon but by two photons being absorbed within a femtosecond timescale, enabling the use of longer wavelength excitation penetrating deeper into samples. In contrast to many organ systems, the migration of immune cells and their dynamic encounters with their surroundings are integral to both the development and the function of the immune system. Naïve T cells are activated/primed by interactions with APC, usually dendritic cells (DCs), presenting a cognate peptide in the context of MHC and providing costimulation. As only very few naïve T cells can recognize any given epitope (the frequency is in the order of magnitude of 10^{-6} to 10^{-7}), mechanisms have to be in place to maximize the number of T cell/DC encounters. This is achieved within secondary lymphoid organs (SLO), especially the lymph nodes. Circulating T (as well as B) cells are continuously recruited to SLO and in the past 5 years TPLSM has allowed us to study single cell dynamics within these structures. Having entered a SLO, T cells move in an amoeboid fashion along a random path [63]. Thereby, the stromal network has been shown to be important in directing lymphocytes that are entering the lymph node to and within the appropriate zones [64]. T cell priming upon interaction with DCs presenting the cognate pMHC seems then to occur in distinct successive phases. Von Andrian et al. [65] found that, during a first phase, CD8+ T cells only interacted shortly with DCs but, although these interactions resembled in frequency and duration the brief random collisions that occur even in the absence of antigen, T cells still clearly got activated. Phase two was characterized by long-lasting (≥ 1 h) T cell/DC interactions, accompanied by full activation of the T cells. Finally, the T cells left their DC partners, proliferated and displayed high motility again. Such a multistage priming of CD8+ as well as CD4+ T cells (although with some slight differences) has also been reported by other groups [66, 67]. TPLSM has also been successfully employed to study single T cell trafficking in other contexts, for example within tumors (e.g. Refs. [68, 69]), or the thymus, the organ where T cells develop (see Refs. [70, 71] for review).

Besides the above-mentioned microscopic techniques employing fluorescent markers, "conventional" intravital microscopy in sufficiently translucent tissues has been applied to study the migration of single leukocytes in blood vessels and tissues in live animals for more than a century and is still a powerful tool to get important insights to leukocyte trafficking (see Ref. [72] for review).

9.5
Other Techniques for Single Cell Analysis

9.5.1
Enzyme-Linked Immunospot Assay

The enzyme-linked immunospot (ELISPOT) assay is a frequently used tool for detecting and analyzing individual cells that secrete a particular protein in vitro. Originally developed for analyzing specific antibody-secreting cells, the assay has meanwhile been adapted for measuring the frequencies of cells that secrete a variety of other molecules, especially cytokines. In comparison to intracellular staining and FACS analysis, the ELISPOT technique has the advantage of higher sensitivity, allowing for the detection of very low frequencies of cells ($>10^{-5}$) that would be below the background threshold in FACS. The principle of this technique is that the plates in which a known number of cells are seeded are coated with a capture antibody that immobilizes the protein in question secreted by the cells during the incubation period. This sequestered protein can then be visualized using a detection antibody that is coupled to an enzyme which catalyzes the conversion of a substrate into a colored product. Thus, secreting cells appear as a "spot" on the membrane and can be counted in an automated manner using an ELISPOT reader.

As CD8+ T cells secrete IFN-γ upon activation, IFN-γ ELISPOT assays have been used to detect pMHC-specific responses of human cells ex vivo. By incubating these cells *in vitro* with APC loaded with candidate peptides, immunogenic peptides can be identified in diseases like autoimmunity [73], cancer [74] and infection [75]. Moreover, assessing the frequency of CD8+ T cells recognizing these disease-associated immunogenic peptides in humans at risk may become an important tool to predicting the likelihood of an individual to develop a disease [76] and to evaluate efficiency of a treatment [77].

9.5.2
In Situ Hybridization

The principle behind in situ hybridization (ISH) is the specific annealing of a labeled nucleic acid probe to complementary DNA or RNA sequences. This technique can be used to locate DNA sequences on chromosomes, to detect RNA or viral DNA/RNA. Today, colorimetric or radioactive ISH has been largely replaced by fluorescence ISH (FISH) in which the probes either contain a fluorescent molecule or an antigenic site that can be recognized with fluorescent antibodies. Although this technique is mainly applied to the analysis of fixed tissue sections, it has also been used for single cell analysis. In addition, FISH can be done in conjunction with confocal microscopy to allow for single cell analysis in three dimensions (3D FISH).

As cells differentiate, their genomes are modified at an epigenetic level. T cells differentiate from bone marrow-derived hematopoietic precursors into CD8+ or CD4+ single positive cells in the thymus. In early stages of differentiation, T cells express neither CD4 nor CD8. In the cortex of the thymus, these double negative cells then begin to form a TCR and to express both CD4 and CD8. At that stage, T cells undergo positive selection, and depending on whether they bind to pMHC I or II they stop expressing CD4 or CD8 and differentiate in CD8 and CD4 single positive T cells, respectively. Using 3D FISH, it has been demonstrated that the stable silencing of CD4 or CD8 co-receptor loci, and thus the lineage commitment, is anticipated by the repositioning of co-receptor alleles to repressive centromeric heterochromatin domains [78].

As mentioned above, expression of a functional TCR is preceded by the rearrangement of V, D and J segments. Thereby, successful rearrangement of one locus suppresses the further rearrangement of the other locus. This allelic exclusion ensures the productive rearrangement of only one of the two respective alleles, which leads to the expression of a single receptor with unique antigen specificity on T cells. The same is true for the receptor of B cells (BCR). Studies on B cells using FISH have shown that one important mechanism for allowing the rearrangement of distal V_H genes and proximal DJ_H segments at the Igh locus is the looping of intermediate domains [79, 80]. This contraction is reversed upon successful Igh recombination at one locus. This "decontraction" physically separates the distal V_H genes from the proximal Igh domain in all subsequent developmental stages, and thus prevents further rearrangement of the second DJ_H-rearranged Igh allele in pre-B cells [79]. Moreover, pre-BCR signaling induces the recruitment of the second DJ_H-rearranged Igh allele to repressive pericentromeric chromatin [79]. Thus, both "decontraction" and repositioning seem to be two mechanisms important in establishing allelic exclusion. Recent evidence suggests that that these mechanisms also contribute to initiation and maintenance of allelic exclusion at the TCRb locus [81].

9.5.3
Electron Microscopy

Electron microscopy is based on the same principles as light microscopy but uses electrons instead of light as electromagnetic radiation to "illuminate" a specimen.

However, as the resolution of a microscope is limited by the wavelength of the radiation used, electron microscopy has a much higher magnification and resolution power. Developed by Max Knoll and Ernst Ruska in 1931, transmission electron microscopy (TEM) has since allowed insights in subcellular structures with unprecedented resolution. In addition, antibodies linked to golden nanoparticles can be used to localize proteins on a subcellular level.

As mentioned above, perforin and granzymes are important effector molecules allowing cytotoxic CD8+ T cells to kill target cells such as virus-infected or tumor cells. Electron microscopy has substantially contributed to our understanding how the "lethal hit" is delivered. The immunogold method has been employed to show that perforin and granzymes are the main cytotoxic components of the electron-dense core of specialized secretory organelles, the cytotoxic granules [82]. After antigen recognition of the target, CTL rapidly polarize their cytotoxic granules toward target cell contact and release their contents at a specialized secretory domain of the IS, which lies next to the cSMAC [83]. Thereby, the cytotoxic granules are transported along microtubules and cluster around the microtubule-organizing center (MTOC) which moves to and contacts the plasma membrane at the cSMAC [84].

9.6
Conclusions and Outlook

The above-mentioned examples show how much scientific progress in the field of immunology is dependent on technological advances in methodology, instrumentation, bioinformatics and the availability of reagents. They also demonstrate how much single cell analysis has contributed in recent years to our understanding of how the immune system works.

We believe that one trend in the near future will be to bring multiplexing to a whole new level, for example by applying genome-wide gene expression analysis at the single cell level. We also envision that new micro- and nanotechnological tools will allow us to apply single immune cells as experimental platforms, that is, as a "laboratory in a cell" that is interfaced with the outside world. We predict that single cell analysis will become an increasingly important part of immunologic research that will enable us to unravel fundamental principles underlying the immune system and thus eventually to find a cure for diseases like cancer, allergy, autoimmune and infectious diseases and to make substantial progress in transplantation medicine.

References

1 Walter, U., Toepfer, T., Dittmar, K.E., Kretschmer, K., Lauber, J., Weiss, S., Servos, G., Lechner, O., Scherbaum, W.A., Bornstein, S.R., Von Boehmer, H. and Buer, J. (2003) Pancreatic NOD beta cells express MHC class II protein and the frequency of I-A(g7) mRNA-expressing beta cells strongly increases during progression to autoimmune diabetes. *Diabetologia*, **46**, 1106–1114.

2 Walter, U., Franzke, A., Sarukhan, A., Zober, C., von Boehmer, H., Buer, J.

and Lechner, O. (2000) Monitoring gene expression of TNFR family members by beta-cells during development of autoimmune diabetes. *European Journal of Immunology*, **30**, 1224–1232.

3 Veiga-Fernandes, H., Walter, U., Bourgeois, C., McLean, A. and Rocha, B. (2000) Response of naive and memory CD8+ T cells to antigen stimulation in vivo. *Nature Immunology*, **1**, 47–53.

4 Maryanski, J.L., Jongeneel, C.V., Bucher, P., Casanova, J.L. and Walker, P.R. (1996) Single-cell PCR analysis of TCR repertoires selected by antigen in vivo: a high magnitude CD8 response is comprised of very few clones. *Immunity*, **4**, 47–55.

5 Turner, S.J., Kedzierska, K., Komodromou, H., La Gruta, N.L., Dunstone, M.A., Webb, A.I., Webby, R., Walden, H., Xie, W., McCluskey, J., Purcell, A.W., Rossjohn, J. and Doherty, P.C. (2005) Lack of prominent peptide-major histocompatibility complex features limits repertoire diversity in virus-specific CD8+ T cell populations. *Nature Immunology*, **6**, 382–389.

6 Turner, S.J., Diaz, G., Cross, R. and Doherty, P.C. (2003) Analysis of clonotype distribution and persistence for an influenza virus-specific CD8+ T cell response. *Immunity*, **18**, 549–559.

7 La Gruta, N.L., Kedzierska, K., Pang, K., Webby, R., Davenport, M., Chen, W., Turner, S.J. and Doherty, P.C. (2006) A virus-specific CD8+ T cell immunodominance hierarchy determined by antigen dose and precursor frequencies. *Proceedings of the National Academy of Sciences of the United States of America*, **103**, 994–999.

8 Aifantis, I., Azogui, O., Feinberg, J., Saint-Ruf, C., Buer, J. and von Boehmer, H. (1998) On the role of the pre-T cell receptor in alphabeta versus gammadelta T lineage commitment. *Immunity*, **9**, 649–655.

9 Aifantis, I., Buer, J., von Boehmer, H. and Azogui, O. (1997) Essential role of the pre-T cell receptor in allelic exclusion of the T cell receptor beta locus. *Immunity*, **7**, 601–607.

10 Lambolez, F., Azogui, O., Joret, A.M., Garcia, C., von Boehmer, H., Di Santo, J., Ezine, S. and Rocha, B. (2002) Characterization of T cell differentiation in the murine gut. *The Journal of Experimental Medicine*, **195**, 437–449.

11 Klein, F., Feldhahn, N., Lee, S., Wang, H., Ciuffi, F., von Elstermann, M., Toribio, M.L., Sauer, H., Wartenberg, M., Barath, V.S., Kronke, M., Wernet, P., Rowley, J.D. and Muschen, M. (2003) T lymphoid differentiation in human bone marrow. *Proceedings of the National Academy of Sciences of the United States of America*, **100**, 6747–6752.

12 Attuil, V., Bucher, P., Rossi, M., Mutin, M. and Maryanski, J.L. (2000) Comparative T cell receptor repertoire selection by antigen after adoptive transfer: a glimpse at an antigen-specific preimmune repertoire. *Proceedings of the National Academy of Sciences of the United States of America*, **97**, 8473–8478.

13 Correia-Neves, M., Waltzinger, C., Mathis, D. and Benoist, C. (2001) The shaping of the T cell repertoire. *Immunity*, **14**, 21–32.

14 Hamrouni, A., Aublin, A., Guillaume, P. and Maryanski, J.L. (2003) T cell receptor gene rearrangement lineage analysis reveals clues for the origin of highly restricted antigen-specific repertoires. *The Journal of Experimental Medicine*, **197**, 601–614.

15 Babbe, H., Roers, A., Waisman, A., Lassmann, H., Goebels, N., Hohlfeld, R., Friese, M., Schroder, R., Deckert, M., Schmidt, S., Ravid, R. and Rajewsky, K. (2000) Clonal expansions of CD8(+) T cells dominate the T cell infiltrate in active multiple sclerosis lesions as shown by micromanipulation and single cell polymerase chain reaction. *The Journal of Experimental Medicine*, **192**, 393–404.

16 Skulina, C., Schmidt, S., Dornmair, K., Babbe, H., Roers, A., Rajewsky, K., Wekerle, H., Hohlfeld, R. and Goebels, N.

(2004) Multiple sclerosis: brain-infiltrating CD8+ T cells persist as clonal expansions in the cerebrospinal fluid and blood. *Proceedings of the National Academy of Sciences of the United States of America*, **101**, 2428–2433.

17 Hofbauer, M., Wiesener, S., Babbe, H., Roers, A., Wekerle, H., Dornmair, K., Hohlfeld, R. and Goebels, N. (2003) Clonal tracking of autoaggressive T cells in polymyositis by combining laser microdissection, single-cell PCR, and CDR3-spectratype analysis. *Proceedings of the National Academy of Sciences of the United States of America*, **100**, 4090–4095.

18 Peixoto, A., Monteiro, M., Rocha, B. and Veiga-Fernandes, H. (2004) Quantification of multiple gene expression in individual cells. *Genome Research*, **14**, 1938–1947.

19 Julius, M.H., Masuda, T. and Herzenberg, L.A. (1972) Demonstration that antigen-binding cells are precursors of antibody-producing cells after purification with a fluorescence-activated cell sorter. *Proceedings of the National Academy of Sciences of the United States of America*, **69**, 1934–1938.

20 Kohler, G. and Milstein, C. (1975) Continuous cultures of fused cells secreting antibody of predefined specificity. *Nature*, **256**, 495–497.

21 Appay, V., Nixon, D.F., Donahoe, S.M., Gillespie, G.M., Dong, T., King, A., Ogg, G.S., Spiegel, H.M., Conlon, C., Spina, C.A., Havlir, D.V., Richman, D.D., Waters, A., Easterbrook, P., McMichael, A.J. and Rowland-Jones, S.L. (2000) HIV-specific CD8(+) T cells produce antiviral cytokines but are impaired in cytolytic function. *The Journal of Experimental Medicine*, **192**, 63–75.

22 Miyahara, N., Swanson, B.J., Takeda, K., Taube, C., Miyahara, S., Kodama, T., Dakhama, A., Ott, V.L. and Gelfand, E.W. (2004) Effector CD8+ T cells mediate inflammation and airway hyper-responsiveness. *Nature Medicine*, **10**, 865–869.

23 Waldrop, S.L., Pitcher, C.J., Peterson, D.M., Maino, V.C. and Picker, L.J. (1997) Determination of antigen-specific memory/effector CD4+ T cell frequencies by flow cytometry: evidence for a novel, antigen-specific homeostatic mechanism in HIV-associated immunodeficiency. *The Journal of Clinical Investigation*, **99**, 1739–1750.

24 Gauduin, M.C., Yu, Y., Barabasz, A., Carville, A., Piatak, M., Lifson, J.D., Desrosiers, R.C. and Johnson, R.P. (2006) Induction of a virus-specific effector-memory CD4+ T cell response by attenuated SIV infection. *The Journal of Experimental Medicine*, **203**, 2661–2672.

25 Perez, O.D. and Nolan, G.P. (2002) Simultaneous measurement of multiple active kinase states using polychromatic flow cytometry. *Nature Biotechnology*, **20**, 155–162.

26 Van De Wiele, C.J., Marino, J.H., Murray, B.W., Vo, S.S., Whetsell, M.E. and Teague, T.K. (2004) Thymocytes between the beta-selection and positive selection checkpoints are nonresponsive to IL-7 as assessed by STAT-5 phosphorylation. *Journal of Immunology (Baltimore, Md: 1950)*, **172**, 4235–4244.

27 Ilangumaran, S., Ramanathan, S., Ning, T., La Rose, J., Reinhart, B., Poussier, P. and Rottapel, R. (2003) Suppressor of cytokine signaling 1 attenuates IL-15 receptor signaling in CD8+ thymocytes. *Blood*, **102**, 4115–4122.

28 Haring, J.S., Corbin, G.A. and Harty, J.T. (2005) Dynamic regulation of IFN-gamma signaling in antigen-specific CD8+ T cells responding to infection. *Journal of Immunology (Baltimore, Md: 1950)*, **174**, 6791–6802.

29 Krutzik, P.O., Hale, M.B. and Nolan, G.P. (2005) Characterization of the murine immunological signaling network with phosphospecific flow cytometry. *Journal of Immunology (Baltimore, Md: 1950)*, **175**, 2366–2373.

30 Altman, J.D., Moss, P.A., Goulder, P.J., Barouch, D.H., McHeyzer-Williams,

M.G., Bell, J.I., McMichael, A.J. and Davis, M.M. (1996) Phenotypic analysis of antigen-specific T lymphocytes. *Science*, **274**, 94–96.

31 Han, B., Serra, P., Amrani, A., Yamanouchi, J., Maree, A.F., Edelstein-Keshet, L. and Santamaria, P. (2005) Prevention of diabetes by manipulation of anti-IGRP autoimmunity: high efficiency of a low-affinity peptide. *Nature Medicine*, **11**, 645–652.

32 Lieberman, S.M., Evans, A.M., Han, B., Takaki, T., Vinnitskaya, Y., Caldwell, J.A., Serreze, D.V., Shabanowitz, J., Hunt, D.F., Nathenson, S.G., Santamaria, P. and DiLorenzo, T.P. (2003) Identification of the beta cell antigen targeted by a prevalent population of pathogenic CD8+ T cells in autoimmune diabetes. *Proceedings of the National Academy of Sciences of the United States of America*, **100**, 8384–8388.

33 Verdaguer, J., Schmidt, D., Amrani, A., Anderson, B., Averill, N. and Santamaria, P. (1997) Spontaneous autoimmune diabetes in monoclonal T cell nonobese diabetic mice. *The Journal of Experimental Medicine*, **186**, 1663–1676.

34 Trudeau, J.D., Kelly-Smith, C., Verchere, C.B., Elliott, J.F., Dutz, J.P., Finegood, D.T., Santamaria, P. and Tan, R. (2003) Prediction of spontaneous autoimmune diabetes in NOD mice by quantification of autoreactive T cells in peripheral blood. *The Journal of Clinical Investigation*, **111**, 217–223.

35 Amrani, A., Verdaguer, J., Serra, P., Tafuro, S., Tan, R. and Santamaria, P. (2000) Progression of autoimmune diabetes driven by avidity maturation of a T-cell population. *Nature*, **406**, 739–742.

36 Irish, J.M., Kotecha, N. and Nolan, G.P. (2006) Mapping normal and cancer cell signalling networks: towards single-cell proteomics. *Nature Reviews. Cancer*, **6**, 146–155.

37 Bruchez, M., Jr., Moronne, M., Gin, P., Weiss, S. and Alivisatos, A.P. (1998) Semiconductor nanocrystals as fluorescent biological labels. *Science*, **281**, 2013–2016.

38 Michalet, X., Pinaud, F.F., Bentolila, L.A., Tsay, J.M., Doose, S., Li, J.J., Sundaresan, G., Wu, A.M., Gambhir, S.S. and Weiss, S. (2005) Quantum dots for live cells, in vivo imaging, and diagnostics. *Science*, **307**, 538–544.

39 Chattopadhyay, P.K., Price, D.A., Harper, T.F., Betts, M.R., Yu, J., Gostick, E., Perfetto, S.P., Goepfert, P., Koup, R.A., De Rosa, S.C., Bruchez, M.P. and Roederer, M. (2006) Quantum dot semiconductor nanocrystals for immunophenotyping by polychromatic flow cytometry. *Nature Medicine*, **12**, 972–977.

40 Tsien, R.Y. (1998) The green fluorescent protein. *Annual Review of Biochemistry*, **67**, 509–544.

41 Lippincott-Schwartz, J. and Patterson, G.H. (2003) Development and use of fluorescent protein markers in living cells. *Science*, **300**, 87–91.

42 Monks, C.R., Freiberg, B.A., Kupfer, H., Sciaky, N. and Kupfer, A. (1998) Three-dimensional segregation of supramolecular activation clusters in T cells. *Nature*, **395**, 82–86.

43 Freiberg, B.A., Kupfer, H., Maslanik, W., Delli, J., Kappler, J., Zaller, D.M. and Kupfer, A. (2002) Staging and resetting T cell activation in SMACs. *Nature Immunology*, **3**, 911–917.

44 Pentcheva-Hoang, T., Egen, J.G., Wojnoonski, K. and Allison, J.P. (2004) B7-1 and B7-2 selectively recruit CTLA-4 and CD28 to the immunological synapse. *Immunity*, **21**, 401–413.

45 Krummel, M.F., Sjaastad, M.D., Wulfing, C. and Davis, M.M. (2000) Differential clustering of CD4 and CD3zeta during T cell recognition. *Science*, **289**, 1349–1352.

46 Lee, K.H., Holdorf, A.D., Dustin, M.L., Chan, A.C., Allen, P.M. and Shaw, A.S. (2002) T cell receptor signaling precedes immunological synapse formation. *Science*, **295**, 1539–1542.

47. Lee, K.H., Dinner, A.R., Tu, C., Campi, G., Raychaudhuri, S., Varma, R., Sims, T.N., Burack, W.R., Wu, H., Wang, J., Kanagawa, O., Markiewicz, M., Allen, P.M., Dustin, M.L., Chakraborty, A.K. and Shaw, A.S. (2003) The immunological synapse balances T cell receptor signaling and degradation. *Science*, **302**, 1218–1222.

48. Purbhoo, M.A., Irvine, D.J., Huppa, J.B. and Davis, M.M. (2004) T cell killing does not require the formation of a stable mature immunological synapse. *Nature Immunology*, **5**, 524–530.

49. Campi, G., Varma, R. and Dustin, M.L. (2005) Actin and agonist MHC-peptide complex-dependent T cell receptor microclusters as scaffolds for signaling. *The Journal of Experimental Medicine*, **202**, 1031–1036.

50. Yokosuka, T., Sakata-Sogawa, K., Kobayashi, W., Hiroshima, M., Hashimoto-Tane, A., Tokunaga, M., Dustin, M.L. and Saito, T. (2005) Newly generated T cell receptor microclusters initiate and sustain T cell activation by recruitment of Zap70 and SLP-76. *Nature Immunology*, **6**, 1253–1262.

51. Varma, R., Campi, G., Yokosuka, T., Saito, T. and Dustin, M.L. (2006) T cell receptor-proximal signals are sustained in peripheral microclusters and terminated in the central supramolecular activation cluster. *Immunity*, **25**, 117–127.

52. Huppa, J.B., Gleimer, M., Sumen, C. and Davis, M.M. (2003) Continuous T cell receptor signaling required for synapse maintenance and full effector potential. *Nature Immunology*, **4**, 749–755.

53. Das, V., Nal, B., Dujeancourt, A., Thoulouze, M.I., Galli, T., Roux, P., Dautry-Varsat, A. and Alcover, A. (2004) Activation-induced polarized recycling targets T cell antigen receptors to the immunological synapse; involvement of SNARE complexes. *Immunity*, **20**, 577–588.

54. Maldonado, R.A., Irvine, D.J., Schreiber, R. and Glimcher, L.H. (2004) A role for the immunological synapse in lineage commitment of CD4 lymphocytes. *Nature*, **431**, 527–532.

55. Irvine, D.J., Purbhoo, M.A., Krogsgaard, M. and Davis, M.M. (2002) Direct observation of ligand recognition by T cells. *Nature*, **419**, 845–849.

56. Krogsgaard, M., Li, Q.J., Sumen, C., Huppa, J.B., Huse, M. and Davis, M.M. (2005) Agonist/endogenous peptide-MHC heterodimers drive T cell activation and sensitivity. *Nature*, **434**, 238–243.

57. Zal, T. and Gascoigne, N.R. (2004) Using live FRET imaging to reveal early protein-protein interactions during T cell activation. *Current Opinion in Immunology*, **16**, 674–683.

58. Zal, T., Zal, M.A. and Gascoigne, N.R. (2002) Inhibition of T cell receptor-coreceptor interactions by antagonist ligands visualized by live FRET imaging of the T-hybridoma immunological synapse. *Immunity*, **16**, 521–534.

59. Yachi, P.P., Ampudia, J., Gascoigne, N.R. and Zal, T. (2005) Nonstimulatory peptides contribute to antigen-induced CD8-T cell receptor interaction at the immunological synapse. *Nature Immunology*, **6**, 785–792.

60. Moldovan, M.C., Sabbagh, L., Breton, G., Sekaly, R.P. and Krummel, M.F. (2006) Triggering of T cell activation via CD4 dimers. *Journal of Immunology (Baltimore, Md: 1950)*, **176**, 5438–5445.

61. Yachi, P.P., Ampudia, J., Zal, T. and Gascoigne, N.R. (2006) Altered peptide ligands induce delayed CD8-T cell receptor interaction – a role for CD8 in distinguishing antigen quality. *Immunity*, **25**, 203–211.

62. Bhatia, S., Edidin, M., Almo, S.C. and Nathenson, S.G. (2005) Different cell surface oligomeric states of B7-1 and B7-2: implications for signalling. *Proceedings of the National Academy of Sciences of the United States of America*, **102**, 15569–15574.

63. Miller, M.J., Wei, S.H., Cahalan, M.D. and Parker, I. (2003) Autonomous T cell

trafficking examined in vivo with intravital two-photon microscopy. *Proceedings of the National Academy of Sciences of the United States of America*, **100**, 2604–2609.
64 Bajenoff, M., Egen, J.G., Koo, L.Y., Laugier, J.P., Brau, F., Glaichenhaus, N. and Germain, R.N. (2006) Stromal cell networks regulate lymphocyte entry, migration, and territoriality in lymph nodes. *Immunity*, **25**, 989–1001.
65 Mempel, T.R., Henrickson, S.E. and Von Andrian, U.H. (2004) T-cell priming by dendritic cells in lymph nodes occurs in three distinct phases. *Nature*, **427**, 154–159.
66 Hugues, S., Fetler, L., Bonifaz, L., Helft, J., Amblard, F. and Amigorena, S. (2004) Distinct T cell dynamics in lymph nodes during the induction of tolerance and immunity. *Nature Immunology*, **5**, 1235–1242.
67 Miller, M.J., Safrina, O., Parker, I. and Cahalan, M.D. (2004) Imaging the single cell dynamics of CD4+ T cell activation by dendritic cells in lymph nodes. *The Journal of Experimental Medicine*, **200**, 847–856.
68 Boissonnas, A., Fetler, L., Zeelenberg, I.S., Hugues, S. and Amigorena, S. (2007) In vivo imaging of cytotoxic T cell infiltration and elimination of a solid tumor. *The Journal of Experimental Medicine*, **204**, 345–356.
69 Mrass, P., Takano, H., Ng, L.G., Daxini, S., Lasaro, M.O., Iparraguirre, A., Cavanagh, L.L., von Andrian, U.H., Ertl, H.C., Haydon, P.G. and Weninger, W. (2006) Random migration precedes stable target cell interactions of tumor-infiltrating T cells. *The Journal of Experimental Medicine*, **203**, 2749–2761.
70 Bousso, P. (2004) Real-time imaging of T-cell development. *Current Opinion in Immunology*, **16**, 400–405.
71 Bhakta, N.R. and Lewis, R.S. (2005) Real-time measurement of signaling and motility during T cell development in the thymus. *Seminars in Immunology*, **17**, 411–420.
72 Mempel, T.R., Scimone, M.L., Mora, J.R. and von Andrian, U.H. (2004) In vivo imaging of leukocyte trafficking in blood vessels and tissues. *Current Opinion in Immunology*, **16**, 406–417.
73 Toma, A., Haddouk, S., Briand, J.P., Camoin, L., Gahery, H., Connan, F., Dubois-Laforgue, D., Caillat-Zucman, S., Guillet, J.G., Carel, J.C., Muller, S., Choppin, J. and Boitard, C. (2005) Recognition of a subregion of human proinsulin by class I-restricted T cells in type 1 diabetic patients. *Proceedings of the National Academy of Sciences of the United States of America*, **102**, 10581–10586.
74 Lennerz, V., Fatho, M., Gentilini, C., Frye, R.A., Lifke, A., Ferel, D., Wolfel, C., Huber, C. and Wolfel, T. (2005) The response of autologous T cells to a human melanoma is dominated by mutated neoantigens. *Proceedings of the National Academy of Sciences of the United States of America*, **102**, 16013–16018.
75 Sacre, K., Carcelain, G., Cassoux, N., Fillet, A.M., Costagliola, D., Vittecoq, D., Salmon, D., Amoura, Z., Katlama, C. and Autran, B. (2005) Repertoire, diversity, and differentiation of specific CD8 T cells are associated with immune protection against human cytomegalovirus disease. *The Journal of Experimental Medicine*, **201**, 1999–2010.
76 Mallone, R., Martinuzzi, E., Blancou, P., Novelli, G., Afonso, G., Dolz, M., Bruno, G., Chaillous, L., Chatenoud, L., Bach, J.M. and van Endert, P. (2007) CD8+ T-cell responses identify beta-cell autoimmunity in human type 1 diabetes. *Diabetes*, **56**, 613–621.
77 Kaufman, H.L., Deraffele, G., Mitcham, J., Moroziewicz, D., Cohen, S.M., Hurst-Wicker, K.S., Cheung, K., Lee, D.S., Divito, J., Voulo, M., Donovan, J., Dolan, K., Manson, K., Panicali, D., Wang, E., Horig, H. and Marincola, F.M. (2005) Targeting the local tumor microenvironment with vaccinia virus expressing B7.1 for the treatment of melanoma. *The Journal of Clinical Investigation*, **115**, 1903–1912.

78 Merkenschlager, M., Amoils, S., Roldan, E., Rahemtulla, A., O'Connor, E., Fisher, A.G. and Brown, K.E. (2004) Centromeric repositioning of coreceptor loci predicts their stable silencing and the CD4/CD8 lineage choice. *The Journal of Experimental Medicine*, **200**, 1437–1444.

79 Roldan, E., Fuxa, M., Chong, W., Martinez, D., Novatchkova, M., Busslinger, M. and Skok, J.A. (2005) Locus "decontraction" and centromeric recruitment contribute to allelic exclusion of the immunoglobulin heavy-chain gene. *Nature Immunology*, **6**, 31–41.

80 Sayegh, C., Jhunjhunwala, S., Riblet, R. and Murre, C. (2005) Visualization of looping involving the immunoglobulin heavy-chain locus in developing B cells. *Genes and Development*, **19**, 322–327.

81 Skok, J.A., Gisler, R., Novatchkova, M., Farmer, D., de Laat, W. and Busslinger, M. (2007) Reversible contraction by looping of the Tcra and Tcrb loci in rearranging thymocytes. *Nature Immunology*, **8**, 378–387.

82 Peters, P.J., Borst, J., Oorschot, V., Fukuda, M., Krahenbuhl, O., Tschopp, J., Slot, J.W. and Geuze, H.J. (1991) Cytotoxic T lymphocyte granules are secretory lysosomes, containing both perforin and granzymes. *The Journal of Experimental Medicine*, **173**, 1099–1109.

83 Stinchcombe, J.C., Bossi, G., Booth, S. and Griffiths, G.M. (2001) The immunological synapse of CTL contains a secretory domain and membrane bridges. *Immunity*, **15**, 751–761.

84 Stinchcombe, J.C., Majorovits, E., Bossi, G., Fuller, S. and Griffiths, G.M. (2006) Centrosome polarization delivers secretory granules to the immunological synapse. *Nature*, **443**, 462–465.

10
Molecular Characterization of Rare Single Tumor Cells
James F. Leary

10.1
Introduction

10.1.1
Importance of Rare Cells

Rare cells can be defined as those of less than 0.1% frequency. Ultra-rare cells can be defined as those of less than 0.001% frequency. These are terms for ease of discussion and are not universally agreed upon. In terms of flow cytometry and cell sorting, rare cell applications not only push the limits of the technology but also require levels of staining specificity beyond those assumed by most biologists. It is important for engineers to understand that good technology can be made irrelevant by bad cell staining and preparation. The greatest difficulty with rare cell applications is not the immediate technological constraints but the requirement that there be no weak links in the experimental methodology anywhere in the process from cell preparation, flow cytometry/cell sorting and data analysis or subsequent analysis of isolated cells. Each and every one of these steps in the methodology must be excellent and the entire process must be thought through to eliminate or deal with the weaker links of a given rare cell application. For reviews on "rare event" analysis techniques, see Refs. [1–3].

There are many rare cell applications of importance to basic or clinical research. The ones discussed in this paper help show the range not only of the applications but also of the technological challenges to engineers working in this field. Some basic research examples that are briefly discussed include: (1) isolation of rare cell clones with specific mutations or transfected genes, (2) isolation of clones with combinatorial libraries of inserted genes and (3) studies of environmentally induced mutations in human cells.

10.1.2
Detection of Rare Tumor Cells

While many biomarkers look very good in non-rare cell applications, few biomarkers look good at rare cell level and very few at the level of ultra-rare cells. It is important to have some idea of the S/N ratio of the marker/probe system being used before wasting considerable energies trying to do an impossible application. Simple mathematical modeling of data using each biomarker can be used to predict the S/N ratio of that probe. In most cases one should expect to use multiple biomarkers to detect rare cells [4]. Ultra-rare cell applications usually require a minimum of three to four parameters and frequently two or more biomarkers designed to minimize the number of required acquisition parameters and to reduce data complexity. Receiver operating characteristic (ROC) [5–8], long familiar to many engineers but only starting to be used in flow cytometry and cell sorting [9–14], can be used to predict the power of each probe, singly or in combination. A biomarker strategy used in this paper is illustrated in Figure 10.1.

ROC analysis can be used to estimate the performance of various cell classifiers both singly and in combination [10, 11, 14], as shown in Figure 10.2. In Figure 10.2 ROC analysis is applied to a rare cell classification problem, in this case the problem of rare metastatic tumor cell analysis and purging (negative sorting to remove these cells) from stem cells to prevent their being co-transplanted back into a patient – a potentially very important future clinical application.

One of the most critical problems of rare event analysis is the correct classification of the cells. To determine the correctness one must have a truth standard, that is one must be able to unequivocally determine the identity of each and every cell in a training set. Obviously in the new test sample there will be incoming data which resides in a classification "gray zone" where different cell subpopulations overlap. But if the training set is done properly, every cell can at least have a probability of assignment calculated on a cell by cell basis. This probability becomes an important

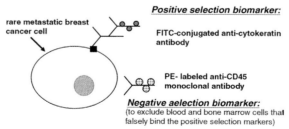

Figure 10.1 A simple biomarker strategy for detecting rare cells using one positive biomarker and one negative biomarker. No cells of interest should stain with the negative biomarker which eliminates most undesired cells.

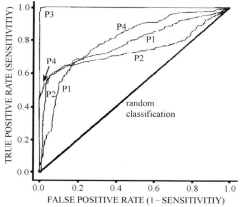

Figure 10.2 In this ROC plot of actual flow cytometric parameters P1-P4, we see that P2 (side scatter) is similar in performance as a classifier to P1 (forward scatter). P3 (a fluorescent monoclonal antibody against antigens found on breast cancer cells) is a very good positive selection probe and classifier. P4 (a fluorescent monoclonal antibody against CD45 found on blood cells but not on breast cancer cells) is a negative selection probe and is needed when the breast cancer cells become very rare in blood.

part of a determination of sort boundaries to allow optimization of sorted cell yield and purity. In the case of cells which truly have partial membership in two or more classes, for example in cases of normally differentiating cells or in the case of hematopoietic cancers such as leukemias, fuzzy logic can be used to provide fuzzy rather than crisp classifiers.

10.2
Finding Rare Event Tumor Cells in Multidimensional Data

In one application, done in collaboration with Drs. Jonathan Ward and Marinel Ammenheuser at the University of Texas Medical Branch, Galveston, Texas, human peripheral blood mononuclear cells were exposed ex vivo to environmental mutagens as an assay development model for environmental carcinogenesis. In this three-color fluorescence listmode data, BUdR uptake was used as a measure of strand breaks and repair and measured by FITC-conjugated anti-BUdR. A mature human T cell subset, singled out for analysis in this assay, were labeled with PE-Cy5 conjugated anti-CD4. Dead or damaged (membrane leaky) cells were identified on the basis of a propidium iodide (PI) exclusion assay.

BUdR was diluted in culture media at various concentrations from $10\,\mu M$ to $1\,mM$ to test for optimal labeling and assuring that the thymidine analog was present in excess of intracellular thymidine pools. 5-fluoro-uracil (5-FU) was included to block endogenous thymidine synthesis. Anti-BUdR binding was optimal at a substitution level of about 25% total thymidines. Too high a substitution rate leads to steric hindrance problems and too low a rate leads to poor signal levels. Growing cells were allowed to incorporate BUdR for 18 h as in the established protocols of the autoradiographic ^{3}H-TdR procedure to which this method was compared. The cells were then washed and incubated with PE-Cy5-conjugated anti-cell surface antigen antibody (initially anti-CD4) and incubated for 30 min at 4 °C. The cells were washed once in PBS and resuspended in 1% paraformaldehyde/0.01% Tween 20 overnight at 4 °C. The cells were then washed in PBS to remove the paraformaldehyde. The cells were then incubated in PBS containing Mg^{2+} and Ca^{2+} with 50 Kunitz units of DNase I and incubated for 30 min at 37 °C. The cells were then washed and suspended in 150 μL of PBS containing 10% bovine serum albumin and 0.5% Tween 20 and 20 μL of FITC-conjugated anti-BUdR antibody for 45 min at room temperature. Finally the cells were washed with PBS, incubated with PE-Cy5-conjugated streptavidin and suspended in 1 mL PBS containing 10 μg/mL PI, ready for flow cytometric analysis.

This is a good example how rare cell subpopulations can sometimes be hidden even if one tries looking at all possible two dimensional projections of the data ("bivariate scattergrams"). No matter how one tries to project this four-dimensional data, one cannot see the rare cells of interest. To deal with this problem we used a special "subtractive clustering" analysis (SCA) of the flow cytometric. A complete description of SCA is beyond the scope of this paper and has been published elsewhere [15, 16]. But briefly SCA, invented by the author, [17] is a way of comparing two or more multidimensional data sets to find the differing subpopulations (in this case a rare cell subpopulation of interest, as shown in Figure 10.3).

10.2.1
Rare Event Sampling Statistics

In the case of limited total sample size, there must be a sufficient number of cells analyzable to obtain statistically or biologically meaningful results. For example, if total sample size is limited and frequencies of rare cells which are analyzable are low, it may not be possible to analyze enough cells to be meaningful at either the statistical or biological level. Since most rare cell experiments are difficult and costly, this should be taken into account before plunging mindlessly into an attempt to analyze or sort rare cells. Mathematical and statistical modeling of the situation is wise to do beforehand as it may change the fundamental approach to the problem. When the number of rare cells is very small it is important to use combinatorial rather than Poisson statistics as Poisson statistics are only an approximation which is not particularly stable below about 50 rare cells [18]. For example, to insure with 95% confidence that the single rare (10^{-6} frequency) cell has been sorted actually requires that a minimum of nearly three million cells must be sampled. Since the probability of successfully isolating the rare cell of interest above false-positive background events becomes more difficult for rare cells, it is important to sample an even higher number than this because it is highly likely that there will be false-positive cells sorted. The times needed to isolate a given number of desired rare cells as a function of sorting rate are illustrated in Table 10.1 [18].

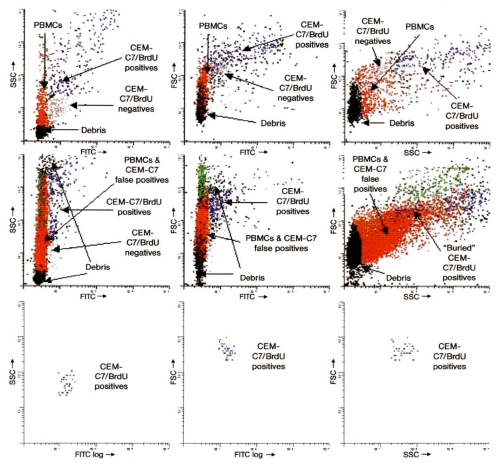

Figure 10.3 Top row shows bivariate displays of FITC, SSC and FSC for a 10 000 event file of the PBMC/CEM-C7 mix that includes debris, peripheral blood mononuclear cells (PBMCs), negative CEM-C7s and positive CEM-C7s. Middle row are bivariate displays for a 100 000 event file showing an additional cluster of debris and how "buried" the true CEM-C7/BrdU positives were under false positives. Bottom row are bivariates of the final result of the multiple subtractions used to pull out the true BrdU positive population at 0.01% from the initial 8.0×10^7 total cells. Row 1: Bivariate displays of FITC, SSC and FSC for 10 000 event file of the PBMC/CEM-C7 mix showing debris, PBMCs, false positive CEM-C7s and true positive CEM-C7 cells. Row 2: Bivariate displays for 100 000 event file showing an additional cluster of debris and also showing how "buried" the true CEM-C7/BrdU positives were to begin with. Row 3: The final result of the "multiple subtractive" process showing the true BrdU positive population at 0.01% from 8×10^7 cells.

If rare single cells are to be isolated for either cloning or for single cell PCR the proper approach is to take these rare cell sampling statistics into account and perform single cell, multitube sorting. An important principle of rare cell sorting is that at very rare cell levels perhaps no cell markers, for example, monoclonal antibodies, will be

10 Molecular Characterization of Rare Single Tumor Cells

Table 10.1 Frequency of cells with selected characteristics = 10 E–06 (0.95 level of assurance).

Desired number of cells with selected properties	Total number of sort decisions	Time (s) required to collect desired cells					
		Sorting rate (sort decisions/s)					
		2500	5000	10 000	20 000	50 000	10 0000
1	2 990 000	1196	598	299	150	60	30
10	15 705 214	6282	3141	1570	785	314	157
100	116 997 126	46 799	23 399	11 700	5850	2340	1170
1000	1 052 577 091	421 031	210 515	105 258	52 629	21 052	10 526

good enough to pull out pure populations of rare cells from a noisy background. But if rare cells are isolated by single cell sorting, 100% pure rare cells can always be isolated. If the S/N ratio is very poor, the number of tubes of single isolated cells becomes impractical. But with modern molecular biology methods such as PCR, even single copy DNA or small numbers of mRNA sequences from a single cell can be quickly expanded into 10^6–10^8 copies [19]. So the principle is to recover a single copy of the DNA or mRNA template from a sorted single cell and then make many, many copies of that template much more rapidly than can be accomplished even by high-speed cell sorting. The basic idea of single cell, multitube cell sorting is shown in Figure 10.4. Obviously cells can be sorted into tube arrays, or multiwell plates. The flow cytometric information associated with each sorted cell can remain associated with all of the subsequent molecular measurements by a method known in the field as "indexed cell sorting" [20]. We have combined flow cytometric indexed cell sorting

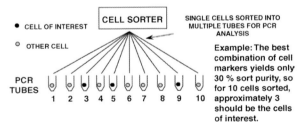

Figure 10.4 The general principle of single cell, multitube PCR analysis of sorted single cells is to obtain one or more copies of a DNA or mRNA template from a single sorted cell. Some of the singly sorted cells will be false positives due to the inability of specific marker probes to have a high enough S/N ratio above false-positive staining background.

data with molecular data in hybrid files which we have then used for subsequent analysis with multivariate statistical packages such as S-Plus for Windows.

Our practical experience agrees quite well with these theoretical predictions. When we have attempted to isolate rare (10^{-6} frequency) cells (labeled with very good probes!) we find that we usually get two to three rare cells successfully isolated per 10 single cell sorts, as depicted in Figure 10.1. Similarly we have had applications where the probes are not very good and the combination of limited sampling statistics due to low numbers of cells available and poor probes have made impossible the successful isolation of less rare cells of a frequency of 10^{-5}. So one must pay careful attention to whether enough cells are available to reach proper sampling statistics and whether the probes are good enough to practically isolate the rare cells of interest.

10.2.2
High-Speed Sorting of Rare Cells

Cells must be isolated fast enough in sufficient quantity and purity and in a condition to allow the subsequent molecular characterization steps. High-speed "enrichment sorting" is one way to diminish the overall time necessary to obtain the most cells of the highest purity. The first sort is done at very high speeds, eliminating most undesired cells, but still containing some nearby cells that are within the sorting interval (typically a two-droplet sorting unit). The sorted sample is then re-sorted a second time at slower speeds to yield the highest possible purity of sorted cells of interest. The concept of "enrichment sorting" [21] is shown in Figure 10.5.

10.2.3
Sorting Speeds must be Fast Enough to be Practical

When the total sample size is large and the rare cell frequency small, there may be some fundamental limits imposed by instrument stability/operation, marker stability and cell viability. Thus there is usually a time window imposed by these variables that should be carefully considered when either designing an instrument for an application or attempting an application on an existing instrument. In most cases this time window will require an analysis speed in excess of 10 000 cells/s for rare cell applications and an analysis speed in excess of 40 000 cells/s for ultra-rare cell applications. In terms of fundamental limitations it is possible to build instruments capable of analysis speeds in excess of 100 000 cells/s, if properly designed. But to do efficient high-speed sorting or rare cells, it is wise to do the fundamental cell classification/sort decision in two stages. The first stage should classify most of the cells with easy decisions so that the second stage can spend more time making the tough decisions. Most of the time we know that the cells are not of interest and the decision is straightforward and can usually be accomplished by some simple Boolean combination of simple gates of the original parameters. But there will always be some potential false positive cells which must be distinguished from the true positive rare cells of interest. To make these tougher choices may require more elaborate gating and perhaps mathematical combinations of parameters (e.g. discriminant functions)

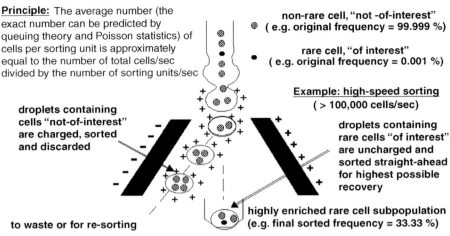

Figure 10.5 Cell sorters are much more efficient at very high throughput speeds in terms of the enrichment factor from the original rare cell concentration to the final sorted rare cell concentration. In this concept, a cell of original frequency of 10^{-5} (one rare cell per 100 000 total cells) can be sorted at 100 000 cells/s with a sorter capable of generating 33 000 droplets/s. A high-speed first-pass sort enrichment of more than 30 000-fold, based on up to three rare parameters and five additional total parameters, can be attained. Sorted cells can then be resorted to any desired purity based on the quality of the selection probes.

which are more complicated and require more time to process the signals. One should also remember that mindless high-speed digitization of signals merely leads to a river of digital data that must be processed at least partially in real time. Otherwise, one would have to deal with the problem of storing and processing of many gigabytes of data per sample.

10.2.4
Limits in Sorting Speeds and Purities

Sorting is fundamentally more difficult and slower than analysis for a number of reasons. First, sorting requires that the rare cells be properly classified, which itself is difficult. Cell sorting should really be considered as real-time data analysis. The calculations and classifications must be done fast enough to reach a sort decision, typically less than a millisecond after excitation of the cell. Fortunately there are new methods available to accomplish this feat [9–13, 22] By using one or more linked, high-speed lookup tables one can perform complex linear and nonlinear "calculations" at memory speeds enabling their use for real-time statistical classifications suitable for cell sorting decisions. The lookup tables are pre-calculated for a range of useful values based on data obtained from an aliquot of the cell sample, as is typically done for any cell sorting decision. The incoming data is then compared to lookup table values which then can be used to issue a sort/no sort decision.

Second, as any flow cytometrist knows, the Achilles heel of a flow cytometer is the flow, that is, the stability of the fluidics is fragile and difficult to maintain. Since high-speed fluidic switching is still under development and has not yet been applied to cell sorting, droplet sorting is presently the method of choice, although this may change in the not-too-distant future. While fluidic switching has problems of its own, it has many potential advantages over droplet sorting, not the least of which is the safety issue of sorting biohazardous materials in a closed system rather than in an open system with aerosol generation. Advances in microfluidics may well change this situation.

Not only is the break-off position of a cell sorter difficult to maintain, but as the number of cells/sec is increased for sorting of rare cells, the stability becomes less reliable. The stickiness of the cells makes clumping a problem at high cell concentrations (e.g. greater than 5×10^7 cells/mL). In terms of the fluidics, the cells themselves start to contribute to the viscosity of the sample stream, making transport difficult and less predictable. As discussed later in more detail, a high-speed cell sorter for rare cells should be able to measure the arrival statistics of the cells to determine if the cell preparation is adequate to permit isolation of these rare cells. As discussed later in this chapter, cell arrival statistics is a critically important measure of cell sorter performance during actual cell sorting of rare cells. If the fluidics are working properly, cell arrival statistics should be random. Cell arrival statistics can be described by simple queuing theory [23] familiar to many engineers and discussed in the flow cytometry/cell sorting field in a number of papers by different groups [1, 24, 25].

10.3
Classification of Rare Tumor Cells

It is possible for a given cell to belong to more than one classification through use of fuzzy set theory (e.g. differentiating cells which are partly a member of two, or more, neighboring differentiation states). But most of the time we use "crisp" rather than fuzzy logic and assign membership of cells (e.g. a tumor cell) to a particularly category. The consequence of using crisp logic is that the classification can be correct if it is a true positive (TP) or if the cell is a true negative (TN). But there is also a given probability that the cell will be incorrectly classified into a false positive (FP) or false negative (FN) as shown conceptually in Figure 10.6. The classification curves need not be Gaussians and usually are not. When a cell is incorrectly classified into a FP or FN, then the purity of the sorted cells is lowered and subsequent molecular characterizations of sorted cells can and will be affected. Since we separate rare cells on the basis of multiple cell classifiers (e.g. multicolor labeling of cells), the classifiers shown in Figure 10.6 must be extended to multiple dimensions of the multiparameter data. Most people use principal component analysis (PCA) to reduce the dimensionality of the data for human visualization. PCA tilts the projection plane onto which the multidimensional data is projected to maximize the statistical variance of the data points. While PCA provides a good first look at multidimensional data, it is actually not the best way of projecting the data in terms of cell classifiers. Discriminant function analysis (DFA) uses a different projection metric which makes the cell subpopulations clusters in

Conditional Probabilities for Classifiers

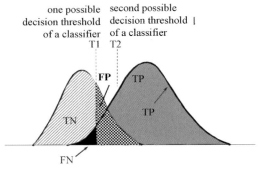

Figure 10.6 Every cell needs to be classified as either desired (e.g. tumor cell) or undesired (e.g. normal blood cell). During this classification process four types of classifications can occur: TP = true positives, TN = true negatives, FP = false positives, FN = false negatives.

multidimensional data-space as tight as possible and maximizes the distance between clusters. DFA classifiers provide a better way to perform statistical classifiers with the least amount of FP and FN misclassifications. Different cutoff scores in DFA can be chosen based on the costs of misclassification (Figure 10.7) [26]. While there are many

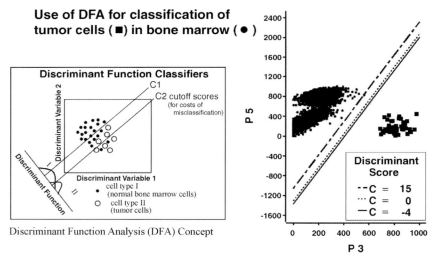

Figure 10.7 In this example stem cells and tumor cells are separated by DFA with different cutoff scores to exclude varying percentages of tumor cells from stem cells. DFA with costs of misclassification for normal human bone marrow and MCF-7 breast cancer cells where discriminant scores are in units of standard deviations.

ways to think and quantitatively assign costs of misclassification, the easiest way is to think of the cost in terms of how many TP cells must be discarded to lower the FP and FN rates to what is acceptable in terms of the application. An example is the simultaneous sorting of stem/progenitor cells from blood or bone marrow for an autologous transplant while simultaneously removing tumor cells that you do not want to accidentally co-transplant back into the patient. This co-transplantation can have serious consequences to an immune-suppressed patient as previously shown [27]. The concept of multidimensional classifiers is shown conceptually in Figure 10.7a with actual data shown in Figure 10.7b [26].

Real-time statistical classifiers, with "flexible sorting" decisions implementing the costs of misclassification, have been implemented in a high-speed cell sorter in the author's laboratory. This system provides for a two-step classification process which results in total throughput speeds of more than 100 000 cells/s and successful detection and sorting of rare tumor cells at frequencies as low as 10^{-6}. This high-speed sorter, shown conceptually in Figure 10.8, has been described previously in many publications and patents and summarized in a recent overview article [3].

Re-weighting of the input variables using embedded algorithms: if additional information indicates that the input variables should be re-weighted, this can be done

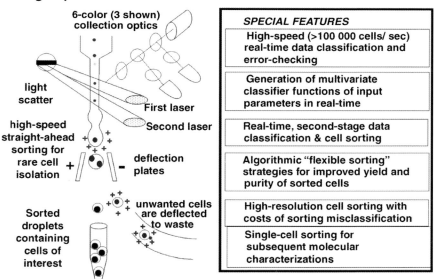

Figure 10.8 Overview of a six-color high-speed flow cytometer with a single layer neural network architecture that allows real-time classification/sort decisions on the basis of multivariate statistical functions. For high recovery of cells sorted at high speeds and for efficient single cell recovery, we often sort cells of interest straight-ahead, achieving sort recoveries of more than 95% even at the single cell level at rare cell frequencies as low as 10^{-6}. The multivariate statistical functions calculated in real-time provide for statistically-based sort boundaries and a measure of the degree of misclassification.

by passing outputs to another linked lookup table. This other information may come from information known about the sample. Or if one form of analysis is more accurate in classification over a subset of the data space, data can be re-mapped to a different lookup table that is more appropriate for that subregion of the input data space. One application of this method is the use of multiple expert systems, whereby the expert is chosen dynamically depending on the sub-region of the input data space. Figure 10.9 shows the use of such an embedded algorithm for sorting to optimize sort yield/purity and to include penalties (or "costs") of misclassification [10].

10.3.1
Using Classifiers to Sort Rare Tumor Cells

The overall system for multivariate statistical sort decisions is sufficiently complicated that a "roadmap" is useful to the reader. Figure 10.10 shows how these methods can be applied to the problem of high-speed human stem cell sorting with simultaneous tumor purging such that cancer cells are removed from a patient's bone marrow before the cells are re-infused into the patient after high-dose chemotherapy as is done in some advanced procedures to treat breast cancer. Gene marking experiments have shown that this may be a serious problem [27].

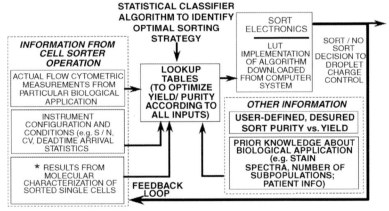

U.S. Patents 5,199,576 (1993) and 5,550,058 (1996)

Figure 10.9 Embedded algorithms allow for the real-time shifting of weighting factors on input variables according to other information known to the researcher or through expert systems analysis of similar data. Importantly, this re-weighting of the weighting factors of the classifying function can be done dynamically such that different expert systems can be applied in real time to different subsets of the incoming data stream. Such a system also can be set up in learning mode such that improvements in classifier/sort decisions can be attained by studying the results of previous sorts. Many of our sort algorithms are first studied using data mixtures and classifier tags so that the accuracy of classification/sort algorithms can be simulated and the best classifier decisions applied to given experiment.

Schematic of Entire Statistical Classifier / Data Mining Proces Applied to High-Speed Stem-Cell Sorting with Tumor Purging

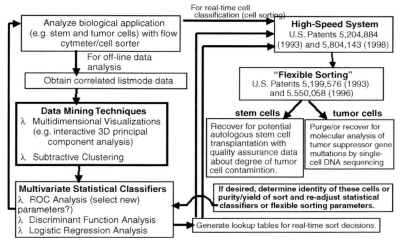

Figure 10.10 A "roadmap" of the entire process used to perform high-throughput system (HTS) analysis and sorting for the separation of human adult stem/progenitor cell subpopulations free of contaminating tumor cells. Special home-built data mining software known as "subtractive clustering" helps find the rare cells of interest. Then multivariate statistical classifiers are produced for sort decisions. These statistical classifiers (e.g. discriminant functions) are then reformulated into LUT format and downloaded into the hardware for subsequent real-time classification and sorting of stem cells and tumor cells.

10.4
Molecular Characterization of Sorted Tumor Cell Cells

10.4.1
Model Cell Systems

A model system was developed to detect mutations in a tumor suppressor gene, PTEN, thought to be an important tyrosine kinase inhibitor responsible for maintaining a normal cell phenotype. PTEN, a putative protein tyrosine phosphatase gene may suppress tumor cell growth by antagonizing protein tyrosine kinases and may regulate tumor cell invasion and metastasis. The mutation and deletion in PTEN gene has been found in two breast cancer cell lines (BT-549, ATCC number HTB-122; MDA-MB-468, ATCC number HTB-132).

10.4.2
Design of PCR Primers to Detect the PTEN Gene Region

We first designed primers to amplify the target region of PTEN gene using the Primer3 software package. This code is available through the Whitehead Institute of

MIT at http://www-genome.wi.mit.edu/genome software/other/primer3.html). In the MDA-MB-468 cell line we expect to find a 44 bp deletion in codon 70 (210 nt). The primers to amplify DNA sequences around this deletion were designed to have a Tm = 60 °C, with a left primer (23-mer; cat caa aga gat cgt tag cag aa) and a right primer (25-mer; ctt gta atg gtt ttt atg ctt tga a). The PCR product from RNA should be 190 bp for the normal gene and 146 bp for the deletion gene. The PCR product from DNA should be 380 bp for the normal gene and 336 bp for the deletion gene [12].

For the BT-549 cell line we expect to find a 1 bp deletion in codon 274 (822 nt). We designed PCR with a Tm = 62–63 °C, a left primer (20-mer; cct cca att cag gac cca ca) and a right primer (20-mer; gga gaa aag tat cgg ttg gc). The PCR product from RNA should be 340 bp for the normal gene and 339 bp for the deletion gene. The PCR product from DNA should be 393 bp for the normal gene and 392 bp for the deletion gene.

After PCR amplification these amplimers were further expanded using TA cloning methods (Shuman, 1994) which involved subcloning the PCR fragments to a TA vector. Subsequently, the PCR-amplified, cloned sequences were sequenced (using a Perkin–Elmer ABI 377XL sequencer with ABI Prism 377 ver. 2.1.1 sequencing software) for the target gene area of PTEN. For studies on sorted single cells, the nested PCR primers (NP for MDA-MB-468; NP2 for BT-549) were used to amplify PTEN target sequences in the single sorted cell [12].

10.4.3
Processing BT-549 Human Breast Cancer Cells

The breast cancer cell line BT-549 was cultured in a T-75 flask and subsequently harvested with trypsin/EDTA digestion and washed with PBS. The cells were then fixed with 1% paraformaldehyde at 4 °C until used, at which time the cells were washed with PBS two times to remove the paraformaldehyde. The cells were resuspended with 100 μL of 0.5% saponin solution at pH 7.5 to permeabilize for intracellular staining of cytokeratins. The cells were then immunofluorescently labeled by adding 40 μL of anti-cytokeratin (CAM 5.2) conjugated with FITC (Becton Dickinson, catalog no. 347653). The reagent was mixed thoroughly and incubated for 1 h at 4 °C. After labeling, the cells were washed twice with PBS and examined by fluorescence microscopy prior to flow cytometric analyses [12].

The cells were filtered with a Steri-Dual filter (Miltenyi Biotec, Inc.) before flow cytometric analysis to remove cell clumps. Single cells were sorted into a PCR tube containing 5 μL of cell lysis buffer. We also sorted samples of 50 cells and 10 cells per tube. We then lysed the sorted cells for subsequent PCR. Then 1 mL (2 mg) of proteinase K (Life Technologies, catalog no. 25530-049) was added to each sample PCR tube and mixed thoroughly. The samples were then incubated at 94 °C for 10 min, followed by incubation at 55 °C for 30 min. Nested PCR amplification of the PTEN gene was then performed. In the first PCR reaction the primer set CGF-3 and CGR-3 is designed for amplification of single by deletion of exon 8 in PTEN gene (the primer sequence was generously provided by Dr. Eng, Harvard University).

Then 6 µL of lysate was added to 1 µL of CGF/R-3, 18 µL of water and 25 µL of PCR master. The PCR reaction was performed beginning with a "pre-PCR" step of 94 °C for 2 min, which is a pre-heating step to prevent the formation of primer dimers. The PCR reaction was then carried out for a total of 40 cycles with the following thermal cycling regimen: 94 °C for 45 s, 50 °C for 1 min, 72 °C for 2 min, 72 °C for 10 min. But this time we not only had to PCR-amplify DNA sequences, we also had to sequence those amplimers to find the mutated tumor suppressor genes [12].

10.5
Detection of Mutated Sequences in Tumor Suppressor Genes

This is an application of rare cell sorting that requires high-speed or ultra high-speed analysis and moderate speed, but very high-precision, single cell sorting. The problem is to look at a statistically meaningful number of cells to find the mutant cell and then have enough power in the sort decision to precisely grab that cell free from a large background. In this example, a rare (10^6 frequency) tumor cell bearing a mutation in a tumor suppressor gene was analyzed and carefully sorted to a single cell level (Figure 10.11). The DNA from the sorted single cell was then amplified with polymerase chain reaction (PCR), cloned into a vector, further grown and then DNA sequenced. The result was the detection of a single base pair mutation in the PTEN tumor suppressor gene thought to be important in the outcome of breast cancer (Figure 10.11) [12].

10.5.1
Detection of Mutations in Breast Cancer Tumor Suppressor Genes by High-Throughput Flow Cytometry, Single Cell Sorting and Single Cell Sequencing

An example of rare tumor cell detection used this methodology [12]. A rare tumor cell clone was mixed with a human T-cell line at a frequency of 10^{-6}. A defined cell mixture of CEM/C7 (human T cells) and MCF-7 cells (human breast carcinoma cells, ATCC no. HTB-22) was constructed to yield a frequency of 10^6 tumor cells for flow cytometry and cell sorting. The cells were then analyzed on the basis of cytokeratin-positive fluorescence and for negative labeling with CD45. We then tested for sort recovery and purity in the case of a defined cell mixture of MCF-7 and CEM/C7 cells. We used DNA HLA-DQalpha typing of the MCF-7 cells (1.2 and 4.0) and the CEM/C7 cells (1.2 and 1.3) to see our efficiency of successful selection of tumor cells. Because the amount of PCR-amplified DNA from a single cell is too small to be seen by conventional ethidium bromide staining of the PCR products on a gel, enzymatic amplification techniques must be used. One such technique we have used is enhanced chemiluminescence (ECL; Amersham Life Science, catalog no. RPN 3021) whereby Southern blotting is performed with enzymatically labeled complementary sequence oligonucleotide ("oligo") probes. An enzymatic reaction gives rise to chemiluminescence that can then be detected on x-ray film, but without the hazards and disposal problems of radioactive probes. ECL provides an enzymatic

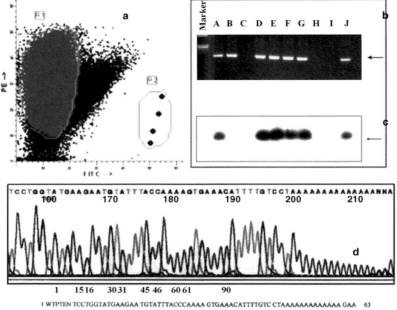

```
1 WTPTEN  TCCTGGTATGAAGAA TGTATTTACCCAAAA GTGAAACATTTTGTC CTAAAAAAAAAAAAA GAA  63
2 BT-549  TCCTGGTATGAAGAA TGTATTTACC- AAAA GTGAAACATTTTGTC CTAAAAAAAAAAAAA GAA  62
```

Figure 10.11 Flow cytometric results from a defined cell mixture of 1025 frequency MCF-7 cells in a major population of human CEM/C7 T cells. Cells were labeled with a phycoerythrin (PE)-conjugated anti-CD45 antibody and a fluorescein isothiocyanate (FITC)-conjugated anti-cytokeratin antibody. A small subpopulation of rare MCF-7 cells was detected in region R2 in an aliquot of the sample. Cells in this region were then sorted at the single cell level for subsequent PCR analysis, TA cloning and DNA sequencing. (a) The four tumor cells, shown in this aliquot of cell sample, have been highlighted as dark enlarged circles in the flow cytometric distribution for easier viewing. (b, c) ECL detection of PCR-amplified sequences from sorted, rare, single tumor cells as shown in (a). (b) Nested PCR product on 2% agarose gel stained with ethidium bromide. (c) The result of Southern blotting with HLA DQ-alpha type 4 probe. The result indicates that the sorting efficiency is 7 of 10; and the Southern blot shows 6 of 10 are MCF-7 cells (lanes A. D, E, F, G, J). (d) The result of ECL Southern blotting with HLA DQ-alpha type 4 probe. The result indicates that the overall sorting efficiency was seven recovered single rare cells out of 10 (a fairly typical recovery over many experiments) and the Southern blot reveals that six of those seven sorted cells were MCF-7 cells (lanes A, D, E, F, G, J), showing that the sort classification was fairly accurate. The alignment of DNA sequencing of the PTEN gene region from a single sorted cell (top) and alignment (bottom) shows that the PTEN mutation consists of a missing single base pair at nucleotide 61 of exon 8. These results show successful detection of a single base pair mutation in a single sorted cell.

amplification factor of approximately 1000 times, thus permitting detection of PCR products from single cells. Single cell DNA sequencing results from TA cloning. As we described previously [12], DNA sequencing of sorted single tumor cells can be performed. We sorted single cells with several mutations, including the BT-549 cells that have a single base pair mutation in the so-called PTEN gene, a tumor suppressor

10.5 Detection of Mutated Sequences in Tumor Suppressor Genes | 213

TA Cloning for Sequencing Analysis of Exon 8 of PTEN Gene in Sorted Single BT-549 Breast Cancer Cell Lines

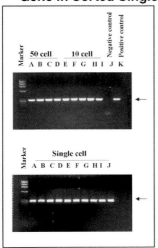
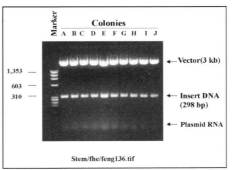

Lanes A to J: 10 white colonies digested with EcoR I to release cloned insert DNA. pGEM-T Easy vector is ~ 3,018 bp and the insert PCR product is ~ 298 bp. Marker: ⌀X174 RF DNA/Hae III fragment.

Figure 10.12 PCR analyses of conventionally sorted MCF-7 cells, with 50 cells, 10 cells and one cell per PCR tube. As can be seen from the results, the overall efficiency of sorting was virtually 100% for single cells.

gene that has been previously described. A clone of BT-549 cells was analyzed by flow cytometry and sorted into single PCR tubes for subsequent analysis by PCR and TA cloning. The DNA from the PTEN gene region of sorted single cells was amplified by PCR and cloned into a TA cloning vector. The vector was then amplified by growth in host *Escherichia coli* bacterial cells and the plasmid DNA was isolated and purified. This purified plasmid DNA containing the PTEN gene sequences from sorted single cells was then analyzed on a DNA sequencing instrument. A typical DNA sequencing result from a single sorted BT-549 cell is shown in Figure 10.12. Cells without the mutated sequence show normal PTEN gene sequences (data not shown). An alignment analysis of this sequence data reveals a single base pair deletion at nucleotide 61 of exon 8 of the PTEN gene, which is indicative of the PTEN mutation present in these BT-549 cells.

To see if we could indeed perform DNA sequencing of sorted single tumor cells with mutations we next sorted single cells with several mutations, including the BT-549 cells which have a single base pair mutation in the so-called PTEN gene, a tumor suppressor gene that has been previously described [28]. A clone of BT-549 cells was analyzed by flow cytometry and sorted into single PCR tubes for subsequent analysis by PCR and TA cloning. The DNA from the PTEN gene region of sorted single cells was amplified by PCR and cloned into a TA cloning vector. The vector was then amplified by growth in host *E. coli* bacterial cells and the plasmid DNA was isolated and purified. This purified plasmid DNA containing the PTEN gene sequences from sorted single cells was then analyzed on a DNA sequencing instrument. A typical DNA sequencing result from a single sorted BT-549 cells is shown in Figure 10.11 [12]. Cells without the mutated sequence show normal PTEN gene sequences (data not shown).

As can be seen in Figure 10.12, while the overall single cell sorting efficiency of this experiment was close to 100% (the average efficiency is approximately 90–95%, taking into account day to day variations in sorting efficiency and stability), the percentage of cells that were sufficiently amplified with nested primers around the HLA-DQ-alpha region for single cell DNA typing was only 70%. However, by using oligo probes for each DNA type, six of the seven sorted single cells detectable by DNA typing were the of the correct DNA type (i.e. MCF-7 tumor cells) indicating that we were able to detect and isolate the rare tumor cells of interest [12].

10.5.2
Single Cell Sorting for Mutational Analysis by PCR

Before attempting to sort the rare BT-549 cells for DNA sequencing, we first tested the sensitivity of the overall system by sorting 50 cells, 10 cells and single cells from a pure mixture of MCF-7 cells (Figure 10.12). As can be seen from the results, the overall efficiency of sorting was virtually 100% for single cells. But in this defined cell mixture experiment, the additional complication involves successful detection of the rare cells using the positive and negative selection markers described in the beginning of this chapter in Figure 10.1. There is also the possibility for contamination of the sorted cell sample during the entire process which we sometimes find in these experiments. Hence DNA typing of each sorted single cell serves as a "truth set" for assessment of sorting purity and accuracy in the case of defined cell mixtures with cells of different DNA types. Unfortunately, in the "real case" of tumor cells in a patient's blood or bone marrow, the DNA types are the same. But for this model system the DNA typing tells us the exact identity of the sorted cell. Depending on the success rate in obtaining the correct rare cells, we must sort additional single cells to insure that we obtain the statistical sampling necessary for detecting a tumor cell subclone of a given frequency.

We then tested for sort recovery and purity in the case of a defined cell mixture of MCF-7 and CEM-C7 cells. We used DNA HLA-DQ-alpha typing of MCF-7 cells and CEM/C7 cells to show our efficiency at successfully selecting tumor cells. Since the amount of PCR-amplified DNA from a single cell is too small to be seen by conventional ethidium bromide staining of the PCR products on a gel, enzymatic amplification techniques must be used. One such technique we have used is enhanced chemiluminescence (ECL; Amersham Life Science, catalog no. RPN 3021) whereby Southern blotting is performed with enzymatically labeled complementary sequence oligonucleotide ("oligo") probes. The enzymatic reaction gives rise to chemiluminescence which can then be detected on X-ray film, but without the hazards and disposal problems of radioactive probes. ECL provides an enzymatic amplification factor of approximately 1000 times, permitting detection of PCR products from single cells.

10.5.3
TA Cloning

"TA Cloning" techniques (Figure 10.13) have been developed and are commercially available. Our laboratory uses the TOPO TA cloning version B kit with the pCRII-TOPO vector (Invitrogen, catalog no. 45-0640). Briefly, this kit provides a highly efficient, 5-min, single step strategy for the direct insertion of Tug polymerase-amplified PCR products into a plasmid vector. No ligase, post-PCR procedures, or PCR primers containing specific sequences are required. The plasmid vector pCR II-TOPO is linearized with a single overhanging 3' deoxythymidine (T) residues. The Taq polymerase has a nontemplate-dependent terminal transferase activity which adds a single deoxyadenosine (A) to the 3' ends of the PCR products. This arrangement exploits the ligation reaction of topoisomerase by providing an "activated",

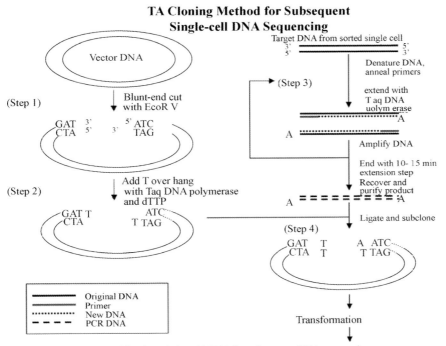

Figure 10.13 TA cloning allows for more faithful expansion (in terms of fidelity of DNA sequences) of the DNA from single sorted cells for subsequent DNA sequencing of particular genes that might be mutated (figure adapted from Invitrogen, Inc.). The overall strategy of "TA cloning" is depicted in this figure. Sorted cells are PCR-amplified with appropriate primers to amplify the sequences of interest. Then these amplimers with A overhang ends, are ligated into a TA cloning vector which contains a region which can be cut by a restriction endonuclease to have a T overhang region. The cloning vector can then be amplified in its host to produce quantities of DNA sequences sufficient for subsequent DNA sequencing [12] and adapted from Invitrogen product literature.

linearized TA vector using proprietary technology (Shuman, 1994). Ligation of the vector with a PCR product containing 3' overhangs is very efficient and occurs spontaneously within 5 min at room temperature (p. 1,TOPO TA cloning version B manual; Invitrogen). The TOPO-cloning reaction can then be transformed into competent cells for subsequent transformation and ultimately plasmid purification. This technique therefore involves PCR amplification of a particular gene of interest (e.g. a tumor suppressor gene) and then ligating these amplimers into a specially designed TA cloning vector.

Sorted cells were PCR-amplified with appropriate primers to amplify the sequences of interest. Then these amplimers. with A-overhang ends are ligated into a TA cloning vector which contains a region which can be cut by a restriction endonuclease to have a T overhang region. The cloning vector can then be amplified in its host to produce quantities of DNA sequences sufficient for subsequent DNA sequencing [12].

10.5.4
Single Cell Analysis of Gene Expression Profiles

The sorting of cell subpopulations for subsequent gene expression microarray analyses pushes the limits of current sorting technology. Because the gene expression profile (GEP) of heterogeneous cell populations will reflect the weighted average of all cell subpopulations present according to their relative frequencies, cells must be sorted to sufficient purity to allow for meaningful results [29]. In addition, to avoid the distortions of the GEP caused by so-called linear amplification technologies (that are never linear!), the number of sorted cells per gene chip to be analyzed needs realistically to be in the order of 100 000 cells and more commonly 100 000 000 cells(Figure 10.14), with more cells being required for oligo-based microarrays than spotted microarrays (Figure 10.15) [30].

Total RNA yield	
10^6 cells:	10µg (3-15µg)
10^3 cells:	10ng (1-10ng)
1 cell:	10pg (0.1-10pg)

T7 amplification	
Total RNA ⇒ aRNA	
1 round:	10^3-fold (1-3 x 10^3)
2 rounds:	10^6-fold (1-5 x 10^6)

Note: T7 amplification only amplifies **mRNA** so the resulting aRNA is 10-20 times less in volume than if it was total RNA.

Figure 10.14 Oligo microarrays (e.g. Affymetrix) require approximately 10^6 copies of gene sequences to be properly read on the chip. Each round of T7 amplification provides approximately a 1000-fold expansion of the gene sequences. One round of T7 amplification of the mRNA provides enough material from 1000 cells for an Affymetrix analysis. Two rounds of T7 amplification provide enough material from a single cell for such an analysis.

Array type:	Required amounts / aray	
	Spotted	Oligo
Total RNA:	1µg (0.5-5µg)	10µg (3-15µg)
mRNA or aRNA:	0.1µg (0.05-1µg)	1µg (0.5-5µg)
Cells wihout amp:	10^5	10^6
Cells, 1 round amp:	$10^2 - 10^5$	$10^3 - 10^6$
Cells, 2 round amp:	$1 - 10^2$	$1 - 10^3$

➢ RNA amplification is necessary below 10^5 cells.

➢ 1 round of T7 amplification is enough for 1000 cells and above.

➢ 2 round of T7 amplification is enough for even a single cell.

Figure 10.15 Oligo-based microarrays (e.g. Affymetrix, Inc.) require approximately 10 times as much RNA as spotted microarrays (e.g. Clontech, Inc.). Approximately 10^6 cells are required for oligo microarrays or 10^5 cells from spotted arrays if RNA amplifications methods are not used.

However, the GEP profile of all well-behaving genes (in terms of their oligo sequences on the Affymetrix chips) can be recovered for 1000 sorted cells with one round of T7 amplification and on a single sorted cell after two rounds of T7 amplification. To obtain meaningful results for even "moderately expressing" genes requires sort purities greater than 70%. Low expressing genes require sort purities greater than 90% and more typically greater than 95% purity [29]. If one is trying to purify cell subpopulations with less than 10% purity, this requires a minimum cell sorting rate faster than 100 000 cells/s. For cell subpopulations with less than 1%, for example, stem/progenitor cells, a more realistic sorting rate would be in excess of 500 000 cells/s. Importantly, it should be remembered that the GEPs of these live cells are rapidly changing, so sort experiment times of 4–10 h are not possible without considerable degradation of the mRNA of these cells. Nevertheless, meaningful if not perfect information can be obtained about the GEP of rare sorted cells and even single cells (Figure 10.16) [30]. An example of results from sorted human stem/progenitor cell lines and cells from cord bloods of newborns are shown in Figure 10.17 [30].

Sorting by brute force for many hours is simply not an option with live cells because the cells themselves are changing faster than this time interval. Stating an obvious but frequently ignored scientific logic, the measuring process must always be much faster than the rate of change of the objects being analyzed! Proteins tend to be a little more stable in terms of half-life. Subsequent mass spectrometry of proteins from sorted cells probably requires on the order of 10^6 to 10^7 cells, depending on the type and sensitivity of the instrumentation. Hence, sorting a 10% cell subpopulation would require analysis of 10^7 to 10^8 total cells, assuming no cell losses. This is doable with current high-speed cell sorters. In contrast, if only 1% of the cells are sorted, we start to get into the realm where higher rates of cell sorting are needed.

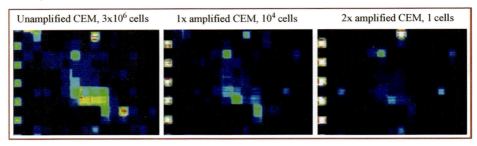

Figure 10.16 Affymetrix array images of linear GEP amplification - CEM cells. GEPs were compared for 10^6 cells unamplified, 10^4 cells after one round and a single cell after two rounds of T7 amplification. Each small square represents the expression level of an individual gene sequence. To visualize the raw image data of microarray analysis, corresponding segments of Affymetrix images were cropped, magnified and pseudo-colored using the Affymetrix Microarray Suite 5.0 software. Even a single cell produced an analyzable GEP. The pattern of expressed genes was similar at each stage.

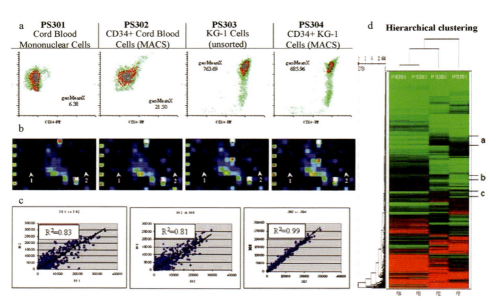

Figure 10.17 Microarray analysis of purified CD34+ cord blood stem/progenitor cells. (a) Flow cytometry scattergrams of cord blood mononuclear cells (CBMCs) magnetic activated cell sorting (MACS)-purified CD34+ cord blood stem/progenitor cells, unsorted KG-1a cells and MACS-sorted KG-1a cells. (b) Affymetrix microarray images of the same four samples. Arrow 1 is pointing at a sequence expressed only in the first two samples. Arrow 2 is pointing at a sequence expressed only in CD34+ cord blood stem/progenitor cells. (c) Analysis of data. (d) Heat map of the same four samples. Samples and genes are ordered by Spotfire hierarchical clustering analysis based on normalized expression levels. Some groups of genes (a, b, c) are differentially expressed in CD34+ cord blood stem/progenitor cells.

10.6
Conclusions and Discussion

Using a variety of rare event staining, detection, sorting and data analysis methods, single rare cells of frequencies as low as 10^{-6} can now be successfully identified and isolated. That is not to say that such experiments are easy to perform. There are challenges at several levels and the major challenge is performing all phases of the experiments well. There are now a variety of molecular characterizations that can be performed at the single cell level to analyze specific DNA sequences to search for mutations related to the outcome of certain human diseases. RNA amplification methods can recover the general GEP of rare cells even at the single cell level.

Acknowledgements

This work was supported by several NIH grants from the US Public Health Service (CA61531, GM38645, EB 00245). The data shown in Figure 10.3 were obtained from cell samples processed by Drs. Jonathan Ward and Marinel Ammenheuser from the University of Texas Medial Branch in Galveston during our collaborations. The data shown in Figures 10.15–10.17 and Table 10.1 were obtained by Dr. Leary's graduate student, Mr. Peter Szaniszlo and published as part of his PhD dissertation in 2007.

References

1 Leary, J.F. (1994) Strategies for rare cell detection and isolation in *Methods in Cell Biology: Flow Cytometry* (eds Z. Darzynkiewicz, J.P. Robinson and H.A. Crissman), vol.42, pp. 331–358.

2 Leary, J.F. (2000) Rare event detection and sorting of rare cells, in Emerging Tools for Cell Analysis: Advances in Optical Measurement Technology (eds G. Durack and J.P. Robinson), pp. 49–72, Wiley-Liss, Inc., Wilmington, DE 2000.

3 Leary, J.F. (2005) Ultra high speed cell sorting. Cytometry Part A. **67**, 76–85.

4 Ryan, D.H., Mitchell, S.J., Hennessy, L.A., Bauer, K.D., Horan, P.K. and Cohen, H.J. (1984) Improved detection of rare CALLA-positive cells in peripheral blood using multiparameter flow cytometry. *Journal of Immunological Methods*, **74**, 115–128.

5 Swets, J.A. and Pickett, R.M. (1982) Evaluation of Diagnostic Systems Methods from Signal Detection Theory, Academic Press in Cognition and Perception.

6 Beck, J.R. and Shultz, E.K. (1986) The use of relative operating characteristic (ROC) curves in test performance evaluation. *Archives of Pathology & Laboratory Medicine*, **110**, 13–20.

7 Hanley, J.A. (1989) Receiver operating characteristic (ROC) methodology: the state of the art. *Critical Reviews in Diagnostic Imaging*, **29** (3), 307–335.

8 Choi, B.C.K. (1998) Slopes of a receiver operating characteristic curve and likelihood ratios for a diagnostic test. *American Journal of Epidemiology*, **148** (11), 1127–1132.

9 Leary, J.F., Hokanson, J.A. and McLaughlin, S.R. (1997) High speed cell classification systems for real-time data classification and cell sorting. SPIE (Journal of the Optical Society of America), **2982**, 342–352.

10 Leary, J.F., McLaughlin, S.R., Hokanson, J.A. and Rosenblatt, J.I. (1998) High speed real-time data classification and cell sorting using discriminant functions and probabilities of misclassification for stem cell enrichment and tumor purging. *SPIE (Journal of the Optical Society of America)*, **3260**, 274–281.

11 Leary, J.F., McLaughlin, S.R., Hokanson, J.A. and Rosenblatt, J.I. (1998) New High Speed Cell Sorting Methods for Stem Cell Sorting and Breast Cancer Cell Purging. *SPIE (Journal of the Optical Society of America)*, **3259**, 114–121.

12 Leary, J.F., He, F. and Reece, L.N. (1999) Detection and isolation of single tumor cells containing mutated DNA sequences. *SPIE Proceedings Systems and Technologies for Clinical Diagnostics, and Drug Discovery II*, **3603**, 93–101.

13 Leary, J.F., McLaughlin, S.R., Reece, L.N., Rosenblatt, J.I. and Hokanson, J.A. (1999) Real-time multivariate statistical classification of cells for flow cytometry and cell sorting: a data mining application for stem cell isolation and tumor purging. *SPIE*, **3604**, 158–169.

14 Hokanson, J.A., Rosenblatt, J.I. and Leary, J.F. (1999) Some theoretical and practical considerations for multivariate statistical cell classification useful in autologous stem cell transplantation and tumor cell purging. *Cytometry*, **36**, 60–70.

15 Leary, J.F., McLaughlin, S.R. and Reece, L.M. (2000) Application of a novel data mining technique to cytometry data. *SPIE*, **3921**, 84–89.

16 Leary, J.F., Jacob Smith, J., Peter Szaniszlo, P., Lisa, M. and Reece, L.M., (2007) Comparison of multidimensional flow cytometric data by a novel data mining technique. *Proceedings of the SPIE*, **6441**, 64410N-1-8.

17 Leary, J.F. (2006) Subtractive Clustering for Use in Analysis of Data (advanced data mining algorithms for comparing two or more large multidimensional data sets) U.S. Patent 7,043,500.

18 Rosenblatt, J.A., Hokanson, J.A., McLaughlin, S.R. and Leary, J.F. (1997) A theoretical basis for sampling statistics appropriate for the detection and isolation of rare cells using flow cytometry and cell sorting. *Cytometry*, **26**, 1–6.

19 Zhang Cui, L.X., Schmitt, K., Hubert, R., Navidi, W., and Arnheim, N. (1992) Whole genome amplification from a single cell: implications for genetic analysis. *Proceedings of the National Academy of Science of the United States of America*, **89**, 5847–5851.

20 Stovel, R.T. and Sweet, R.G. (1979) Individual cell sorting. *Journal of Histochemistry and Cytochemistry*, **27**, 284–288.

21 Leary, J.F., Szaniszlo, P., Prow, T., Reece, L.M., Wang, N. and Asmuth, D.M., (2002) The importance of high-throughput cell separation technologies for genomics/proteomics-based clinical therapeutics. *Proceedings of the SPIE*, **4625**, 1–8.

22 Corio, M.A. and Leary, J.F. (1999) System for Flexibly Sorting Particles U.S. Patents 5,199,576(1993) and 5,550,058 (1996) and 5,998,212.

23 Gross, D. and Harris, C.M. (1985) *Fundamentals of Queuing Theory*, 2nd edn, John Wiley and Sons, New York.

24 Lindmo, T. and Fundingsrund, K. (1981) Measurement of the distribution of time intervals between cell passages in flow cytometry as a method for the evaluation of sample preparation procedures. *Cytometry*, **2**, 151–154.

25 van Rotterdam, A., Keij, J. and Visser, J.W. (1992) Models for the electronic processing of flow cytometric data at high particle rates. *Cytometry*, **13**, 149–154.

26 Leary, J.F., Hokanson, J.A., Rosenblatt, J.I. and Reece, L.N., (2001) Real-time decision-making for high-throughput screening applications. *Proceedings of the SPIE*, **4260**, 219–225.

27 Brenner, M.K., Rill, D.R., Moen, R.C., Krance, R.A., Mirroe, J., Jr, Anderson, W.F. and Ihle, J.N. (1993) Gene marking to trace origin of relapse after autologous bone marrow transplantation. *The Lancet*, **341**, 85–86.

28 Li, J., Yen, C., Liaw, D., Podsypanina, K., Bose, S., Wang, S.1., Puc, J., Miliaresis, C., Rodgers, L., McCombie, R., Bigner, S.H., Giovanella, B.C., Ittmann, M., Tycko, B., Hibshoosh, H., Wigler, M.H. and Parsons, R. 1997 PTEN, a putative protein tyrosine phosphatase gene mutated in human brain, breast, and prostate cancer. *Science*, **275** (5308), 1943–1947.

29 Szaniszlo, P., Wang, N., Sinha, M., Reece, L.M., Van Hook, J.W., Luxon, B.A. and Leary, J.F. (2004) Getting the right cells to the array: gene expression microarray analysis of cell mixtures and sorted cells. *Cytometry*, **59**, 191–202.

30 Szaniszlo, P. Gene Expression Profile Analysis of Purified Human Stem/Progenitor Blood Cell Subsets. Ph.D. Dissertation 2007 University of Texas Medical Branch, Galveston, Texas (Supervised by Dr. James F. Leary).

31 Shuman, S. (1994) Novel Approach to Molecular Cloning and Polynucleotide Synthesis Using Vaccinia DNA Topoisomerase. *Journal of Biological Chemistry*, **269**, 32678–32684.

11
Single Cell Heterogeneity
Edgar A. Arriaga

11.1
Introduction

One of the wonders in nature is the process of embryogenesis in which one single fertilized egg develops into a multicellular organism. During this process, cells replicate multiple times and, at the same time, differentiate into different cell types with a variety of properties and functions. Neurons, T cells, hepatocytes, skeletal muscle fibers and bone marrow cells are just a few representative examples of how diverse cell types that originate from a single cell might be. Cleary the process of cellular differentiation is an important source of cellular diversity in complex organisms. Although this diversity is a manifestation of cellular heterogeneity, this form of heterogeneity is not discussed here, as it is presently straightforward to analyze and compare different cell types due to their unique morphologies. Instead this chapter preferentially focuses on heterogeneities among cells that are isogenic (from the same origin) and that are, in practical terms, morphologically indistinguishable.

Within a given cell type or in an ensemble of isogenic cells, cells may show variations in their morphology, biochemical properties, composition, function and behavior. These variations may be subtle in comparison to the dramatic differences among cell types, but their relevance to functional status of the organism cannot be underscored. As an example, beta cells in the pancreas are responsible for the production of insulin in response to elevation in glucose levels and the global failure of this function leads to diabetes type II onset. These cells do not lose their ability to release insulin all at once, but display sporadic periods in which they are inactive in releasing this hormone. These periods occur at higher frequencies as the disease onsets. The progression of the disease is initially unnoticed because the remaining functional beta cells can cover the insulin demands from the body. Some dysfunctional beta cells may be replaced, somewhat alleviating the shortage of insulin, but as the number of dysfunctional cells increase, the insulin demand from the body are not fulfilled. This is an example of a heterogeneous dynamic population defined by one cellular function: the ability to release insulin. Understanding the heterogeneous nature of insulin release by single beta cells would be of tremendous benefit in defining the role of this heterogeneity

Single Cell Analysis: Technologies and Applications. Edited by Dario Anselmetti
Copyright © 2009 WILEY-VCH Verlag GmbH & Co. KGaA, Weinheim
ISBN: 978-3-527-31864-3

underlying this terrible disease. Fortunately single cell analysis techniques capable of monitoring insulin release are becoming available, thereby facilitating the description of cellular heterogeneity [1]. Other examples of single cell analysis techniques that may be used in single cell heterogeneity studies are described in the next section.

11.2
Measuring Heterogeneity using Single Cell Techniques

In principle any of the single cell analysis techniques that are described in this book are suitable for investigating the heterogeneous nature of single cells within a given cell type or isogenic group. Examples of such techniques are listed in Table 11.1 and a brief commentary on each of them is presented below.

Table 11.1 Examples of single cell analysis techniques useful to investigate heterogeneity.

Technique	Single cell measurement	Cellular system	References
Flow cytometry	Noise in abundances of GFP-fusion proteins	*Sacharomyces cereviciae*	[2]
Optical well arrays	Kinetics of *recA and lacZ* gene expression using GFP reporter	*Escherichia coli*	[3]
Fluorescence microscopy	Intracellular calcium release to identify subpopulations differing in expression of P2 receptors.	Human osteoblasts	[10]
Electrochemical detection	Time profile of bursts of insulin secretion	Rat and human pancreatic beta cells	[1]
Raman microspectroscopy	Changes in Raman spectra indicating coexisting cell types	*Clostridium beijerinckii* in an acetone-butanol fermentation reactor	[4]
CE –biomolecules	Two dimensional separation of proteins that are fluorescently labeled on-line	Cultured MC3T3-E1 osteoprogenitor and MCF-7 breast cancer cells	[6]
CE – organelles	Separation and detection of mitochondria labeled with DsRed2	Cultured 143B osteosarcoma cells	[7]
MALDI-MS	Identification of neuropeptides	Neuron cells isolated from *Aplysia californica*	[5]
LCM and cDNA microarray analysis	Gene expression profiling pointing to two subpopulations	CA1 neurons from rat hypocampus	[8]
Multiplexed, real-time RT-PCR	Quantification of 20 different mRNAs	Human small intestine cells selected by FACS	[9]

GFP: Green fluorescent protein; CE: capillary electrophoresis; MALDI-MS: matrix assisted laser desorption/ionization mass spectrometry; LCM: laser capture microdissection; RT-PCR: reverse transcriptase polymerase chain reaction; FACS: fluorescence-activated cell sorting.

High-throughput techniques, such as flow cytometry and optical well arrays, are ideal to rapidly measure large collections of single cells, but they report only a limited number of parameters per cell [2, 3]. Real-time monitoring techniques, such as optical well arrays and electrochemical detection, are ideal to monitor kinetics and time profiles as long as the single cells contain adequate reporters.

Information-rich techniques, such as Raman microspectroscopy and mass spectrometry, are still not widely used but hold great potential [4, 5]. Raman microspectroscopy, although very sensitive to changes in the chemical composition of single cells, has the disadvantage that the interpretation of the spectra may not be straightforward. Mass spectrometry has been successfully used in large single cells such as the neurons of the sea snail *Aplysia californica* but useful determinations in picoliter-sized cells has not been reported so far [5].

Separation techniques, such as capillary electrophoresis, have been successfully used to monitor fluorescent reporters found in picoliter-sized single cells. Both fluorescently labeled molecules that are free in solution or that are bound to organelles have been reported [6, 7]. Presently the capillary electrophoresis approaches is limited by their relatively low throughput.

Techniques for gene expression of single cells have also appeared [8, 9]. These techniques have excellent multiplexing capabilities, but require prior isolation and selection of single cells using techniques such as laser capture microdissection or fluorescent activated cell sorting.

Two emerging techniques among those found in Table 11.1 are described below. They are: (i) optical well arrays to monitor the time profiles of gene expression and (ii) capillary electrophoresis (CE) analysis of organelles released from single cells.

11.2.1
Optical Well Arrays

An optical well array is fabricated by etching a bundle of imaging fibers to ensure that the end of each fiber in the bundle has a diameter of 3.1 μm [3]. Figure 11.1 shows an image of such bundle in which *Escherichia coli* cells have been deposited. Cells expressing the reporter green fluorescent protein (GFP) upon stimulation with mitomycin C are deposited on the wells and the GFP fluorescence is monitored over time. Aproximately 200 cells, each collected in a separate well, are being monitored in Figure 11.1; the insets are images of part of the array containing some cells that are expressing GFP after 0 min (left) and 80 min (right) of mitomycin C treatment. This is a promising emerging technique because the kinetic information on gene activation can be obtained in high throughput.

11.2.2
Capillary Electrophoresis Analysis of Organelles Released from Single Cells

The second example illustrates the use of CE with laser-induced fluorescence (LIF) detection to detect mito-DsRed2 in individual mitochondria released from single cells [7]. Figure 11.2a is a schematic of the steps resulting in the release of organelles

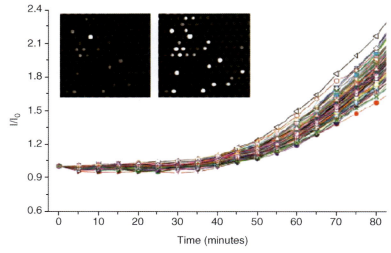

Figure 11.1 Monitoring the expression of the GFP reporter of the gene recA in an optical well array. Wells containing E. coli cells display GFP fluorescence that increases with time after treatment with mitomycin C. The left and right insets are partial images of the array at 0 and 80 min after treatment, respectively. Reprinted with permission from Ref. [3].

within the separation CE capillary. First the single cell is sandwiched between two plugs of a solution containing trypsin and digitonin (Figure 11.2a, part iii). Electrophoretic mobilization of the cell to one of the plugs causes disruption of the plasma membrane and then release of the organelles (Figure 11.2a, parts iv and v, respectively). Upon release, the organelles are electrophoretically separated and detected individually by a LIF detector resulting an electropherogram in which most spikes correspond to individual organelles (Figure 11.2b). The electrophoretic mobility of all detected mitochondria and their respective mito-DsRed2 contents are plotted to represent heterogeneity (Figure 11.2c and d, respectively). Thus every single cell has a distribution of electrophoretic mobilities and mito-DsRed2 contents of its individual mitochondria. This technique is unique in that it provides a description of heterogeneities among the organelles of single cells. That is, it describes how the fluorescent reporter mito-DsRed2 is distributed among the mitochondria from a single cell. Furthermore it suggests that mitochondria within the cell differ in regards to their surface charge, as inferred from the electrophoretic mobility measurements.

In summary, there are multiple single cell techniques capable of providing valuable information about the heterogeneity of cellular properties. Depending on the application and the techniques available, an investigator interested in exploring cellular heterogeneity may select a single cell technique based on its throughput, sensitivity, temporal or subcellular resolution, reporters (e.g. electrochemically active versus fluorescent reporters), destructive nature and molecular identification capabilities (e.g. peptide masses or gene expression).

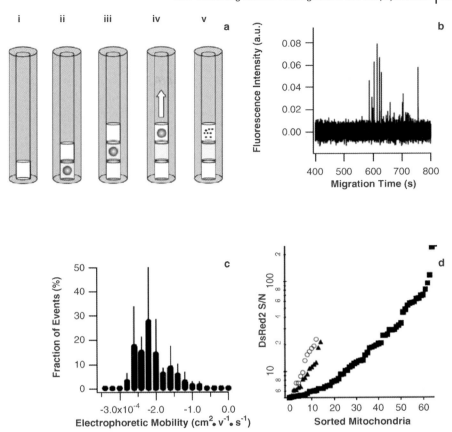

Figure 11.2 Capillary electrophoresis analysis of individual mitochondria released from single cells. (a) Schematic of loading disruption solution (i), a single cell (ii), and disruption solution (iii), the cell is then moved electrokinetically to the disruption solution (iv) and incubated for 5 min (v). The released organelles are then analyzed by capillary electrophoresis with laser-induced fluorescence detection. (b) Electropherogram consisting of individual mitochondria containing mito-DsRed2. (c) Distribution of electrophoretic mobilities of mitochondria contained in single cells. (d) Mito-DsRed2 contents of individual mitochondria released from single cells sorted from low to high contents. Reprinted with permission from Ref. [7].

11.3
Describing Cellular Heterogeneities and Subpopulations

Single cell measurements indicate that some cellular properties are defined by an on–off response or a response that may take many possible outcomes. An example of an on–off response is the cytokine mRNA expression in T-cells upon antigen presentation [11]. In such a case, individual T-cells can be distinguished by the presence or absence of cytokine mRNA. An example of a complex response is the intracellular Ca^{2+} release upon stimulation of mucosal mast cells with dinitrophenyl [12]. This response is characterized by bursts of Ca^{2+} release that vary in intensity, frequency and duration.

Regardless of the nature of the response, binary or complex, a large number of single cell measurements and their interpretation are usually needed in order to establish the boundaries and nature of such responses and to describe their heterogeneity. This is why high-throughput techniques (e.g. flow cytometry) are more appealing when investigating cellular heterogeneity. In addition, these investigations are usually carried out using cultured immortal cell lines or unicellular organisms because the environment in which the cells reside can be controlled with high accuracy thereby eliminating environmental effects.

If the boundaries and nature of cellular responses are known, it should be feasible to explore the effect of external parameters (e.g. cytokines, growth factors, nutrients) on the heterogeneity of cellular responses and properties. In addition, it should also be feasible to uncover cell subpopulations or even detect foreign cells that belong to a different cell type. For example, Openshaw and coworkers relied on immunofluorescence measurements to define that there are two subpopulations of $Cd4+$ T cell helpers, Th1 and Th2, in a cell culture that is exposed to an antigen [13]. These studies, whether on on–off responses, complex cellular behaviors, cellular subpopulations, or mixed cell types would not be possible when conducting measurements done in bulk preparations containing millions, because that would report only an average value.

11.4
Origins of Cellular Heterogeneity

Single cell analysis techniques make it possible to describe the heterogeneity in a cell ensemble. However uncovering and explaining the molecular origins of such heterogeneities are far from trivial and is still an open field. This section summarizes some of the current thoughts on the origins of cellular heterogeneities.

A cell may be viewed as complex network of molecular processes in which information is encoded and transmitted to ultimately produce a specific cell response, function or behavior that is measured using a single cell technique. For the sake of simplicity, one may envision one flow chart of one line of information in such a complex network. In this flow chart (Figure 11.3), the outcome of one molecular process affects the input of the next one and, consequently, the output of other downstream molecular processes. As expected in any natural process, the outcome of each molecular process fluctuates. This fluctuation may be binary, a series of bursts, the accumulation of a product, or the degradation of a substrate. It may depend on the subcellular location or have a temporal response. Logically the fluctuation in the outcome of a process would affect the next process in the flow chart. Activation of the next process may require an on-response (e.g. the transcription of a gene), a certain frequency of bursts (e.g. Ca^{2+} release), or a total number of product molecules above a threshold from the previous process. Extending this concept to the whole flow chart, the fluctuations in the measured property would result from the

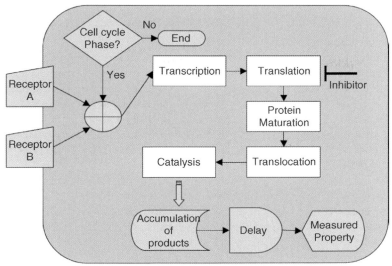

Figure 11.3 An oversimplified flow chart outlining molecular processes that may contribute to the fluctuations in a property measured using a single cell technique. The receptors A and B suggest that other environmental factors would also contribute to these fluctuations. Other processes are implicit. For example, chromatin unfolding, which is a main noise contributor, is included under *Transcription*.

fluctuations in the property itself and those propagated through the network. The reader is referred to the works of Barkai and coworkers for an example of mathematical models describing the propagation of fluctuations through a cellular network [14].

The fluctuations associated to molecular processes that lead to heterogeneities observed when measuring single cell properties is termed biological noise. It is postulated that there are three types of biological noise: stochastic, intrinsic and extrinsic. However when the type of noise is not well defined, it is simply called biological noise. It is important to notice that instrumental noise needs to be taken into consideration and corrected for in studies dealing with biological noise.

Stochastic noise associated with every one of the molecular processes in a flow line of information is the most fundamental reason for observing heterogeneity in a cellular property. If this noise is represented as a standard deviation, it is generally assumed that the variance, that is the (squared) standard deviation, is proportional to average of a stochastic distribution. Importantly the stochastic noise associated with each molecular process of a flow chart does not contribute equally to the observed heterogeneity.

Intrinsic noise refers to fluctuations originating from molecular processes, for example those processes explicitly shown or implicitly included in Figure 11.3 have an intrinsic noise. In a recent study [2], Newman and coworkers determined that the proximity of genes in a given chromosome, mRNA half-life and participation in protein–protein interactions do not have a strong correlation with intrinsic noise, even when the opposite is generally assumed to be true. They found that some proteins involved in the regulation of chromatin remodeling and histone modification have high intrinsic noise, that is SAGA, SWI/SNF, Ino80, Swr1 and Swr2. They postulate that high intrinsic noise levels may result from the binary output of a slow transcription process that leads to production of mRNA in bursts. In contrast, they found that the abundance of proteins involved in translation, proteolysis and acidification of vesicles in the secretory pathway show low noise, suggesting that these processes are highly regulated. A particularly interesting observation was that proteins involved in oxidative phosphorylation, heat shock, stress and amino acid synthesis show high noise. The fact that these proteins are associated with organelles that are distributed to during cell mitosis suggests that subcellular localization is another contributor to intrinsic biological noise.

Extrinsic noise, also known as global noise, refers to other sources of fluctuation not associated with individual molecular processes within the cell. A dramatic example is the differentiation of hematopoietic stem cells as a result changes in the microenvironment [15]. At the molecular level, events such as a ligand binding to receptors in Figure 11.3 represent environmental effects that would be considered extrinsic noise. Other environmental effects contributing to extrinsic noise are: cell to cell and cell to extracellular matrix contacts, mechanical stress and thermal stress. Finally extrinsic noise may lead to fluctuations in cell size, cell cycle frequency or status, and cell morphology, all of which must be taken into account when dissecting the origins of cellular heterogeneity.

11.5
Identifying Extrinsic and Intrinsic Noise Sources

Multiple investigators have laid the path to identify the sources of biological noise, that is the origins cellular heterogeneity. This section summarizes one approach, based on flow cytometry, to investigate biological noise. It is important to point out that other single cell analysis techniques could also be useful to investigate the origins of biological noise, that is cellular heterogeneity.

In order to investigate the origins of biological noise, Newman and coworkers measure the products of genes expressed as fusion green fluorescent proteins in single yeast cells using flow cytometry [2]. Their strategy can be summarized in four steps: (i) validation of the flow cytometry measurements, (ii) dissection of the observed fluctuations into instrumental, extrinsic and intrinsic noise, (iii) identification of gene products deviating from the noise-abundance trends, (iv) correlation of deviant gene products with parameters that contribute to noise.

11.5.1
Validation of the Flow Cytometry Measurements

It is important to establish that the fluctuations associated with the fluorescence intensities are indeed a representation of the cellular properties. Figure 11.4a shows the histogram distributions of GFP fluorescence intensities of single cells with two

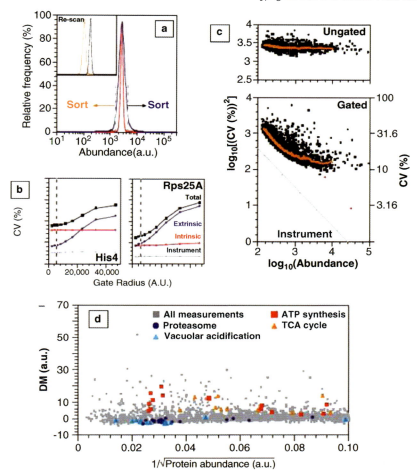

Figure 11.4 Using flow cytometry to investigate the origins of cellular heterogeneity. (a) Two genes expressing GFP-fusion products with similar protein abundace are analyzed by flow cytometry and each of them have different CVs. After sorting and re-analysis, their CVs remain unchanged suggesting that flow cytometry is a suitable technique to investigate heterogeneity in protein abundance. (b) Examples of the use of gating centered on the average forward and side scattering to estimate CV variations associated with intrinsic and extrinsic noise; left and right panels correspond to His4 and Rps25A genes, respectively; instrumental noise is estimated in a separate experiment; the vertical line indicates the selected gate ratio that makes possible to measure the CV associated with intrinsic noise. (c) Dependence of intrinsic noise on protein abundance. (d) Deviations (DMs) from the average CV values plotted in (c) versus the square root of protein abundance. Reprinted with permission from Ref. [2].

different GFP fusion proteins (Rpl35A-GFP and NOp8-GFP). They show coefficients of variation (CV) of 11.8% and 38.6%, respectively. When the cells are sorted and re-measured they retain the CVs of the original distributions, suggesting that flow cytometry is adequate to reproducibly measure the cellular properties of the system.

11.5.2
Noise Dissection

Gating is based on forward scattering (FSC), which is a measurement of size, and side scattering (SSC), which is a measurement of cellular granularity and is used to identify extrinsic noise. The CV is calculated for a given gate radius, centered at the average FSC and SSC. As the gate radius is reduced (right to left in Figure 11.4b) the CV decreases until it practically flattens out. The flat CV value suggests that extrinsic noise is basically eliminated and that the remaining fluctuations represent the sum of the instrument and intrinsic noise. This flat CV value can then be used to define a gate radius value (vertical dashed line in Figure 11.4b) that can be used to investigate intrinsic noise. The intrinsic noise is determined from the difference between the CV at the selected gate and the instrument noise, determined from the analysis of standard particles. In Figure 11.4b, the intrinsic noise is associated with the GFP fusion product of the His4 gene (left) and the RPS25A gene (right). Both the intrinsic and instrumental noise are uncorrelated and thus they can be represented as flat lines in the CV versus gate radius plot, as shown in Figure 11.4b. By subtracting the instrument noise and the intrinsic noise from the total noise over the entire gate range, one obtains the extrinsic noise.

11.5.3
Identification of Deviant Gene Products

The third step focuses on identifying gene products that deviate from the average trend observed when plotting variance versus abundance. A plot of $\log_{10} [CV^2]$ versus $\log_{10}$ [abundance] is shown in Figure 11.4c for all the gene products that were investigated in the above-mentioned study [2]. In this figure, CV^2 shows no correlation with abundance for ungated results. However the trend is that CV^2 is inversely proportional to protein abundance for gate radii determined, as indicated in Figure 11.4b. It is interesting to note that, for the least abundant gene products, Figure 11.4c indicates that the corresponding CVs are around 30%.

CV deviations from the average trend (DM) are plotted versus a function of protein abundance (the inverse of the square root of protein abundance) in Figure 11.4d. This plot is useful to identify genes with associated intrinsic noise deviating significantly from the running average shown in Figure 11.4c. These genes, with either positive or negative DMs, are highly relevant to identify intrinsic sources of cellular heterogeneity.

11.5.4
Correlation of Gene Products with Potential Sources of Noise

While gene products displaying an average trend in Figure 11.4c are mainly associated with stochastic noise, gene products displaying deviations from the average trend in Figure 11.4c can be correlated with other parameters that shed light on the origins of intrinsic noise. Some properties strongly correlated with the observed variation are: protein copy number, mRNA copy number and mRNA variation. Properties weakly correlated are: gene proximity, protein interactions and mRNA half-life. In addition, by correlating CVs of genes with high DMs products with gene ontology (GO) terms, it is possible to identify those GO terms that are associated with low or high intrinsic noise. Some of the gene products correlated to GO terms were already mentioned in Section 11.4, where origins of cellular heterogeneity were described. One of these gene products, SAGA, is a protein involved with chromatin remodeling and histone modification that shows high intrinsic noise. In the study by Newman and coworkers, SAGA has the strongest correlation with a GO term that includes 230 gene products regulated by this protein. A second example of this study is the strong correlation of a GO term including gene products involved in translation and ribosomal processing with gene products showing low noise.

Although not explicitly discussed in the report by Newman and coworkers, similar studies to investigate the effects extrinsic noise ought to be possible. For example single cell measurements of cells grown under different experimental conditions could provide the data to identify how these global factors affect cellular heterogeneity.

11.6
Concluding Remarks

This section stresses the importance of single cell analysis as a tool to describe cellular heterogeneity within a cell type or an isogenic group of cells. The origin of such heterogeneity is explained in terms of fluctuations associated with intrinsic and extrinsic noise. There has been some progress in identifying the molecular bases of both noise types, mainly by using high-throughput techniques such as flow cyometry, but most current explanations need further verification.

A frequent question from experimentalists seeking to define the origins and effects cellular heterogeneity is: how many cells are necessary to define the heterogeneity associated with a given property? This is particularly critical when using low-throughput single cell analysis techniques. From the example discussed in the previous section, the limiting number of cells should be sufficient for selection of a threshold (i.e. the gate radius) defining the CV representing the intrinsic noise. Sufficient cells must remain when using this threshold, for example 10 cells when using a low-throughput single cell analysis technique. However at least twice that number (20 cells) must be used when a threshold is large. In practice, the experimentalist using a low-throughput single cell technique may want to implement an iterative process in which: (i) 20 cells are initially analyzed, (ii) the CV is plotted as a function of the threshold (i.e. the gate radius), (iii) the low end of the threshold range is monitored to determine whether a flat CV is observed and (iv) if the CV values do not plateau, more cells are analyzed and the overall process is repeated.

Indeed, an alternative to the tedious use of low-throughput user-intensive single cell analysis techniques would be the further development of the existing techniques to make their use compatible with acquisition of large sets of single cell data. Other chapters in this book highlight some of the impressive development that may, in the future, enable a clear understanding of the origins of cellular heterogeneity.

Acknowledgements

The author thanks the National Institutes of Health for a Career Award (K02 AG021453).

References

1 Huang, L., Shen, H., Atkinson, M.A. and Kennedy, R.T. (1995) *Proceedings of the National Academy of Sciences of the United States of America*, **92**, 9608–9612.
2 Newman, J.R., Ghaemmaghami, S., Ihmels, J., Breslow, D.K., Noble, M., DeRisi, J.L. and Weissman, J.S. (2006) *Nature*, **441**, 840–846.
3 Kuang, Y., Biran, I. and Walt, D.R. (2004) *Analytical Chemistry*, **76**, 6282–6286.
4 Schuster, K.C., Urlaub, E. and Gapes, J.R. (2000) *Journal of Microbiological Methods*, **42**, 29–38.
5 Li, L., Romanova, E.V., Rubakhin, S.S., Alexeeva, V., Weiss, K.R., Vilim, F.S. and Sweedler, J.V. (2000) *Analytical Chemistry*, **72**, 3867–3874.
6 Hu, S., Michels, D.A., Fazal, M.A., Ratisoontorn, C., Cunningham, M.L. and Dovichi, N.J. (2004) *Analytical Chemistry*, **76**, 4044–4049.
7 Johnson, R.D., Navratil, M., Poe, B.G., Xiong, G., Olson, K.J., Ahmadzadeh, H., Andreyev, D., Duffy, C.F. and Arriaga, E.A. (2007) *Analytical and Bioanalytical Chemistry*, **387**, 107–118.
8 Kamme, F., Salunga, R., Yu, J., Tran, D.T., Zhu, J., Luo, L., Bittner, A., Guo, H.Q., Miller, N., Wan, J. and Erlander, M. (2003) *The Journal of Neuroscience*, **23**, 3607–3615.
9 Peixoto, A., Monteiro, M., Rocha, B. and Veiga-Fernandes, H. (2004) *Genome Research*, **14**, 1938–1947.
10 Dixon, C.J., Bowler, W.B., Walsh, C.A. and Gallagher, J.A. (1997) *British Journal of Pharmacology*, **120**, 777–780.
11 Bucy, R.P., Panoskaltsis-Mortari, A., Huang, G.Q., Li, J., Karr, L., Ross, M., Russell, J.H., Murphy, K.M. and Weaver, C.T. (1994) *The Journal of Experimental Medicine*, **180**, 1251–1262.
12 Kuchtey, J. and Fewtrell, C. (1996) *Journal of Cellular Physiology*, **166**, 643–652.
13 Openshaw, P., Murphy, E.E., Hosken, N.A., Maino, V., Davis, K., Murphy, K. and O'Garra, A. (1995) *The Journal of Experimental Medicine*, **182**, 1357–1367.
14 Bar-Even, A., Paulsson, J., Maheshri, N., Carmi, M., O'Shea, E., Pilpel, Y. and Barkai, N. (2006) *Nature Genetics*, **38**, 636–643.
15 Roeder, I. and Loeffler, M. (2002) *Experimental Hematology*, **30**, 853–861.

12
Genome and Transcriptome Analysis of Single Tumor Cells
Bernhard Polzer, Claudia H. Hartmann, and Christoph A. Klein

12.1
Introduction

In the European Union each year about one million people die from various types of cancer [1]. Over recent decades, molecular diagnosis has advanced greatly, but progress in therapy has been disappointingly small. This indicates that our understanding of the disease is still rather incomplete. One aspect of the disease is its remarkable cellular heterogeneity. A wealth of data suggests that there are no two individual cancer cells in a patient that are identical [2]. This excessive heterogeneity questions the usual approach of studying cancer biology with samples from pooled cells. It seems much more adequate to address the process of selection and mutation by analyzing individual cancer cells [3].

This perspective is justified when we consider the most fatal process of the disease, metastasis, which accounts for the majority of cancer-related deaths. Important steps during the metastatic process include dissemination of cells from the primary tumor, intravasation, survival in the circulation, extravasation, resistance to anoikis at the distant site and colonization of the ectopic organ (Figure 12.1). Selection for these steps acts on individual cells and our understanding of the cellular mechanisms that enable cancers to adapt to the ectopic environment will only increase if we manage to study individual tumor cells directly.

12.2
Detection and Malignant Origin of Disseminated Cancer Cells

For epithelial cancers, the most common approach for early detection of disseminated tumor cells is based on the expression of epithelial markers such as cytokeratins or the epithelial cell adhesion molecule (EpCAM) in mesenchymal organs like blood, lymph node, or bone marrow [4]. Cells positive for these markers are found at a frequency of one or two disseminated cancer cells per million bone marrow cells in 20–50% of cancer patients that present to the

Single Cell Analysis: Technologies and Applications. Edited by Dario Anselmetti
Copyright © 2009 WILEY-VCH Verlag GmbH & Co. KGaA, Weinheim
ISBN: 978-3-527-31864-3

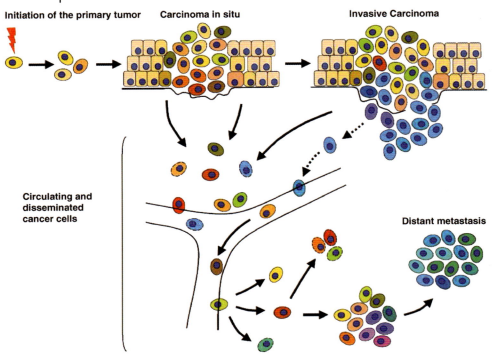

Figure 12.1 Systemic progression of cancer. The primary lesion progresses from a noninvasive *in situ* carcinoma to an invasive tumor – a process that is paralleled by the acquisition of genomic aberrations (upper part). Disseminating tumor cells may migrate from early lesions and not necessarily resemble late stage tumor cells, which are depicted in blue. The disseminating cancer cells, which form a heterogeneous cell population, enter the circulation and move with the blood or lymph to a distant site, where they leave the vessel and infiltrate the microenvironment. Even after long latency periods, selected tumor cells may grow into manifest metastases.

physician without manifest metastasis. Since the late 1980s, detection of cytokeratin-positive cells in bone marrow or EpCAM-positive cells in lymph nodes of patients suffering from various types of epithelial cancers has been demonstrated to be an important risk factor associated with reduced overall and disease-free survival [5], the most convincing data being available for breast cancer [6]. The clinical evidence that accumulated during recent decades has turned disseminated cancer cells defined by these markers into prime candidates for being direct metastatic precursor cells.

Although clinical studies investigating the prognostic impact of disseminated cancer cells in patients without manifest metastasis were initiated in the early 1990s, molecular characterization was hampered for many years due to the fact that only one or two cells per sample were available for analysis. Therefore, only techniques could be applied that used a marker antibody to detect the tumor cells plus one additional molecule, like an antigen of interest or specific DNA probes. Such double-labeling approaches demonstrated either the histogenetic [5, 7] or the malignant origin of cytokeratin-positive cells from bone marrow. A FISH study in prostate cancer could reveal chromosomal abnormalities in cytokeratin-positive cells in bone marrow [8]. In addition to these descriptive studies, a cell line could be established from micrometastatic cells. A lymph node from an esophageal cancer patient without clinical evidence of lymph node metastasis was taken into culture and the esophageal cancer cell line derived thereof formed tumors in immunocompromised mice [9]. Today, this remains the only cell line from disseminated tumor cells that can be generated without genetic manipulation.

Taken together, to address the molecular biology of these extremely rare metastatic precursor cells, methods were needed that enable comprehensive molecular characterization. Not surprisingly, PCR techniques opened the first gate to obtain comprehensive information from disseminated cancer cells.

12.3
Methods for Amplifying Genomic DNA of Single Cells

When trying to amplify the whole genome of a single cell, one is confronted with the challenge of amplifying single copy sequences and the risk of losing one of the two copies in an diploid genome ("allelic loss"). This phenomenon is thought to be caused by varying distributions of reagents at the target sites, such that not all components necessary for the PCR reaction are available [10]. Template quality may also become a limiting factor in particular when clinical material is under study.

PCR-based protocols for DNA amplification of a few or even single cells have been published from the early 1990s and a selection is presented in Table 12.1. The first widely used method employed degenerative oligonucleotide primers for PCR (DOP). It relies on a mixture of defined 5′ and 3′ bases and six random core nucleotides which enable annealing of the primers throughout the DNA template [11]. Another approach is primer extension pre-amplification (PEP) PCR, which uses highly degenerated oligonucleotides to prime template DNA [12–14]. However, all methods employing degenerate oligonucleotides suffer from the limitations that not all genomic regions are equally primed and that amplification conditions for the 4096 (DOP) or 10^9 (PEP) primers within the reaction vary considerably. Yet, an improved procedure (I-PEP) was successfully used for single cell analysis of defined genomic loci [15]. Some problems inherent to DOP or PEP protocols were overcome by the highly processive polymerase phi29 and the development of multiple strand displacement (MSD) amplification [16]. The method generates relatively long DNA fragments during amplification and has successfully been used for diluted DNA samples [17] as well as single cells [18].

While all the methods mentioned so far are suited for the analysis of predefined genetic loci, they failed to convincingly demonstrate a homogenous amplification of a diploid single cell-genome. As a consequence, comparative genomic hybridization (CGH, a technique that enables genome-wide screening for DNA copy number variations [19, 20]) cannot be applied to single cells. To this end we developed an

Table 12.1 Methods for amplifying limited amounts of genomic DNA.

Method	Advantages	Limitations	Demonstrated applicability for single cell array CGH	References
DOP-PCR	Direct PCR on template DNA	No reliable single cell metaphase CGH	No	[11, 43]
PEP-PCR	Direct PCR on template DNA	Not tested for single cell metaphase CGH	No	[12–14]
Adapter–linker PCR (SCOMP)	Excellent for metaphase CGH. Only 5–10% of allelic drop out. Good performance with minute amounts of damaged DNA from paraffin-embedded samples	Complex procedure, requiring digestion of template DNA and adapter-ligation before PCR	Yes, detection of aberrant regions as small as 1–2 Mb	[21, 44, 49]
MSD-PCR	Direct amplification of template DNA	Not tested for single cell metaphase CGH	Yes, but no improvement in resolution compared to metaphase CGH	[16–18]
Combination of DOP- and PEP-PCR	Direct PCR on template DNA	25–33% allelic drop out. High resolution not shown for all genomic regions	Yes, but no improvement in resolution compared to metaphase CGH	[26] www.sigmaaldrich.com

adapter–linker PCR strategy that amplifies the whole genome [21], does not introduce a bias and, for the first time, enables reproducible application of metaphase comparative genomic hybridization (CGH). We therefore called the procedure single cell *comp*arative hybridization (SCOMP). The resulting single cell genome representation can be used not only for genome-wide screens but also as a template for multiple sequence-specific PCR.

12.4
Studying the Genome of Single Disseminated Cancer Cells

We used SCOMP for the analysis of single disseminated cancer cells. During validation we applied the method to single cells with known karyotypes (such as single normal cells or cells from donors with trisomy 21) and obtained the expected results [21]. In contrast, no prior knowledge was available for single tumor cells isolated from bone marrow or lymph nodes of patients. Therefore, we reasoned that several individual cells from one patient should be similar (i.e. assuming clonal relatedness) which would serve as internal control. We therefore screened over 500 bone marrow and lymph node samples of various carcinomas and detected more than one disseminated cancer cell in 71 samples (14%). As expected, individual disseminated cancer cells from the same patient displayed a high degree of similarity in their chromosomal imbalances, but to our surprise only when the patient was in the stage of manifest metastasis (stage M1). As long as the patients were clinically free of metastases (i.e. in clinical stage M0) we observed an excessive genomic heterogeneity of tumor cells from the same individual. Thus, the data strongly suggest that, during progression from M0 to M1, very aggressive cancer cells are selected from a very high number of variant cells which clonally expand at a later time point. Disseminated tumor cells from M1 patients display characteristic chromosomal aberrations such as amplification of 8q and 17q or loss of 17p that are rarely detected in cytokeratin-positive cells isolated from M0 patients. Interestingly these aberrations belong to a set of five gains or losses that contain the information whether the cytokeratin-positive cell is obtained from a patient with or without manifest metastasis. Thus CGH analysis of a single disseminated cancer cell can predict the metastatic stage of disease of a breast cancer patient with 85% accuracy [22].

While it is interesting that a single cytokeratin-positive cell from bone marrow of a breast cancer patient contains the information about the stage of disease, this observation is diagnostically irrelevant as there are standard imaging procedures to detect metastases. Likewise the finding of an excessive genetic heterogeneity during the latency period prior to manifest metastasis is important to understand the evolutionary dynamics of systemic cancer spread but is less helpful to identify genetically activated therapy target genes that are shared among disseminated cancer cells. Such genes are desperately needed for the development of drugs that interfere with oncogenic mechanisms in order to prevent lethal metastasis. Taking into account that, in many cases, individual tumor cells from one patient often did not share a single chromosomal aberration, we must conclude: (i) clonal relatedness at this stage cannot be determined through molecular karyotyping, (ii) shared subchromosomal changes or point mutations must be identified in these cells in order to call breast cancer a monoclonal disease, and (iii) if cancer is a monoclonal disease, subchromosomal changes shared with tumor cells at the primary lesion may indicate the earliest genetic changes occurring during transformation of epithelial tissues and thus may reveal the causes of sporadic cancers.

In order to address these questions we tried to increase resolution of our single cell analysis. We applied polymorphic markers and quantitative PCR (using genomic DNA as template) for the detection of allelic losses and gene amplification, respectively. In order to identify early lesions we concentrated on cells that – to our great surprise – did not display any karyotypic aberrations. The finding of these cells was unexpected because karyotypic abnormalities are present early in primary tumors. However, 60% of the isolated cytokeratin-positive cells did not display any genomic rearrangements detectable by metaphase CGH, but showed frequent loss of subchromosomal regions when loss of heterozygosity (LOH) analysis was performed. Some of the changes were shared with the matched primary tumors and therefore not only identified these cells as being derived from the neoplastic mammary growth but also identified genetic lesions that were present before dissemination. Thus the finding of subchromosomal aberrations in disseminated cancer cells with a normal karyotype points to an early time point of dissemination before the onset of chromosomal instability [23]. Recently, we could confirm the early time-point of dissemination in mouse models and breast cancer patients with ductal carcinoma in situ (DCIS) [50]. Of clinical relevance, we found in about 20% of early disseminated breast cancer cells a gain of the well known oncogene HER2, which renders the patient eligible for therapy with Trastuzumab (herceptin), an antibody directed against the oncogene. This finding was surprising because most of the primary tumors of these patients did not display a HER2 amplification [23] and would therefore not receive adjuvant HER2 therapy. Clinical studies are under way to address the question whether or not patients with HER2-amplified disseminated cancer cells (in the absence of a Her2-amplification in the primary tumor) would benefit from trastuzumab treatment. In addition, we found in esophageal cancer patients, that Her2 amplifications in disseminated tumor cells but not in the primary tumor were prognostically relevant. In disseminated tumor cells from patients with adenocarcinoma of the esophagus amplifications of the Her2 gene were found even more frequent than in disseminated tumor cells of breast cancer patients. In future therapeutic trials these patients should be stratified by Her2 gene amplifications of their disseminated tumor cells [51].

12.5
The Need for Higher Resolution: Array CGH of Single Cells

The LOH study demonstrated that the analysis of a single cell genome by several hundred of gene specific PCR is feasible after global amplification; however it is extremely laborious and not comprehensive. Thus there is a need to apply genome-wide techniques that provide higher resolution than metaphase CGH, such as array

CGH. Array CGH uses oligonucleotides or cloned genomic DNA (e.g. bacterial artificial chromosomes, BACs) as hybridization matrix instead of metaphase chromosomes. So far, several hundred nanograms of cellular DNA are needed to exploit the advantages provided by this method. Attempts to apply the method to single cells have not yet resulted in an increase of resolution beyond metaphase CGH, as the signals of several BAC clones have to be integrated for noise reduction [24, 25]. For example single cell array CGH using DOP- or MSD-based amplification protocols reach a resolution of 34–60 Mb [25]. Recently, a method combining DOP and PEP PCR for single cell genome amplification (GenomePlex single cell whole genome amplification kit) was tested on a genome-wide array comprising more than 30 000 BAC clones. The authors analyzed single epithelial cancer cells from cell lines and uncovered an amplification of 8.3 Mb [26]. While changes of this size are in the range of metaphase CGH, it was clearly demonstrated that single cell array CGH is able to resolve aberrations smaller than chromosome arms. The failure to increase resolution further likely results from stochastic amplification of single cell DNA. Possibly the allelic drop-out rate of 25–30% of the GenomePlex kit is too high to achieve further improvements. In this respect it is noteworthy that SCOMP has an allelic drop-out of only 5–10% [23]. Furthermore, recently we could show that our customized SCOMP array CGH is able to detect genomic aberrations down to a resolution of 1–2 Mb in single TD47 breast cancer cells and can be applied for the analysis of single disseminated tumor cells [49].

12.6
Studying the Gene Expression of Single Disseminated Cancer Cells

Single cells contain about 1–6 pg of mRNA, with most transcripts being present in only 10–15 copies per cell [27]. Techniques to study the gene expression of single cells therefore need to reach an exquisite sensitivity, which is achieved by gene-specific reverse transcription PCR that can be performed on a selected target mRNA. In cancer research this application is sometimes used to detect individual circulating tumor cells in blood samples [28]. However if the cells of interest are isolated and individual transcripts are to be directly quantified, the maximum number of PCR reactions is limited to about 10 without prior amplification of the target cDNA. This limitation results from the fact that, in 10 aliquots of the cellular cDNA, most transcripts are present as single copy molecules. Thus in any experimental setting, diluting or splitting the sample to perform more than one PCR reaction inevitably results in a stochastic bias.

Thus for a comprehensive transcriptional profiling of disseminated cancer cells, gene-specific approaches are less suited. For gene expression analysis on microarrays, which enable to study the expression of all known transcripts in parallel, several micrograms of RNA are usually required. Consequently, the single cell transcriptome needs to be amplified for this application. This is achieved by T7-based *in vitro* transcription [29, 30], a method that enables linear amplification of target sequences, or by PCR-based amplification, which results in an exponential increase of amplicons [31, 32], as summarized in Table 12.2. The PCR-based methods

Table 12.2 Methods for amplifying limited amounts of mRNA.

Method	Principle	Advantages	Limitations	References
T7-based amplification (Eberwine)	Linear amplification by T7-RNA polymerase	Widely used, well documented, kits available	Sensitivity for comprehensive single cell gene expression analysis not convincingly demonstrated	[30, 45]
PCR-based amplification (Brady)	Exponential amplification using dT/dA primers	Relatively simple protocol, widely used Higher amplification efficiency than linear methods	cDNA synthesis under limiting conditions Limited sensitivity due to oligo-dT/dA primer in PCR	[32, 41, 46]
PCR-based amplification (Kurimoto)	Exponential amplification using dT/dA primers	Higher amplification efficiency than linear methods Higher sensitivity than Brady (two rounds of PCR plus T7 amplification) Applicable to oligonucleotide arrays using T7 technology	cDNA synthesis under limiting conditions Limited sensitivity due to oligo-dT/dA primer in PCR complex protocol	[40]
PCR-based amplification (SCAGE)	Exponential amplification using dC/dG primers	Relatively simple protocol, robust, higher amplification efficiency compared to linear methods and dT/dA–PCR protocols Can be combined with SCOMP for analysis of genomic DNA from same cell (Table 12.1)	Limited sensitivity for very low abundant transcripts in single murine cells	[31]
PCR-based amplification (SCAGE)	Exponential amplification using dC/dG primers, optimized mRNA extraction	Improved mRNA isolation and cDNA synthesis compared previous SCAGE protocol Demonstrated sensitivity for rare transcripts in murine cells	Use of T7-based hybridization techniques and Affymetrix platforms only after modification of the amplification products	[33]

TALPAT	Combination of T7 and PCR amplification	Applicable to oligonucleotide arrays Can be combined with SCOMP for analysis of genomic DNA from same cell (Table 12.1) Higher amplification efficiency compared to T7-based linear amplification alone Applicable to oligonucleotide arrays using T7 technology	Sensitivity for single cells not convincingly demonstrated multistep protocol	[47]
RACE	Amplification of 3′ sequences of RNAs	Simple, fast	Mostly 3′-UTR amplified Currently limited to 40 genes	[48]

are believed to be less complicated and more sensitive, while it is often assumed that only linear amplification methods are quantitative. However we could demonstrate by quantitative PCR that our PCR-based method preserves the relative transcript ratios [33]. For the T7 method [30], it is still unclear whether the original proportions of mRNAs are preserved, because a sequence specific bias was observed during amplification at least in one of three studies [34–36]. For PCR-based approaches, it has been proposed that the amplification reaction should be stopped before reaching the plateau phase or that certain sequences are amplified preferentially [37–39]; however neither we nor the group of Norman Iscove observed such an effect [32].

While no fully validated method for comprehensive transcriptional profiling of single cells has been published using linear amplification, several PCR-based protocols for single cell gene expression profiling on large-scale arrays have been published [33, 40, 41]. Tietjen and coworkers used the Brady protocol for single cell amplification and hybridized the PCR products onto an Affymetrix platform. However, several aspects in this study escape critical evaluation, such as the control for efficient and specific binding of the DNA to the short Affymetrix oligos that preferentially bind RNA; also no comprehensive gene expression analysis was performed. In principle the protocol suffers from the limitations of the original Brady protocol, which combines conditions of limited processivity for the enzymes used in cDNA synthesis and primers of imperfect amplification efficiency (see supplemental information in Ref. [31]). The protocol by Kurimoto *et al.* consists of two rounds of PCR amplification. During the second PCR, a T7 promoter sequence is introduced into the cDNA fragments to allow for the subsequent T7-based amplification, labeling and hybridization of the PCR products. Using qPCR, the authors could demonstrate small inter-experimental variations of diluted total RNA especially for genes with higher expression (i.e. 20 copies per cell). Since the reliability and reproducibility was only assessed with diluted RNA, the number of transcripts in single cells by this protocol and thus its sensitivity are unknown. The method was applied to morphologically indistinguishable single cells from mouse blastocysts and it was found that these cells consist of two individual cell populations (primitive endoderm- and epiblast-like) by embryonic day 3.5.

High sensitivity and reproducibility for single cell analysis was recently achieved by slight modification of our protocol (*s*ingle *c*ell *a*nalysis of *g*ene *e*xpression, SCAGE [31]) which enabled retrieval of the correct histogenetic origin of cells and monitoring of the cellular differentiation and pathway activation in individual cells [33]. On microarrays that comprised 17 000 transcripts epithelial cells, hematopoietic stem cells, immature and mature dendritic cells were all correctly classified, providing the most comprehensive gene expression study on large-scale arrays so far. Maturation of dendritic cells after stimulation with LPS could be monitored in single cells by analysis of the Tlr4 pathway that is activated after LPS stimulation.

The protocol is based on a solid-phase capturing of the cellular mRNA, its reverse transcription into cDNA and PCR amplification. The size of the cDNA is restricted by usage of random primers during cDNA synthesis to a size optimal for PCR. The first primer binding site (a poly-dC stretch) for PCR is a flanking 5′ region in all cDNA primers that are elongated during reverse transcription. The second poly-dG-containing primer binding site is generated by a tailing reaction, enabling amplification of all fragments with a single poly-dC primer. Making all sequences equally CG-rich reduces sequence-dependent variation of amplification efficiency and maintains the relative transcript ratios. The quantitative nature of the amplification was demonstrated by qPCR experiments comparing the expression levels of specific genes from amplified and non-amplified single cells [33]. Moreover a quantitative analysis of transcripts from single heart cells isolated from young versus old animals identified loss of transcriptional control of gene expression as a mechanism that is closely associated with ageing [42].

After primary amplification, we used a secondary amplification in the presence of labeled nucleotides to label the PCR products for subsequent microarray analysis. This labeling method yielded best results. Upon hybridization on microarrays, we noted that contaminating bacterial sequences in the amplified sample may interfere with reliable measurements. Such contaminating sequences originate from the various recombinant enzyme preparations (mostly DNA binding enzymes) which are used for primary amplification and are inevitably coamplified and labeled. They result in false-positive hybridization signals if complementary sequences are also present on the array. Thus using arrays made of cloned cDNAs is often unsuccessful. In contrast, oligonucleotide arrays are free from bacterial or plasmid sequences and,

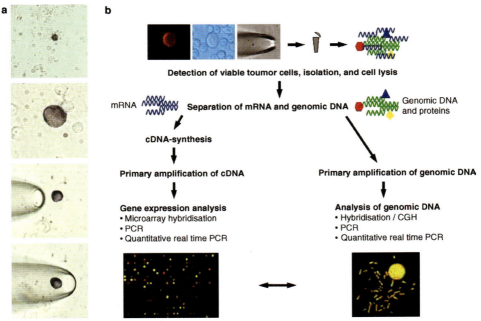

Figure 12.2 Techniques for analyzing single disseminated cancer cells. (a) Cancer cells of epithelial origin (blue; stained with an antibody against cytokeratin) can be isolated from bone marrow of carcinoma patients; (b) Combined genome and transcriptome analysis of single disseminated tumor cells [31, 33].

although hybridization conditions for single cells differ from standard procedures for total RNA, robust experimental conditions can be identified. Our data indicate that single cells express transcripts binding to 25–30% of all genes on the array, suggesting that low-abundance transcripts are also detected with high reliability. The correlation coefficient of technical replicates for both amplification and hybridization was found to range from 88% to 91%, similar to that observed for samples from several micrograms of total RNA.

12.7
Combined Genome and Transcriptome Analysis of Single Disseminated Cancer Cells

The analysis of single disseminated cancer cells isolated from the bone marrow of breast cancer patients in disease state M0 is currently under way and will certainly reveal important insights into the biology of these cells. Future studies of tumor cells isolated from different disease stages and from various cancer types will hopefully lead to an improved understanding of the systemic progression of cancer. Since SCAGE and SCOMP can be combined to analyze the mRNA and genomic DNA of the same cell (Figure 12.2), we might identify not only signaling pathways involved in ectopic survival, tumor dormancy and drug resistance but also the genetic basis of the abnormal gene expression. Hopefully this knowledge may then be translated into therapies that prevent the growth of lethal metastases.

References

1 Boyle, P., d'Onofrio, A., Maisonneuve, P., Severi, G., Robertson, C., Tubiana, M. and Veronesi, U. (2003) Measuring progress against cancer in Europe: has the 15% decline targeted for 2000 come about? *Annals of Oncology*, **14**, 1312–1325.

2 Bielas, J.H., Loeb, K.R., Rubin, B.P., True, L.D. and Loeb, L.A. (2006) Human cancers express a mutator phenotype. *Proceedings of the National Academy of Sciences of the United States, of America*, **103**, 18238–18242.

3 Klein, C.A. (2003) The systemic progression of human cancer: a focus on the individual disseminated cancer cell-- the unit of selection. *Advances in Cancer Research*, **89**, 35–67.

4 Schlimok, G., Funke, I., Holzmann, B., Gottlinger, G., Schmidt, G., Hauser, H., Swierkot, S., Warnecke, H.H., Schneider, B., Koprowski, H. et al. (1987) Micrometastatic cancer cells in bone marrow: in vitro detection with anti-cytokeratin and in vivo labeling with anti-17-1A monoclonal antibodies. *Proceedings of the National Academy of Sciences of the United States, of America*, **84**, 8672–8676.

5 Pantel, K., Cote, R.J. and Fodstad, O. (1999) Detection and clinical importance of micrometastatic disease. *Journal of the National Cancer Institute*, **91**, 1113–1124.

6 Braun, S., Vogl, F.D., Naume, B., Janni, W., Osborne, M.P., Coombes, R.C., Schlimok, G., Diel, I.J., Gerber, B., Gebauer, G., Pierga, J.Y., Marth, C., Oruzio, D., Wiedswang, G., Solomayer, E.F., Kundt, G., Strobl, B., Fehm, T.,

Wong, G.Y., Bliss, J., Vincent-Salomon, A. and Pantel, K. (2005) A pooled analysis of bone marrow micrometastasis in breast cancer. *The New England Journal of Medicine*, **353**, 793–802.

7 Oberneder, R., Riesenberg, R., Kriegmair, M., Bitzer, U., Klammert, R., Schneede, P., Hofstetter, A., Riethmuller, G. and Pantel, K. (1994) Immunocytochemical detection and phenotypic characterization of micrometastatic tumour cells in bone marrow of patients with prostate cancer. *Urological Research*, **22**, 3–8.

8 Mueller, P., Carroll, P., Bowers, E., Moore, D., 2nd, Cher, M., Presti, J., Wessman, M. and Pallavicini, M.G. (1998) Low frequency epithelial cells in bone marrow aspirates from prostate carcinoma patients are cytogenetically aberrant. *Cancer*, **83**, 538–546.

9 Hosch, S., Kraus, J., Scheunemann, P., Izbicki, J.R., Schneider, C., Schumacher, U., Witter, K., Speicher, M.R. and Pantel, K. (2000) Malignant potential and cytogenetic characteristics of occult disseminated tumor cells in esophageal cancer. *Cancer Research*, **60**, 6836–6840.

10 Findlay, I., Ray, P., Quirke, P., Rutherford, A. and Lilford, R. (1995) Allelic drop-out and preferential amplification in single cells and human blastomeres: implications for preimplantation diagnosis of sex and cystic fibrosis. *Human Reproduction (Oxford, England)*, **10**, 1609–1618.

11 Telenius, H., Carter, N.P., Bebb, C.E., Nordenskjold, M., Ponder, B.A. and Tunnacliffe, A. (1992) Degenerate oligonucleotide-primed PCR: general amplification of target DNA by a single degenerate primer. *Genomics*, **13**, 718–725.

12 Barrett, M.T., Reid, B.J. and Joslyn, G. (1995) Genotypic analysis of multiple loci in somatic cells by whole genome amplification. *Nucleic Acids Research*, **23**, 3488–3492.

13 Snabes, M.C., Chong, S.S., Subramanian, S.B., Kristjansson, K., DiSepio, D. and Hughes, M.R. (1994) Preimplantation single-cell analysis of multiple genetic loci by whole-genome amplification. *Proceedings of the National Academy of Sciences of the United States, of America*, **91**, 6181–6185.

14 Zhang, L., Cui, X., Schmitt, K., Hubert, R., Navidi, W. and Arnheim, N. (1992) Whole genome amplification from a single cell: implications for genetic analysis. *Proceedings of the National Academy of Sciences of the United States, of America*, **89**, 5847–5851.

15 Dietmaier, W., Hartmann, A., Wallinger, S., Heinmoller, E., Kerner, T., Endl, E., Jauch, K.W., Hofstadter, F. and Ruschoff, J. (1999) Multiple mutation analyses in single tumor cells with improved whole genome amplification. *The American Journal of Pathology*, **154**, 83–95.

16 Dean, F.B., Nelson, J.R., Giesler, T.L. and Lasken, R.S. (2001) Rapid amplification of plasmid and phage DNA using Phi 29 DNA polymerase and multiply-primed rolling circle amplification. *Genome Research*, **11**, 1095–1099.

17 Paul, P. and Apgar, J. (2005) Single-molecule dilution and multiple displacement amplification for molecular haplotyping. *Biotechniques*, **38**, 553–544, 556, 558–559.

18 Hellani, A., Coskun, S., Benkhalifa, M., Tbakhi, A., Sakati, N., Al-Odaib, A. and Ozand, P. (2004) Multiple displacement amplification on single cell and possible PGD applications. *Molecular Human Reproduction*, **10**, 847–852.

19 du Manoir, S., Speicher, M.R., Joos, S., Schrock, E., Popp, S., Dohner, H., Kovacs, G., Robert-Nicoud, M., Lichter, P. and Cremer, T. (1993) Detection of complete and partial chromosome gains and losses by comparative genomic in situ hybridization. *Human Genetics*, **90**, 590–610.

20 Kallioniemi, A., Kallioniemi, O.P., Sudar, D., Rutovitz, D., Gray, J.W., Waldman, F. and Pinkel, D. (1992) Comparative genomic hybridization for molecular cytogenetic analysis of solid tumors. *Science*, **258**, 818–821.

21 Klein, C.A., Schmidt-Kittler, O., Schardt, J.A., Pantel, K., Speicher, M.R. and Riethmuller, G. (1999) Comparative genomic hybridization, loss of heterozygosity, and DNA sequence analysis of single cells. *Proceedings of the National Academy of Sciences of the United States, of America*, **96**, 4494–4499.

22 Schmidt-Kittler, O., Ragg, T., Daskalakis, A., Granzow, M., Ahr, A., Blankenstein, T.J., Kaufmann, M., Diebold, J., Arnholdt, H., Muller, P., Bischoff, J., Harich, D., Schlimok, G., Riethmuller, G., Eils, R. and Klein, C.A. (2003) From latent disseminated cells to overt metastasis: genetic analysis of systemic breast cancer progression. *Proceedings of the National Academy of Sciences of the United States, of America*, **100**, 7737–7742.

23 Schardt, J.A., Meyer, M., Hartmann, C.H., Schubert, F., Schmidt-Kittler, O., Fuhrmann, C., Polzer, B., Petronio, M., Eils, R. and Klein, C.A. (2005) Genomic analysis of single cytokeratin-positive cells from bone marrow reveals early mutational events in breast cancer. *Cancer Cell*, **8**, 227–239.

24 Hu, D.G., Webb, G. and Hussey, N. (2004) Aneuploidy detection in single cells using DNA array-based comparative genomic hybridization. *Molecular Human Reproduction*, **10**, 283–289.

25 Le Caignec, C., Spits, C., Sermon, K., De Rycke, M., Thienpont, B., Debrock, S., Staessen, C., Moreau, Y., Fryns, J.P., Van Steirteghem, A., Liebaers, I. and Vermeesch, J.R. 2006 Single-cell chromosomal imbalances detection by array CGH. *Nucleic Acids Research*, **34**, e68.

26 Fiegler, H., Geigl, J.B., Langer, S., Rigler, D., Porter, K., Unger, K., Carter, N.P. and Speicher, M.R. (2007) High resolution array-CGH analysis of single cells. *Nucleic Acids Research*, **35**, e15.

27 Alberts, B., Johnson, A., Lewis, J. *et al.* (2002) *Molecular Biology of the Cell*, 4th edn, Garland Publishing, New York.

28 Xenidis, N., Perraki, M., Kafousi, M., Apostolaki, S., Bolonaki, I., Stathopoulou, A., Kalbakis, K., Androulakis, N., Kouroussis, C., Pallis, T., Christophylakis, C., Argyraki, K., Lianidou, E.S., Stathopoulos, S., Georgoulias, V. and Mavroudis, D. (2006) Predictive and prognostic value of peripheral blood cytokeratin-19 mRNA-positive cells detected by real-time polymerase chain reaction in node-negative breast cancer patients. *Journal of Clinical Oncology*, **24**, 3756–3762.

29 Baugh, L.R., Hill, A.A., Brown, E.L. and Hunter, C.P. (2001) Quantitative analysis of mRNA amplification by in vitro transcription. *Nucleic Acids Research*, **29**, E29.

30 Van Gelder, R.N., von Zastrow, M.E., Yool, A., Dement, W.C., Barchas, J.D. and Eberwine, J.H. (1990) Amplified RNA synthesized from limited quantities of heterogeneous cDNA. *Proceedings of the National Academy of Sciences of the United States, of America*, **87**, 1663–1667.

31 Klein, C.A., Seidl, S., Petat-Dutter, K., Offner, S., Geigl, J.B., Schmidt-Kittler, O., Wendler, N., Passlick, B., Huber, R.M., Schlimok, G., Baeuerle, P.A. and Riethmuller, G. (2002) Combined transcriptome and genome analysis of single micrometastatic cells. *Nature Biotechnology*, **20**, 387–392.

32 Iscove, N.N., Barbara, M., Gu, M., Gibson, M., Modi, C. and Winegarden, N. (2002) Representation is faithfully preserved in global cDNA amplified exponentially from sub-picogram quantities of mRNA. *Nature Biotechnology*, **20**, 940–943.

33 Hartmann, C.H. and Klein, C.A. (2006) Gene expression profiling of single cells on large-scale oligonucleotide arrays. *Nucleic Acids Research*, **34**, e143.

34 Feldman, A.L., Costouros, N.G., Wang, E., Qian, M., Marincola, F.M., Alexander, H.R. and Libutti, S.K. (2002) Advantages of mRNA amplification for microarray analysis. *Biotechniques*, **33**, 906–912, 914.

35 Heil, S.G., Kluijtmans, L.A., Spiegelstein, O., Finnell, R.H. and Blom, H.J. (2003) Gene-specific monitoring of T7-based RNA amplification by real-time

quantitative PCR. *Biotechniques*, **35**, 502–504, 506–508.

36 Li, J., Adams, L., Schwartz, S.M. and Bumgarner, R.E. (2003) RNA amplification, fidelity and reproducibility of expression profiling. *Comptes Rendus Biologies*, **326**, 1021–1030.

37 Dixon, A.K., Richardson, P.J., Pinnock, R.D. and Lee, K. (2000) Gene-expression analysis at the single-cell level. *Trends in Pharmacological Sciences*, **21**, 65–70.

38 Freeman, T.C., Lee, K. and Richardson, P.J. (1999) Analysis of gene expression in single cells. *Current Opinion in Biotechnology*, **10**, 579–582.

39 Saghizadeh, M., Brown, D.J., Tajbakhsh, J., Chen, Z., Kenney, M.C., Farber, D.B. and Nelson, S.F. (2003) Evaluation of techniques using amplified nucleic acid probes for gene expression profiling. *Biomolecular Engineering*, **20**, 97–106.

40 Kurimoto, K., Yabuta, Y., Ohinata, Y., Ono, Y., Uno, K.D., Yamada, R.G., Ueda, H.R. and Saitou, M. (2006) An improved single-cell cDNA amplification method for efficient high-density oligonucleotide microarray analysis. *Nucleic Acids Research*, **34**, e42.

41 Tietjen, I., Rihel, J.M., Cao, Y., Koentges, G., Zakhary, L. and Dulac, C. (2003) Single-cell transcriptional analysis of neuronal progenitors. *Neuron*, **38**, 161–175.

42 Bahar, R., Hartmann, C.H., Rodriguez, K.A., Denny, A.D., Busuttil, R.A., Dolle, M.E., Calder, R.B., Chisholm, G.B., Pollock, B.H., Klein, C.A. and Vijg, J. (2006) Increased cell-to-cell variation in gene expression in ageing mouse heart. *Nature*, **441**, 1011–1014.

43 Gribble, S., Ng, B.L., Prigmore, E., Burford, D.C. and Carter, N.P. (2004) Chromosome paints from single copies of chromosomes. Chromosome Research: An International Journal on the Molecular, Supramolecular and Evolutionary Aspects of Chromosome Biology **12**, 143–151.

44 Stoecklein, N.H., Erbersdobler, A., Schmidt-Kittler, O., Diebold, J., Schardt, J.A., Izbicki, J.R. and Klein, C.A. (2002) SCOMP is superior to degenerated oligonucleotide primed-polymerase chain reaction for global amplification of minute amounts of DNA from microdissected archival tissue samples. *The American Journal of Pathology*, **161**, 43–51.

45 Eberwine, J., Yeh, H., Miyashiro, K., Cac, Y., Nair, S., Finnell, R., Zettel, M. and Coleman, P. (1992) Analysis of gene expression in single live neurons. *Proceedings of the National Academy of Sciences of the United States, of America*, **89**, 3010–3014.

46 Brady, G. and Iscove, N.N. (1993) Construction of cDNA libraries from single cells. *Methods in Enzymology*, **225**, 611–623.

47 Aoyagi, K., Tatsuta, T., Nishigaki, M., Akimoto, S., Tanabe, C., Omoto, Y., Hayashi, S., Sakamoto, H., Sakamoto, M., Yoshida, T., Terada, M. and Sasaki, H. (2003) A faithful method for PCR-mediated global mRNA amplification and its integration into microarray analysis on laser-captured cells. *Biochemical and Biophysical Research Communications*, **300**, 915–920.

48 Fink, L., Kohlhoff, S., Stein, M.M., Hanze, J., Weissmann, N., Rose, F., Akkayagil, E., Manz, D., Grimminger, F., Seeger, W. and Bohle, R.M. (2002) cDNA array hybridization after laser-assisted microdissection from nonneoplastic tissue. *The American Journal of Pathology*, **160**, 81–90.

49 Fuhrmann, C., Schmidt-Kittler, O., Stoecklein, N.H. *et al.* (2008) High resolution array comparative genomic hybridization of single micrometastatic tumor cells. *Nucleic Acids Research*, **36**, e39.

50 Husemann, Y., Geigl, J.B., Schubert, F. *et al.* (2008) Systemic spread is an early step in breast cancer. *Cancer Cell*, **13**, 56–58.

51 Stoecklein, N.H., Hosch, S.B., Bezler, M. *et al.* (2008) Direct genetic analysis of single disseminated cancer cells for prediction of outcome and therapy selection in esophageal cancer. *Cancer Cell*, **13**, 441–453.

Index

a

acceptor photobleaching 186
achilles heel 205
acousto-optical tunable filter (AOTF) 5
activated LuxR-type regulators 30
Aequorea victoria 183
Agrobacterium tumefaciens 30
AHL-free buffer solution 31
algebraic reconstruction techniques (ART) 56
alkaline phosphatase 84
allelic exclusion 188
allelic loss 237
amino acids 74, 143
– aromatic 74
analysis speed 203
antigen recognition 177
Anton van Leeuwenhoek 109
AOTF driver 4
Aplysia californica 121, 224
aromatic amino acids 74
array-based techniques 144
arrayed single cell culture 154
atomic force microscopy (AFM) 19, 20, 21, 25, 26, 29, 30
– analysis 24
– cantilever 20
– classification 23
– high-resolution imaging 21, 24
– nano-dissection analysis 21
– schematic representation 20
autofluorescent proteins 6
– conjugates 1
autoimmune diseases 182
autoimmune disorders 179
automated microscopy techniques 141

b

Bacillus anthracis 113

bacterial artificial chromosomes (BAC) 240
BAMS mass spectrometers 113
bioaerosol mass spectrometry (BAMS) 113
bioanalytical sensors 21
biological molecules 19
– antibodies 19
– chaperones 19
– cytochromes 19
– proteoglycans 19
– selectins 19
biological sample 110
– advantage of 110
biological techniques 21
– PCR 21
– SDS-PAGE 21
– sequencing 21
biomarkers 198
biomolecular interactions 19
bio/nanotechnological application 20, 21
blood vessels 187
box-like sequence (TATAGTACATGT) 30
breast cancer 211
Brownian motion 12, 162
buffer solution 26, 27
bulk techniques 138
– western blots 138

c

Caenorhabditis elegans zygote 79
cancer 161, 206
cancerous tissue 173
capillary electrophoresis (CE) 224
capillary based separation devices 92
capillary electrophoresis 78, 84, 114
capillary liquid chromatography column 72
capillary sieving electrophoresis 78
– column 80
– electropherograms 77, 79
– experiment 83

carbonaceous compounds   40
cell
 – adult mesenchymal stem cells   161
 – arrival statistics   205
 – biology   161
 – deformation   171
 – eukaryotic   162
 – keratinocytes   172
 – noncancerous   171
cell-cell contact   143, 144
cell-cell interactions   109
cell-cell peptide heterogeneity   117
cell-cell signaling molecules   109
cell fluorescence microscopy   183
cell heterogeneity   69
cell lysis   92, 95
cell microfluidic analysis   94
cell proteomics   69
cell trapping principles   103
 – cell suction   103
 – dielectrophoresis   103
 – hydrodynamic trapping   103
 – microfluidic constrictions   103
cellular components   92, 119, 141
cellular division   135
cellular heterogeneity   223
cellular homogenate   83
cellular lipids   125
central nervous system   69
central supramolecular activation cluster, (cSMAC)   184
changing trap geometry   143
charged coupled device (CCD)   3
 – camera   4, 53, 163
chemical reactions   136
chromatic aberration   5
 – axial   5
 – lateral   5
chromatin network   10
chromatography column   72
chromium mask   100
cluster of differentiation (CD)   181
coefficients of variation (CV)   232
collision-induced dissociation (CID) fragmentation   118
comparative genomic hybridization (CGH)   237
competent cells   216
confocal microscopy   184
cord blood mononuclear cells (CBMCs)   218
cDNA (complementary DNA)   177, 188
computer-aided design (CAD)   44
computer-assisted tomography   41
 – technique   56

computer-controlled valve   73
continuous wave   1
 – genetically engineered   1
contrast transfer function (CTF)   49
conventional techniques   14
 – FCS/photobleaching   14
Corynebacterium glutamicum   20, 21, 23
 – AFM images   22
 – mutant   25
 – paracrystalline cell surface layers   20
 – regulator   24
 – s-layers   22
 – strains   23, 24
cryo-electron micrograph   42
cryo-electron microscopy of vitrified sections (CEMOVIS)   48
cryo-electron tomography (CET)   40, 43
cryo-transmission electron microscopes   40
Cy5-labeled RNPs   10
 – photobleaching   10
cytometry stains   77
cytoskeletal/nuclear matrix fractions   78
cytotoxic T lymphocytes (CTL)   177

d

degenerative oligonucleotide primers (DOP)   237
dendritic cells (DCs)   186
derivatization reaction   72
derivatization technology   74
derivatizing reagent   74
desorption/ionization on porous silicon (DIOS)   113
detection method   92
diabetes   173, 223
dichromatic beam splitters   4, 5
Dictyostelium discoideum   42, 56
dielectric/transparent material   164
dielectrophoretic techniques   145
diffusion coefficient   13
digital image   2
digitonin-permeabilized cells   8
discriminant function analysis (DFA)   205
2D gel electrophoresis   79, 83
DIOS substrate   113
DNA affinity purification assay   24
DNA electrophoresis   75
DNA fragments   26, 29
DNA-protein complexes   28
DNA sequences   74, 188
DNA target sequences   28, 29
DNA transcription/replication   6
2D projection images   41
double-stranded oligonucleotides   77

dynamic biological systems 183
dynamic force spectroscopy 28

e

electrical connection 81
electrochemical detection 92
electromagnetic radiation 39
electromotive force 146
electron microscope 39, 41, 42, 175, 188
– constant electron bombardment 39
– practical perspective 42–48
– ultra-high vacuum 39
electron multiplying (EM) camera 3
electronic/optical devices 21
electron tomography 51
electroosmotic flow 101
electrophoresis/detection system 77
– performance 77
electrophoretic migration 94
electrophoretic mobility 75
electrospray conditions 114
electrospray ionization (ESI) 112
ELISPOT reader 187
ELISPOT technique 187
EM grid 47
emerging techniques 147
endocrine cells 115
endogenous enzymes 119
engineering system 157
enhanced chemiluminescence 211
environmental SEM image 122
enzyme-linked immunospot (ELISPOT) assay 187
epithelial cell adhesion molecule (EpCAM) 235
epitopes 179
epoxy resin 47
Escherichia coli 213
ESI Fourier transform-ion cyclotron resonance 114
eukaryotic cell 135
eukaryotic RNA transcripts 7
excitation light 5
exogenous enzymes 119
ExpR-DNA interaction 30

f

false negative (FN) 205
false positive (FP) 205
fast timescale measurements 144
FCS/FRAP measurements 15
field emission gun (FEG) 49
FITC-conjugated anti-BUdR antibody 199, 200

fluid droplets 93
fluid manipulation 93
fluorescence
– activated cell sorter (FACS) 161
– emissions 183
– lifetime imaging microscopy (FLIM) 186
– microscopy 224
– proteins 183
forward scattering (FSC) 232
fluorescence-activated cell sorting (FACS) 140, 180, 181
– machines 161
fluorescence bands 4
fluorescence correlation spectroscopy (FCS) 6
fluorescence detectors 76
fluorescence image 3, 4
fluorescence intensity 148
fluorescence interrogation 143
fluorescence lifetime imaging microscopy (FLIM) 186
fluorescence microscope 77
fluorescence microscopic techniques 1
– quantitative 1
– visualization 2
fluorescence signal 180
fluorescence techniques 1
fluorescent dye 140, 152
fluorescent fusion protein 93, 94
fluorescent markers 141, 161
fluorescent microscope 82
fluorescent probe molecule 5
fluorescent proteins 183
fluorescent quantum dots 6
fluorogenic reagents 74, 75
fluorogenic substrates 140
Fourier space 56
Fourier transform-ion cyclotron resonance (FT-ICR) 71, 72
– instruments 71, 72
Fourier transform (FT) mass spectrometers 118
FQ derivatizing reagent 75
freeze-fracture methodologies 124
freeze-fracture preparation methods 123
fused green fluorescent protein (GFP) 84
fusion protein 93
fuzzy set theory 205

g

Gal4/ GFP fusion 84
– expression 84
Gaussian distribution 26
Gaussian function 3

Gaussian illumination pattern 3
Gaussian intensity 167
Gaussian intensity profile 167
gene 175
– expression profile 143, 175, 216
– marking 208
– ontology (GO) terms 233
genomic DNA 180
geometric factor 172
global noise, See for Extrinsic noise
genetic engineering 83
genomic DNA 179
geometrical cell trap 94
green fluorescent protein (GFP) 84, 85, 95, 98
– excitation 8
– protein 138, 140
glow discharging 45
glutamicum strains 21
glycerol stabilization 121
glycerol treatment 116
gradient force 168
granzyme B 180

h

HeLa cells 8, 148, 154
hematopoietic stem cell 230
Herzenberg group 180
heterogeneous systems 14
heterogeneous tissues 119
high-content techniques 147
– implementation 147
high-content microscopy-based techniques 140
high-content screening (HCS) 141
high-content techniques 157
high-precision mass measurement 117
high-pressure freezing (HPF) 46, 47
– device 47
high-resolution imaging techniques 22
high-resolution methods 1
high-resolution topographs 21
high-speed cell sorter 205
high-speed enrichment sorting 203
High-throughput system (HTS) 209
high-speed lookup tables 204
high-speed video microscope 6, 14
high-throughput experimentation 146
high-throughput fluorescence-based analysis 144
high-throughput serial process 140
high-throughput technique 141
high resolution imaging 121
holographic optical traps 103, 146
hormone 177
hostile factors 20
HT-29 human adenocarcinoma cell 76
hydrophilic glass beads/tissue pieces 115

i

ice-embedded samples 40
illumination field size 3
illumination source 49
– field emission 49
– thermionic 49
immunofluorescence 140
in situ hybridization 175, 188
incoming/probing electron beam 46
inorganic compounds 113
inorganic crystals 182
inorganic salt concentration 124
interchromatin granule clusters 7
interconnected molecular interactions (signaling pathways) 136
internal reflection fluorescence microscopy 185
intra-cellular single molecule 2
intracellular membranes 56
– segmentation of 56
intracellular proteins 146, 181
– dephosphorylation of 181
– high-content spatial arrangements 146
intracellular signal transduction pathway 96
intracellular viscosity 6
intramolecular forces 19
intranuclear binding processes 6
intranuclear components 7
– DNA 7
– RNA 7
– protein complexes 7
intranuclear splicing factor tracking 8–10
intranuclear transport pathways 6
ionization methods 112, 113
Islets of Langerhans 177
isotope- coded affinity tag (ICAT) 71

j

jump distance distribution 11, 12
– analysis 12
jump distance histograms 13

l

label-free fluorescence detection 92, 99
laminar flow 169
large-scale cellular homogenate 78
laser beams 4
laser capture microdissection (LCM) 115
laser desorption ionization 112
laser-induced fluorescence (LIF) 74, 225

– approaches 99
– detection 92, 98
laser light sources 2
laser microprobe mass analysis (LAMMA) technique 112
– capability 113
laser scanning cytometry (LSC) 140, 161
laser scanning microscopy 183
– confocal 183
LDI approaches 113
ligand-receptor molecules 20
ligand-receptor pairs 20
ligand-receptor systems 20
light microscopy 1
light transmission 2
– optimization 2
lipid membrane analysis 124
lipophilic small molecules 124
liquid chromatographic separation 78
liquid chromatography/capillary electrophoresis systems 81
lithographic masks 21
loss of heterozygosity (LOH) 240
low-abundance proteins 92, 98
low-contrast electron micrographs 45
– alignment 45
low-dose imaging techniques 40
lysine residues 74, 75
lysis reagents 93

m

machine-based segmentation 56
macromolecular environment 54
magnetic activated cell sorting (MACS) 218
magnetic bead rheology 163
magnetic resonance imaging (MRI) 41, 50
Magnetospirillum griphiswaldense 43
magnetotactic microorganism 43
major histocompatibility complex (MHC) 176, 186
– class I molecule 177, 182
malignant cells 171
mammalian cells 48, 70, 73, 115
mammalian oocytes 72
mammalian system 71
mammary carcinoma cells 119
mass spectrometry 71, 72, 103, 110, 111, 121
massively parallel sample preparation process 116
– schematic representation 116
matrix-assisted laser desorption/ionization (MALDI) 71, 112, 118, 125
– advancement of 121
– detection 124

– experiments 117
– imaging 120, 121
– matrix 115, 123
– plate 72, 115
matrix-free methods 113
mean square displacements (MSD) 11
mesenchymal stem cells 173
metastatic transformation 123
metastatic tumor cell analysis 198
micellar electrophoresis dimension 81
microarray
– oligo-based 216
– spotted 216
microbial genomic databases 117
microchannel walls 147
microfabricated cantilever 19
microfabricated chambers 143
microfluidic cell array 137
microfluidic channel 142
microfluidic chip 164
microfluidic device 94, 96, 149
– scheme of 94
microfluidic devices 92, 93
microfluidic environments 92
microfluidic flow chamber 164
microfluidic format 93
microfluidic platform 143
microfluidic single cell analysis 93, 142
– approach 93
– arrays 148
microfluidic system 94, 97
microfluidic techniques 142, 143
microfluidics-based dynamic single cell culture array 154
microinjection system 5
micropipette aspiration 163
microplate method 162
microtubules 162
molecular ions 111
molecular isoforms 111
molecular resolution tomography 48
molecular screening methods 111
molecule signals 14
monomer-oligomer 151
motor action 110
– whisker movement 110
mRNA molecules 23, 178
mRNA splicing 6, 8
– dynamics of 8
mRNA transcription 11
MS technologies 120
multicomponent nuclear RNP complexes 7
multidimensional protein identification technology (MudPIT) 71

multifocal microscopy   103
multiple mass analyzers   111
multiplex PCR   178
Mycobacterium tuberculosis   113

n

N-acyl homoserine lactones (AHLs)   29
Navier–Stokes equation   170
nervous tissue   116
neurons   120
neuropeptides   119
nonconfocal video images   10
notch filters   3
NPY secretion   125
nuclear speckles   7
nucleoplasmic staining   7

o

oligonucleotides   2
optical density   3
optical gradient force   146
optical system   3
– magnification of   3
optical trapping   146
optical tweezers (OT)   94
organ systems   186

p

Parkinson's diseases   173
PDMS device fabrication   99
polymerase chain reaction (PCR)   178, 179
poly(dimethylsiloxane)   101, 102, 103
– elastomer-based microfluidic systems   102
– fabrication   100
– fluorescence   102
– microfluidic chip   103
– microfluidic devices   102
– microfluidic structure   101
– mold   101
– slab   101
– surface   102
– technology   101
PEG-bound ligands   28
peptide diversity   111
peptide-MHC complexes   185
peptide/protein characterization   112
pericentromeric chromatin   188
perichromatin fibrils   7
phase-contrast microscope   163, 164
phosphate-buffered saline   76
phosphatidylcholine lipids   124
phospho-specific antibodies   185
phospholipid signals   123

photobleaching/photoactivation techniques   14
photodetector   83
photomultiplier tube   73
photomultiplier tubes   180
Phymatodocis nordstedtiana   120
physiological processes   11
– splicing   11
– transcription   11
pipette-like channels   142
Plunge freezing instrumentation   44
– shematic presentation   44
pMHC ligand   185
polyethylene glycol (PEG)   147
polyethylene oxide   76
polytropic montage   56
pore-forming event   151
post-transcriptional modifications   7
post-translational modifications (PTMs)   111
PTMs   111
– characterization   111
post-translational proteolytic processing   111
pre-filled well array   144
pre-messenger RNAs   7
– reaction   10
– splicing   7
pressure-driven sampling   120
primary ions   114
principal component analysis (PCA)   205
probing signals   39
– electrons   39
– photons   39
– X-ray quanta   39
projection-slice theorem   55
prokaryotic cell envelopes   20
proteins   6, 80, 92, 99, 102
– information   20
– label-free LIF detection   99
– LIF detection   92
– morphology   20
– two-dimensional capillary electrophoresis   80
– mRNA content data   147
– DNA complexes   25
– DNA interactions   20, 29, 31
– DNA system   30
– analysis   74
– expression   69, 70, 103
– levels   71
protein electropherograms   77
protein formation/degradation   1
pulsed lasers   1
– LIF detection   102
putative binding sites   10

q

quantum dots 182
- nanocrystals 182
quartz microfluidic chip 99
quasi-deterministic mathematical models 136
quorum sensing (QS) systems 29

r

radical polymerization 76
Raleigh scattered light 3
Raman microspectroscopy 224
rapid vaporization 113
rare cell applications 197
ray optics regime 165
receiver operating characteristic (ROC) 198
reconstruction technique 56
red blood cells 119
refractive index inhomogeneity 82
reverse transcription (RT)-PCR 177
RGD-integrin interaction 146
ribonucleoprotein particles 2
ribosomal ribonucleic acid (rRNA) analysis 72
RNA polymerase 7
RNA processing machinery 7
RNA sequences 188
RNA splicing 14
RNP dynamics 8
RNP trajectories 8
RT-PCR 178

s

S-layer gene expression 23
S-layer proteins 22, 23
Saccharomyces cerevisiae 71
scanning tunneling microscope (STM) 19
scattering force 168
second-order kinetics 74
secondary ion mass spectrometry (SIMS) 114
self-assembly processes 21
separation methods 119
- capillary electrophoresis (CE) 119
- chromatography 119
several alternative methods 93
SFM measurements 163
shake-off method 73
sheath-flow cuvette 73
sieving electrophoresis fraction 81
signal to noise ratio (SNR) 4, 53
signaling pathways 136
signaling processes 110
simple microfluidic device 93
SIMS 120, 123
SIMS imaging 123, 124
single base pair mutation 211
single cell analysis 1, 103, 110, 152
- approach 103, 175
single cell comparative hybridization (SCOMP) 239
single cell heterogeneity 223
single cell sequencing 211
single cell sorting 211
single cell (multiplex) RT-PCR 175
single cell deformation 171
single cell electropherogram 76
- contamination 76
single cell electropherograms 78, 97
- approaches 97
single cell electrophoretic separation 96
single cell mass spectrometry (MS) 109
single cell microfluidic 98
single cell protein analysis 72
single cell protein electropherograms 92
single cell proteomics 69, 70, 71
single cell RT-PCR 179
single cell trapping arrays 153
single fluorescent molecules 3
- microscopic imaging 3
single mammalian cell 72
single molecule force spectroscopy 20, 25
single molecule microscope 2, 4
- architecture 4
- optical setup 2
single molecule microscopy 13, 15
- perspectives, 13–15
- visualization 13
single molecule signals 2, 3
- identification/tracking 2
Sinorhizobium meliloti 25
SLO 152, 186
- monomers 149
smooth transitions 152
Snell's Law 166
snRNP mobility 11
sodium dodecyl sulfate (SDS) PAGE 75, 76, 78, 99
- electropherogram 79
- treated protein 79
- treatment 84
soft lithography 99, 101
solid-phase extraction (SPE) collection technique 124
- extraction 125
- material 124
southern blotting 214
speckles events 10

spectrometric approaches 114
Spiroplasma melliferum 56
splicing factors 1, 7, 9
– gray-scale image 9
– intranuclear dynamics 1
stabilization treatments 115
staining techniques 109
standard deviation 229
stem cell characterization 173
stochastic model 148
stochastic protein expression 143
Stokes–Einstein law 13
streptavidin/avidin-biotin interactions 19
stretch-activated ion channels 135
subcellular profiling approach 120
subcellular protein surfaces 25
submicellar buffer systems 83
submicellar electrophoresis 79
subnanometer scale 21
supramolecular systems 19
surface analysis technique 124
surface antigens 181

t

taq polymerase 215
T cell receptor (TCR) 184
– MHC interactions 177
– microclusters 184
– repertoire 178
– signaling 181, 185
– transgenic model 178
TEM grid 45
tetrahymena cells 124
three-dimensional (3D) object 40
tilt series data acquisition scheme 41
– single axis 41
time-dependent kinetic studies 143
time-dependent transitions 140
TIRFM 185
TOF mass spectrometer 72
tomographic acquisition technique 41
total internal reflection fluorescence microscopy 185
transmembrane pore 151

transmission electron microscope (TEM) 41, 49
– study 48
transmitted light 183
tube arrays 202
tubular capillary chromatography 72
tumor protein expression 69
tube arrays 202
tumor suppressor gene 209
two-dimensional capillary electrophoresis 80
– instrument 81
two-dimensional distribution maps/ion images 121
two-dimensional electropherogram 82
– intensity image 82
two-dimensional gel electrophoresis technology 80
two-photon laser scanning microscopy (TPLSM) 186

u

U-shaped hydrodynamic trapping structures 154
ultrafiltration membranes 21
ultrasensitive LIF detector 80
University of Alberta 73
UV-LIF 99
– detection 101
UV-transparent microfluidic channels 99
UV spectral 92
UV/infrared (IR) laser 113

v

vaporization/ionization processes 111
vitrification process 54

x

Xenopus laevis oocytes 124

y

yeast cell 85
yeast genome 74
yellow fluorescent protein (YFP) 96
– fusion protein 98